# SUSTAINABLE AGRICULTURE RESEARCH AND EDUCATION IN THE FIELD
## A PROCEEDINGS

Board on Agriculture
National Research Council

NATIONAL ACADEMY PRESS
Washington, D.C. 1991

Support for the workshop and the proceedings was provided by the U.S. Department of Agriculture, Cooperative State Research Service, under Agreement No. 90-COOP-2-5028. Any opinions, findings, conclusions, or recommendations expressed in this publication are those of the authors and do not necessarily reflect the views of the U.S. Department of Agriculture.

Library of Congress Catalog Card No. 91-62492
ISBN 0-309-04578-9

A limited number of copies are available from:
Board on Agriculture
National Research Council
2101 Constitution Avenue
Washington, DC 20418

Additional copies are available for sale from:
National Academy Press
2101 Constitution Avenue
Washington, DC 20418

S-417

Printed in the United States of America

# Preface

With the Office of Science and Education, U.S. Department of Agriculture (USDA), the Board on Agriculture cosponsored the workshop, Sustainable Agriculture Research and Education in the Field—1990, April 3–4, 1990. It included research reports from around the country that summarized recent progress in understanding the scientific and technological basis of sustainable and profitable agricultural production systems. The regional panels demonstrated the vastly different challenges facing farmers in different regions of the country. In addition, the workshop explored gaps in ongoing research efforts, common themes and approaches in successful research programs, and areas in need of additional effort or new research strategies.

A mix of nationally recognized experts, including academic scientists, farmer-innovators, and agribusiness leaders, placed into perspective the scientific and technology challenges facing the nation in striving toward sustainable, profitable, and resource-conserving production systems.

The workshop provided a forum for discussing field research results since the inception of the Low-Input Sustainable Agriculture (LISA) program in 1988. In the 1990 farm bill, the LISA program was renamed by Congress to be Sustainable Agriculture Research and Education. During its first 3 years of operation, the program has supported more than 100 projects on many topics essential to the development and profitable adoption of sustainable farming systems. Only a few of the projects were selected for presentation at the workshop. Other sustainable agriculture projects funded by USDA's Agricultural Research Service (ARS) and INFORM, a private nonprofit environmental research organization, were also selected for presentation.

The workshop fostered dialogue about and understanding of the new research priorities and directions emerging across the country, not only in land-grant universities, but also in ARS and private organizations. Information and suggestions during the workshop will assist in setting research priorities in future years.

The introduction to the report of the proceedings is followed by six parts. Part one presents an overview of general information and issues, including USDA's commitment to sustainable agriculture, the background and status of the program, challenges and rewards of research and education efforts, and economic considerations, as they pertain to midwestern farmers. Parts two through five present the findings of research projects in the western, southern, north central, and northeastern regions, respectively, along with reactors' comments. Part six is a summary of the workshop followed by appendixes that contain two poster session papers and a special topic paper.

The contributions of several individuals warrant special mention. Paul O'Connell, deputy administrator of USDA's Cooperative State Research Service; Charles Benbrook, former executive director of the Board on Agriculture; and Patrick Madden, consultant, were instrumental in organizing the workshop and the report of the proceedings. All conference speakers and reactors are gratefully acknowledged for the timely submission of their contributions.

THEODORE L. HULLAR, *Chairman*
Board on Agriculture

# Contents

## PART FIVE: RESEARCH AND EDUCATION IN THE NORTHEASTERN REGION

## PART SIX: SUMMARY

## APPENDIXES

# Introduction

*Charles M. Benbrook*

These proceedings are based on a workshop that brought together scientists, farmer-innovators, policymakers, and interested members of the public for a progress report on sustainable agriculture research and education efforts across the United States. The workshop, which was held on April 3 and 4, 1990, in Washington, D.C., was sponsored by the Office of Science and Education of the U.S. Department of Agriculture and the Board on Agriculture of the National Research Council. The encouraging new science discussed there should convince nearly everyone of two facts.

First, the natural resource, economic, and food safety problems facing U.S. agriculture are diverse, dynamic, and often complex. Second, a common set of biological and ecological principles—when systematically embodied in cropping and livestock management systems—can bring improved economic and environmental performance within the reach of innovative farmers. Some people contend that this result is not a realistic expectation for U.S. agriculture. The evidence presented here does not support such a pessimistic assessment.

The report of the Board on Agriculture entitled *Alternative Agriculture* (National Research Council, 1989a) challenged everyone to rethink key components of conventional wisdom and contemporary scientific dogma. That report has provided encouragement and direction to those individuals and organizations striving toward more sustainable production systems, and it has provoked skeptics to articulate why they feel U.S. agriculture cannot—some even say should not—seriously contemplate the need for such change. The debate has been spirited and generally constructive.

Scholars, activists, professional critics, and analysts have participated in this debate by writing papers and books, conducting research, and offering opinions about alternative and sustainable agriculture for over 10 years. Over the past decade, many terms and concepts have come and gone. Most people—and unfortunately, many farmers—have not gone very far beyond the confusion, frustration, and occasional demagoguery that swirls around the different definitions of alternative, low-input, organic, and sustainable agriculture.

Fortunately, though, beginning in late 1989, a broad cross-section of people has grown comfortable with the term *sustainable agriculture*. The May 21, 1990, issue of *Time* magazine, in an article on sustainable agriculture entitled "It's Ugly, But It Works" includes the following passage:

> [A] growing corps of experts [are] urging farmers to adopt a new approach called sustainable agriculture. Once the term was synonymous with the dreaded O word—a farm-belt euphemism for trendy organic farming that uses no synthetic chemicals. But sustainable agriculture has blossomed into an effort to curb erosion by modifying plowing techniques and to protect water supplies by minimizing, if not eliminating, artificial fertilizers and pest controls.

Concern and ridicule in farm publications and during agribusiness meetings over the philosophical roots of low-input, sustainable, or organic farming have given way to more thoughtful appraisals of the ecological and biological foundations of practical, profitable, and sustainable farming systems. While consensus clearly does not yet exist on how to "fix" agriculture's contemporary problems, a constructive dialogue is now under way among a broad cross-section of individuals, both practitioners and technicians involved in a wide variety of specialties.

This new dialogue is powerful because of the people and ideas it is connecting. Change will come slowly, however. Critical comments in some farm magazines will persist, and research and on-farm experimentation will not always lead to the hoped for insights or breakthroughs. Some systems that now appear to be sustainable will encounter unexpected production problems. Nonetheless, progress will be made.

The Board on Agriculture believes that over the next several decades significant progress can and will be made toward more profitable, resource-conserving, and environmentally prudent farming systems. Rural areas of the United States could become safer, more diverse, and aesthetically pleasing places to live. Farming could, as a result, become a more rewarding profession, both economically and through stewardship of the nation's soil and water resources. Change will be made possible; and it will be driven by new scientific knowledge, novel on-farm management tools and approaches, and economic necessity. The policy reforms adopted in the 1990 farm bill, and ongoing efforts to incorporate environmental objectives into farm policy, may also in time make a significant difference in reshaping the economic environment in which on-farm management decisions are made.

This volume presents an array of new knowledge and insight about the functioning of agricultural systems that will provide the managerial and technological foundations for improved farming practices and systems. Examples of the research projects under way around the country are described. Through exploration of the practical experiences, recent findings, and insights of these researchers, the papers and discussions presented in this volume should demonstrate the value of field- and farm-level systems-based research that is designed and conducted with ongoing input from farmer-innovators.

Some discussion of the basic concepts that guide sustainable agriculture research and education activities may be useful. Definitions of key terms, such as sustainable agriculture, alternative agriculture, and low-input sustainable agriculture, are drawn from *Alternative Agriculture* and a recent paper (Benbrook and Cook, 1990).

## BASIC CONCEPTS AND OPERATIONAL DEFINITIONS

### Basic Concepts

Sustainable agriculture, which is a goal rather than a distinct set of practices, is a system of food and fiber production that

• improves the underlying productivity of natural resources and cropping systems so that farmers can meet increasing levels of demand in concert with population and economic growth;

• produces food that is safe, wholesome, and nutritious and that promotes human well-being;

• ensures an adequate net farm income to support an acceptable standard of living for farmers while also underwriting the annual investments needed to improve progressively the productivity of soil, water, and other resources; and

• complies with community norms and meets social expectations.

Other similar definitions could be cited, but there is now a general consensus regarding the essential elements of sustainable agriculture. Various definitions place differing degrees of emphasis on certain aspects, but a common set of core features is now found in nearly all definitions.

While sustainable agriculture is an inherently dynamic concept, alternative agriculture is the process of on-farm innovation that strives toward the goal of sustainable agriculture. Alternative agriculture encompasses efforts by farmers to develop more efficient production systems, as well as efforts by researchers to explore the biological and ecological foundations of agricultural productivity.

The challenges inherent in striving toward sustainability are clearly dynamic. The production of adequate food on a sustainable basis will become more difficult if demographers are correct in their estimates that the global population will not stabilize before it reaches 11 billion or 12 billion in the middle of the twenty-first century. The sustainability challenge and what must be done to meet it range in nature from a single farm field, to the scale of an individual farm as an enterprise, to the food and fiber needs of a region or country, and finally to the world as a whole.

A comprehensive definition of sustainability must include physical, biological, and socioeconomic components. The continued viability of a farming system can be threatened by problems that arise within any one of these components. Farmers are often confronted with choices and sacrifices because of seemingly unavoidable trade-offs—an investment in a conservation system may improve soil and water quality but may sacrifice near-term economic performance. Diversification may increase the efficiency of resource use and bring within reach certain biological benefits, yet it may require additional machinery and a more stable and versatile labor supply. Indeed, agricultural researchers and those who design and administer farm policy must seek ways to alleviate seemingly unwelcome trade-offs by developing new knowledge and technology and, when warranted, new policies.

## Operational Definitions

Sustainable agriculture is the production of food and fiber using a system that increases the inherent productive capacity of natural and biological resources in step with demand. At the same time, it must allow farmers to earn adequate profits, provide consumers with wholesome, safe food, and minimize adverse impacts on the environment.

As defined in our report, alternative agriculture is any system of food or fiber production that systematically pursues the following goals (National Research Council, 1989a):

• more thorough incorporation of natural processes such as nutrient cycling, nitrogen fixation, and beneficial pest-predator relationships into the agricultural production process;

• reduction in the use of off-farm inputs with the greatest potential to harm the environment or the health of farmers and consumers;

• productive use of the biological and genetic potential of plant and animal species;

• improvement in the match between cropping patterns and the productive potential and physical limitations of agricultural lands; and

• profitable and efficient production with emphasis on improved farm management, prevention of animal disease, optimal integration of livestock and cropping enterprises, and conservation of soil, water, energy, and biological resources.

*Conventional agriculture* is the predominant farming practices, methods, and systems used in a region. Conventional agriculture varies over time and according to soil, climatic, and other environmental factors. Moreover, many conventional practices and methods are fully sustainable when pursued or applied properly and will continue to play integral roles in future farming systems.

*Low-input sustainable agriculture* (LISA) systems strive to achieve sustainability by incorporating biologically based practices that indirectly result in lessened reliance on purchased agrichemical inputs. The goal of LISA systems is improved profitability and environmental performance through systems that reduce pest pressure, efficiently manage nutrients, and comprehensively conserve resources.

Successful LISA systems are founded on practices that enhance the efficiency of resource use and limit pest pressures in a sustainable way. The operational goal of LISA should not, as a matter of first principles, be viewed as a reduction in the use of pesticides and fertilizers. Higher yields, lower per unit production costs, and lessened reliance on agrichemicals in intensive agricultural systems are, however, often among the positive outcomes of the successful adoption of LISA systems. But in much of the Third World an increased level of certain agrichemical and fertilizer inputs will be very helpful if not essential to achieve sustainability. For example, the phosphorous-starved pastures in the humid tropics will continue to suffer severe erosion and degradation in soil physical properties until soil fertility levels are restored and more vigorous plant growth provides protection from rain and sun.

Farmers are continuously modifying farming systems whenever opportunities arise for increasing productivity or profits. Management decisions are not made just in the context of one goal or concern but in the context of the overall performance of the farm and take into account many variables: prices, policy, available resources, climatic conditions, and implications for risk and uncertainty.

A necessary step in carrying out comparative assessments of conventional and alternative farming systems is to understand the differences between farming practices, farming methods, and farming systems. It is somewhat easier, then, to determine what a conventional practice, method, or system is and how an alternative or sustainable practice, method, or system might or should differ from a conventional one. The following definitions are drawn from the Glossary of *Alternative Agriculture* (National Research Council, 1989a).

A *farming practice* is a way of carrying out a discrete farming task such as a tillage operation, particular pesticide application technology, or single conservation practice. Most important farming operations—preparing a seedbed, controlling weeds and erosion, or maintaining soil fertility, for example—require a combination of practices, or a *method*. Most farming operations can be carried out by different methods, each of which can be accomplished by several unique combinations of different practices. The manner in which a practice is carried out—the speed and depth of a tillage operation, for example—can markedly alter its consequences.

A *farming method* is a systematic way to accomplish a specific farming objective by integrating a number of practices. A discrete method is needed for each essential farming task, such as preparing a seedbed and planting a crop, sustaining soil fertility, managing irrigation, collecting and disposing of manure, controlling pests, and preventing animal diseases.

A *farming system* is the overall approach used in crop or livestock production, often derived from a farmer's goals, values, knowledge, available technologies, and economic opportunities. A farming system influences, and is in turn defined by, the choice of methods and practices used to produce a crop or care for animals.

In practice, farmers are constantly adjusting cropping systems in an effort to improve a farm's performance. Changes in management practices generally lead to a complex set of results—some positive, others negative—all of which occur over different time scales.

The transition to more sustainable agriculture systems may, for many farmers, require some short-term sacrifices in economic performance in order to prepare the physical resource and biological ecosystem base needed for long-term improvement in both economic and environmental performance. As a result, some say that practices essential to progress toward sustainable agriculture are not economically viable and are unlikely to take hold on the farm (Marten, 1989). Their contention may prove correct, given current farm policies and the contemporary inclination to accept contemporary, short-term economic challenges as inviolate. Nonetheless, one question lingers: What is the alternative to sustainable agriculture?

## PUBLIC POLICY AND RESEARCH IN
## SUSTAINABLE AGRICULTURE

Farmers, conservationists, consumers, and political leaders share an intense interest in the sustainability of agricultural production systems. This interest is heightened by growing recognition of the successes achieved by innovative farmers across the country who are discovering alternative agriculture practices and methods that improve a farm's economic and environmental performance. Ongoing experimental efforts on the farm, by no means universally successful, are being subjected to rigorous scientific investigation. New insights should help farmers become even more effective stewards of natural resources and produce food that is consistently free of man-made or natural contaminants that may pose health risks.

The major challenge for U.S. agriculture in the 1990s will be to strike a balance between near-term economic performance and long-term ecological and food safety imperatives. As recommended in *Alternative Agriculture* (National Research Council, 1989a), public policies in the 1990s should, at a minimum, no longer penalize farmers who are committed to resource protection or those who are trying to make progress toward sustainability. Sustainability will always remain a goal to strive toward, and alternative agriculture systems will continuously evolve as a means to this end. Policy can and must play an integral role in this process.

If sustainability emerges as a principal farm and environmental policy goal, the design and assessment of agricultural policies will become more complex. Trade-offs, and hence choices, will become more explicit between near-term economic performance and enhancement of the long-term biological and physical factors that can contribute to soil and water resource productivity.

Drawing on expertise in several disciplines, policy analysts will be compelled to assess more insightfully the complex interactions that link a farm's economic, ecological, and environmental performance. It is hoped that political leaders will, as a result, recognize the importance of unraveling conflicts among policy goals and more aggressively seizing opportunities to advance the productivity and sustainability of U.S. agriculture.

A few examples may help clarify how adopting the concept of sustainability as a policy goal complicates the identification of cause-and-effect relationships and, hence, the design of remedial policies.

When a farmer is pushed toward bankruptcy by falling crop prices, a farm operation can become financially unsustainable. When crop losses mount because of pest pressure or a lack of soil nutrients, however, the farming system still becomes unsustainable financially, but for a different reason. In the former example, economic forces beyond any individual farmer's control are the clear cause; in the latter case the underlying cause is rooted in the biological management and performance of the farming system.

The biological and economic performance of a farming system can, in turn, unravel for several different reasons. Consider an example involving a particular farm that is enrolled each year in the U.S. Department of Agriculture's commodity price support programs. To maintain eligibility for government subsidies on a continuing basis, the farmer understands the importance of growing a certain minimum (base) acreage of the same crop each year. Hence, the cropping pattern on this farm is likely to lead to a buildup in soilborne pathogens that attack plant roots and reduce yields. As a result, the farmer might resort to the use of a fumigant to control the pathogens, but the pesticide might become ineffective because of steadily worsening microbial degradation of the fumigant, or a pesticide-resistant pathogen may emerge.

A solution to these new problems might be to speed up the registration of another pesticide that could be used, or relax regulatory standards so more new products can get registered, or both. Consider another possibility. A regulatory agency may cancel use of a fumigant a farmer has been relying upon because of food safety, water quality, or concerns about it effect on wildlife. The farmer might then seek a change in grading standards or an increase in commodity prices or program benefits if alternative pesticides are more costly.

Each of these problems is distinctive when viewed in isolation and could be attacked through a number of changes in policy. The most cost-effective solution, however, will prove elusive unless the biology of the whole system is perceptively evaluated. For this reason, in the policy arena, just as on the farm, it is critical to know what the problem is that warrants intervention and what the root causes of the problem really are.

## Research Challenges

In thinking through agricultural research priorities, it should be acknowledged that the crossroads where the sciences of agriculture and ecology meet remain largely undefined, yet clearly promising. There is too little information to specify in detail the features of a truly sustainable agriculture system, yet there is enough information to recognize the merit in striving toward sustainability in a more systematic way.

The capacity of current research programs and institutions to carry out such work is suspect (see *Investing in Research* [National Research Council, 1989b]). It also remains uncertain whether current policies and programs that were designed in the 1930s or earlier to serve a different set of farmer needs can effectively bring about the types of changes needed to improve ecological management on the modern farm.

In the 1980s, the research community reached consensus on the diagnosis of many of agriculture's contemporary ills; it may take most of the 1990s to agree on cures, and it will take at least another decade to get them into place. Those who are eager for a quick fix or who are just impatient are bound to be chronically frustrated by the slow rate of change.

Another important caution deserves emphasis. The "silver bullet" approach to solving agricultural production problems offers little promise for providing an understanding of the ecological and biological bases of sustainable agriculture. The one-on-one syndrome seeks to discover a new pesticide for each pest, a new plant variety when a new strain of rust evolves, or a new nitrogen management method when nitrate contamination of drinking water becomes a pressing social concern. This reductionist approach reflects the inclination in the past to focus scientific and technological attention on products and outcomes rather than processes and on overcoming symptoms rather than eliminating causes. This must be changed if research aimed at making agriculture more sustainable is to move ahead at the rate possible given the new tools available to agricultural scientists.

One area of research in particular—biotechnology—will benefit from a shift in focus toward understanding the biology and ecology underlying agricultural systems. Biotechnology research tools make possible powerful new approaches in unraveling biological interactions and other natural processes at the molecular and cellular levels, thus shedding vital new light on ecological interactions with a degree of precision previously unimagined in the biological sciences. However, rather than using these new tools to advance knowledge about the functioning of systems as a first order of priority, emphasis is increasingly placed on discovering products to solve specific production problems or elucidating the mode of action of specific products.

This is regrettable for several reasons. A chance to decipher the physiological basis of sustainable agriculture systems is being put off. The payoff from focusing on products is also likely to be disappointing. The current widespread pattern of failure and consolidation within the agricultural biotechnology industry suggests that biotechnology is not yet mature enough as a science to reliably discover, refine, and commercialize product-based technologies. Products from biotechnology are inevitable, but a necessary first step must be to generate more in-depth understanding of biological processes, cycles, and interactions.

Perhaps the greatest potential of biotechnology lies in the design and on-farm application of more efficient, stable, and profitable cropping and livestock management systems. For farmers to use such systems successfully, they will need access to a range of new information and diagnostic and analytical techniques that can be used on a real-time basis to make agronomic and animal husbandry judgments about how to optimize the efficiencies of the processes and interactions that underlie plant and animal growth.

Knowledge, in combination with both conventional and novel inputs, will be deployed much more systematically to avoid soil nutrient or animal nutrition-related limits on growth; to ensure that diseases and pests do not become serious enough to warrant the excessive use of costly or hazardous pesticides; to increase the realistically attainable annual level of energy flows independent of purchased inputs within agroecosystems; and to maximize a range of functional symbiotic relationships between soil micro-

and macrofauna, plants, and animals. Discrete goals will include pathogen-suppressive soils, enhanced rotation effects, pest suppression by populations of plant-associated microorganisms, nutrient cycling and renewal, the optimization of general resistance mechanisms in plants by cultural practices, and much more effective soil and water conservation systems that benefit from changes in the stability of soil aggregates and the capacity of soils to absorb and hold moisture.

Because of the profound changes needed to create and instill this new knowledge and skills on the farm, the recommendations in *Alternative Agriculture* (National Research Council, 1989a) emphasize the need to expand systems-based applied research, on-farm experimentation utilizing farmers as research collaborators, and novel extension education strategies—the very goals of the U.S. Department of Agriculture's LISA program.

Future research efforts—and not just those funded through LISA—should place a premium on the application of ecological principles in the multidisciplinary study of farming system performance. A diversity of approaches in researching and designing innovative farming systems will ensure broad-based progress, particularly if farmers are actively engaged in the research enterprise.

## REFERENCES

Benbrook, C., and J. Cook. 1990. Striving toward sustainability: A framework to guide on-farm innovation, research, and policy analysis. Speech presented at the 1990 Pacific Northwest Symposium on Sustainable Agriculture, March 2.

Marten, J. 1989. Commentary: Will low-input rotations sustain your income? Farm Journal, Dec. 6.

National Research Council. 1989a. Alternative Agriculture. Washington, D.C.: National Academy Press.

National Research Council. 1989b. Investing in Research: A Proposal to Strengthen the Agricultural, Food, and Environmental System. Washington, D.C.: National Academy Press.

# PART ONE
## Overview

# 1

# The U.S. Department of Agriculture Commitment to Sustainable Agriculture

*Charles E. Hess*

Significant progress has been made in the past 5 years in the acceptance of the concept of sustainable agriculture. The U.S. Department of Agriculture (USDA) low-input sustainable agriculture (LISA) programs, the Leopold Center at Iowa State University (Ames), a long-term ecological research program at Michigan State University (East Lansing), and a state-wide sustainable agriculture program in California are examples. Michael Jacobson, executive director of the Center for Science in the Public Interest, recognized one aspect of progress when he said, "Even USDA is uttering the 'O' word [organic] and not choking."

Overall, today's agriculture is being challenged to operate in an environmentally responsible fashion while at the same time continuing to produce abundant supplies of food and fiber both economically and profitably. The scientific community is responding positively and assertively to the challenge.

There is increasing interest in the development and adoption of sustainable land use systems for two very basic reasons: (1) a need to bring about fundamental improvements in the global environment, and (2) an ever-expanding need to provide economically produced food and fiber for a growing world population.

Through technology, the United States has developed an efficient, highly productive food and fiber system that is the envy of the world. Of all the people in the world, consumers in the United States currently spend the lowest percentage of their incomes on food—an incredible 11.8 percent. It is now recognized, however, that current technology has had some costs that were not fully anticipated at the time of its introduction. Scientists are looking more closely at its possible social, environmental, and health impacts. Clearly, the issues—both perceived and real—that are raised by current technology must be addressed.

## THE USDA APPROACH TO SUSTAINABLE AGRICULTURE

The term *sustainable agriculture* means different things to different people. The term itself is not important. What is important is that farmers around the country are closing their conventional farming cookbooks and carefully crafting new recipes for what might be called "smart and considerate farming." Rather than providing yet another definition, this chapter provides a look at the approach used at USDA.

It is the department's responsibility to provide farmers with a range of options that can best fit their economic and environmental situations. The choices range from the optimal use of fertilizers, pesticides, and other off-farm purchases in conjunction with the best management practices, to operations that actively seek to minimize their off-farm purchases and emphasize crop rotation, integration of livestock and crop production, and mechanical or biological weed control. The thing that they have in common is integrated resource management—a systems management approach that looks at the farm as a whole.

To some, this seems a return to the 1930s and "low-tech" production methods. This is not at all the basis of sustainable agriculture, however. It does not mean a return to hoes, hard labor, and low output.

*Low input* is not an exactly appropriate term because it carries the wrong connotation, that something can be achieved for nothing. In fact, the preferred designation is *sustainable agriculture*. This means the use of the very best technology in a balanced, well-managed, and environmentally responsible system. It relies on skilled management, scientific know-how, and on-farm resources.

It should be stressed again that the emphasis is not to eliminate the use of important chemicals and fertilizers. In many instances, such chemicals and fertilizers are absolutely necessary to the farmer. The emphasis is, however, to seek ways to reduce their use and increase their effectiveness to improve and maintain environmental and economic sustainability.

The appropriate measure of a system's productivity and efficiency is not how much it produces but, rather, the relative value of what it produces compared with what went into producing it. Environmental impacts must now be included in the cost-benefit equation; this has not always been considered. Contributions will be needed from all the agricultural sciences to develop sustainability models with sound management practices and techniques for food and fiber production systems.

## The Road Ahead

It must be made absolutely clear that those involved in the U.S. agriculture system care about the environment. It is one of USDA's top priorities, and this is certainly evident in the proposals in the 1990 farm bill and the 1991 federal budget (both of which are discussed below).

Agriculture has always tried to be a careful steward of the nation's land and water resources, but that effort is now receiving renewed emphasis. For example, an excellent summary of data, case studies, and recommendations was presented in *Alternative Agriculture* (National Research Council, 1989a), which has received a great deal of attention.

Since its publication, many people have commended the National Research Council for producing such a comprehensive assessment at such a critical time. Other readers, however, say that it overstates the economic feasibility and the benefits of adopting alternative agriculture practices.

The principles laid out in that report are well worth thoughtful study and can point the way to change. A recent issue of *Chemical and Engineering News* (March 5, 1990) contains a good analysis of the issues involved, and the Board on Agriculture of the National Research Council will soon provide a response to some of the comments that *Alternative Agriculture* has generated.

In fact, some of the reactions miss the point of that report. It was never intended to prove that one kind of agriculture is superior to another but, rather, to help provide an understanding of the kinds of agriculture systems being used on U.S. farms and ranches and to encourage research not only to determine the most environmentally and economically beneficial kinds of farming but to guide development as well. The current scarcity of hard evidence on either side of the issue can only invite unfounded and unhelpful assertions.

## The Need for Hard Data

There must be an effort to gain more hard data so that informed decisions can be made based on science rather than on emotion.

It is human nature to want to know everything without having to wait for it. People want to know immediately what does and does not work and why. These kinds of questions take time to answer, and time is needed to gather the evidence that will eventually lead to conclusions.

## CONSTRUCTIVE APPROACHES THAT ARE UNDER WAY

The following are six concrete examples of current research.

• Under the President's Initiative on Water Quality, research will help to provide a better sense of real versus perceived progress on the issue of water quality. This initiative will determine what agricultural practices adversely affect water quality and then develop alternatives to them. The Cooperative Extension Service and the Soil Conservation Service will extend the existing knowledge of the best management practices.

• On February 9, 1990, USDA announced the establishment of eight water quality demonstration projects to show new ways to minimize the effects of agricultural nutrients and pesticides on water quality. The Soil Conservation Service and Extension Service will provide joint leadership for the on-farm demonstration projects. Five USDA agencies have committed $3.3 million to the projects in 1990. The projects are located in California, Florida, Maryland, Minnesota, Nebraska, North Carolina, Texas, and Wisconsin.

• In the 1990 field season, the Agricultural Stabilization and Conservation Service will test a cost-sharing program for reducing chemical use. The trial program is designed to encourage the adoption of integrated pest and fertilizer management practices. It will be limited to 20 farms in each of five counties per state in all 50 states. Participants must enroll at least 40 acres of small grains, forage, hay, or row crops and follow a written integrated crop management plan that seeks to reduce pesticide or fertilizer use by at least 20 percent.

• Research in integrated pest management will also be continued. Integrated pest management is the study of biological controls and management practices that aid in the more precise use of pesticides and in judicious reductions in the amounts that are used. The goal is to avoid adverse effects on the environment and beneficial organisms. Yet, at the same time, care must be taken so that, in the enthusiasm to remove toxic compounds, conditions are not created in which naturally occurring toxic substances (such as aflatoxins) are able to increase.

• In July 1989, R. Dean Plowman, administrator of USDA's Agricultural Research Service, and I participated in the dedication of a new $11.9 million soil tilth laboratory on the Iowa State University campus. This laboratory will study the effects of a variety of agricultural practices on soil structure, organic matter, microorganisms, and movement of nutrients.

• The Alternative Farming Systems Information Center at the National Agricultural Library is another way that the transfer of knowledge is being increased. As part of the team working with sustainable agriculture, this information center focuses human expertise on the specialized subject area of sustainable agriculture. This center inventories and coordinates data from many sources and plays an important role in meeting the information needs of researchers and producers.

## The Roles of Universities and Farmers

Universities will play a vital role in the future of sustainable agriculture.

In the endeavor to create management systems that combine knowledge from a variety of areas, universities will want to create internal mechanisms to facilitate multidisciplinary approaches to research. It takes cooperative interactions among members of many disciplines for the development of stable systems.

The widespread awareness of the need for economical and environmentally sound ways of farming has not always been matched by the availability of reliable and practical information on what, in fact, can be done. Innovative farmers and researchers have generated considerable new information, but it has not always been shared with and tested by others to the extent that it should. Extension certainly has an historic and very current role in meeting this need.

A group called the Practical Farmers of Iowa (PFI) has as its goal "profitable and environmentally sound farming—pure and simple. It's got to sustain the land, the soil, the people, the communities, and the pocket-book." This group places strong emphasis on action. Its members are involved in a number of demonstration projects that pair customary practices with alternative methods.

For instance, ridge-till farmers have compared chemical weed control with nonchemical weed control in soybean and corn demonstration projects. In 11 soybean field trials in 1989, participating PFI farmers applied no herbicides and substituted nonchemical weed control such as cultivation. They saved an average of $11.12 an acre on cultivation and labor costs as well as on the cost of the herbicides, which had already been reduced to a small, economical level. In five corn crop trials that same year, PFI farmers saved $7.00 an acre by using little or no herbicides. Yields were not affected in either case.

New ways of sharing such information must continue to be examined. Every ounce of careful management and efficient technology that can be mustered is needed to continue to maintain competitiveness in a tough global marketplace and, at the same time, to have an environmentally sensitive agriculture system.

## THE NEED FOR A PROACTIVE EFFORT

It is essential that policymakers, researchers, and farmers join together to take an assertive, proactive approach in dealing with environmental issues. To say that there are no problems or that public concern is completely the product of misinformation is not a productive approach, neither for the future of agriculture nor for the restoration of public confidence.

The public is growing more and more concerned about the impact of agriculture on the environment, particularly its potential effect on water quality. There are recent data that give some credence to that fear. A U.S.

Geological Survey report (1989) showed that in a sampling of surface water in 10 midwestern states, 90 percent of the samples showed the presence of some agricultural chemicals.

The issue is not limited to the United States. In England, there are suits pending against water companies citing the high levels of nitrogen in drinking water, and legislation is being proposed that would regulate the amount of fertilizer an English farmer can use. The legislation proposes that the amount be based on the nitrate content of the region's well water.

If such restrictive legislation is to be avoided in the United States, a positive response to these issues must be made. It is time to be proactive rather than defensive. To do otherwise is to invite legislation and regulation that may remove farmers' decision-making powers and constrain their flexibility in adapting management practices that best fit each farming situation.

## LEGISLATIVE AND FUNDING INITIATIVES

### Farm Bill

Emerging environmental concerns were strongly reflected in the Food Security Act of 1985. I predict that they will be even more strongly present in the current debates over farm legislation. The proposals of the farm bill should be mentioned here because one of its three basic goals is to deal with environmental concerns. The administration is seeking an assertive role in shaping the nation's sustainable agriculture policy in the years to come. The 1990 farm bill will go far in this direction.

The bill proposes the enhancement of resource stewardship of U.S. farmers by giving them greater flexibility in their planting, crop use, crop rotation, and marketing; incentives to change their resource use in environmentally sensitive areas; and lastly, greater research and technical assistance—especially in farming in an environmentally aware way.

In the 1990 farm bill, the administration encourages changes in commodity programs to ensure that the farmers who participate in those programs will not be penalized for adopting sustainable agriculture practices. Currently, commodity programs reward farmers for growing as much of the program crop as they can on their eligible base acres. They would lose that base, and, therefore, future price support payments, if they used it to grow rotation crops, even though those crops could increase environmental and economic sustainability. The administration's flexibility proposals would allow farmers to incorporate rotation crops without having to make that sacrifice.

### Request for LISA Funding

In addition, the administration has requested $4.45 million for USDA's

LISA research and education program in 1991. Furthermore, if Congress funds the proposed $100 million Initiative for Research on Agriculture, Food, and the Environment (see National Research Council, 1989b), another $1 million would be expected to be added to USDA's support for sustainable agriculture research.

LISA is highly favored by some because it provides opportunities for users of the research to have direct input into the decision-making process of selecting the projects that should be funded. Unfortunately, it has sometimes engendered skepticism as well as enthusiastic support—in part, because it differs from traditional research and education. For example, some people are suspicious of the results of studies that put farmers and others in the middle of the research and education process.

The purpose of USDA's LISA program is to help develop and disseminate to farmers practical, reliable information on sustainable farming practices. Now in its third year, the program has supported up to 90 projects ranging from experimental research to the development of educational materials. Most of the projects reviewed in this volume have been funded partially by the LISA program.

The benefits of this effort include more than information for farmers. The program is a catalyst. It is helping to stimulate sustainable agriculture research and education in many universities and other research organizations.

The LISA program is just a start, however. For one thing, it is currently limited to farm-level research and education. As noted by the National Research Council (1989b), very little research is being done on what implications the adoption of sustainable agriculture might have for the structure of agriculture, environmental quality, and rural communities, as well as for national and global food production.

This is not to say that people should ignore the question: How can the world have a clean environment and enough to eat? The pervasive negativism is that the world cannot have both and, therefore, that LISA is a false hope. Regrettably, that conclusion overlooks the other side of the equation, namely, what will be the outcome for future generations if continued reliance is placed on highly specialized, capital-intensive, chemical-intensive ways of farming? Thoughtful research is needed on this fundamental issue.

## Commitment to Research

The ability to offer the farmer a broad range of practices and to tap the full potential of technology depends on a reservoir of knowledge in the basic sciences essential to agriculture. Fortunately, Secretary of Agriculture Clayton Yeutter has a deep appreciation for the role of research, and

President Bush has indicated that "investing in the future for a better America" is one of his major commitments.

This commitment was obvious when the president presented his budget to the Congress. He announced a $100 million Initiative for Research on Agriculture, Food and the Environment, with $50 million to be added annually in the subsequent years to reach at least $300 million, and possibly $500 million. This USDA initiative is based on the National Research Council report *Investing in Agriculture: A Proposal to Strengthen the Agricultural, Food, and Environmental System* (National Research Council, 1989b).

The president has proposed the following levels of funding for the first year in four major areas: plant systems, $50 million; animal systems, $30 million; natural resources and the environment, $15 million; and nutrition, food quality, and health, $5 million. The $15 million for natural resources and the environment does not seem like a significant investment, but nearly one-third of the USDA science and education budget of $1.3 billion is related to the environment.

Also, other areas in the budget are directly related to the goal of a sustainable agriculture system. For example, $15 million is proposed for the plant genome study. The goal is to determine those genes that regulate agriculturally important traits, such as disease and insect resistance.

There is also a department-wide water quality initiative with proposed funding of $207 million (an increase of $52 million) and a global change initiative with funding of $47.4 million (up from $21.2 million in 1990).

The first challenge was to get the initiative into the budget. Now, perhaps an even tougher challenge is to get it through the Congress. This is where the administration needs help and support. In addition, the Office of Management and Budget is watching carefully to see whether the administration can get the initiative passed by the Congress unimpaired by earmarking of funds for special interest purposes. It is clearly in the best interest of everyone to resist the urge to carve up the initiative. It would be killing the goose that laid the golden egg.

## CONCLUSION

The quest for agricultural sustainability in the United States and abroad bears more than a casual resemblance to the astonishing events that have been taking place in Eastern Europe, the Soviet Union, and elsewhere around the world. Both phenomena have caught some by surprise but have captured the imagination of everyone.

The similarities do not stop there. The pursuit of an environmentally and economically sustainable agriculture system, no less than the drive for freedom, involves a deep questioning of the status quo and an intense commitment to help build a better world in the future. Both the search for sustainability and the parting of the Iron Curtain have been brewing for decades, and they are now bursting forth.

I applaud the participants of the workshop on which this volume is based for making an effort to tell others about the important work that is being done and to join in learning about the information provided by this research.

Agriculture in the United States is facing major challenges, some of which may appear to be in conflict. On one hand, agriculture needs to be highly efficient and internationally competitive in order to be economically viable. On the other hand, it needs a system of production that is environmentally sensitive and sustainable and whose products are viewed as safe. Both goals are achievable.

Sustainable agriculture is a direction that makes remarkable sense for farmers and for the rest of U.S. society. It is a direction that must be faced with a spirit of openness and willingness to change for the better.

## REFERENCES

National Research Council. 1989a. Alternative Agriculture. Washington, D.C.: National Academy Press.

National Research Council. 1989b. Investing in Research: A Proposal to Strengthen the Agricultural, Food, and Environmental System. Washington, D.C.: National Academy Press.

U.S. Geological Survey. 1989. Reconnaissance for triazine herbicides in surface waters in agricultural areas of the upper midwestern United States. October 26. Unpublished.

# 2

# Background and Status of the Low-Input Sustainable Agriculture Program

*Neill Schaller*

The roots of the low-input sustainable agriculture (LISA) program go far back in time. It is the product of growing concerns of the public and farmers over unforseen high costs of conventional agriculture. Indeed, the highly specialized, capital-intensive, and chemical-intensive conventional farming methods, while boosting farm output to higher and higher levels, have had a myriad of adverse side effects on natural resources, environmental quality, human health, and food quality and safety.

LISA is but one of several names used to describe a form of agriculture that will not only be productive and profitable for generations to come but will also conserve resources, protect the environment, and enhance the health and safety of the citizenry. Other versions of the same ideal or different paths to it, are known by names such as organic, regenerative, biological, ecological, biodynamic, sustainable, low-input, reduced-input, and alternative agriculture.

## BACKGROUND

"Low input" was added to LISA's name after the LISA program was first authorized to give more meaning to the rather general connotation of sustainability and to head off a possible misinterpretation that the real purpose of the program was eventually to eliminate the use of all purchased agricultural chemicals on farms and ranches as an end in itself. Low input had the advantage that it could include, but not be limited to, the chemical-free path to sustainability. The relevant question implied here is which path is the most environmentally and economically sustainable, not just which path allows farmers to use fewer chemicals.

A discussion of LISA might not have occurred if it were not for important turning points in the support for agricultural sustainability. One such turning point was the farm financial crisis of the 1980s. It has long been known that conventional farming might not be environmentally sustainable. It took the financial tragedy that struck farm families early in the 1980s, however, which was the result of declining exports of U.S. farm products and plunging farmland values, to show that U.S. agriculture might be economically unsustainable as well.

Farmers who survived that economic crisis saw the urgency of farming in ways that would lower production costs and debts. Their discovery helps to explain why so much of the search for a more sustainable agriculture system has focused on reducing the use of purchased chemical fertilizers and pesticides. Not only were those inputs known to be potentially harmful to the environment, but farmers' extreme dependence on them has also been seen as weakening agriculture's economic sustainability.

Another development fostering interest in agricultural sustainability has been the growing realization among environmentalists that (1) environmentally sound farming practices must be profitable if farmers are going to adopt them, and (2) it might not always be essential—or even sustainable—to try to eliminate all synthetic chemicals from farming practices, as was assumed by some of the environmentally concerned critics of conventional agriculture.

## LEGISLATION

Despite the important developments described above, in the early 1980s the U.S. Congress failed to pass legislation supporting sustainable agriculture research and education. Two reasons for this stand out. First, the initial attempts emphasized or were identified with organic farming, which lacked wide support. Some in the agriculture community saw organic farming only as a way to meet a special market niche. Most considered it, incorrectly, as a "cause" rather than as an alternative that had the potential to equal the productivity and profitability of conventional agriculture.

Another possible reason for failure to enact legislation was that early proposals supporting sustainable farming research and education included the establishment of new centers or procedures that departed from and even threatened the traditional structure and practices of the land-grant system and the U.S. Department of Agriculture (USDA).

Because the reasons for interest in a more sustainable agriculture system persisted, however, it was only a matter of time before supportive policies would be enacted (see Table 2-1). That happened in 1985 with passage of

**TABLE 2-1** LISA Program History

| Event | Fiscal Year | | | | | | |
|---|---|---|---|---|---|---|---|
| | 1985 | 1986 | 1987 | 1988 | 1989 | 1990 | 1991 |
| Congress Legislation | Food Security Act of 1985 authorizes program | | | | | To be re-authorized in farm bill | |
| Appropriation | None | None | None | $3.9 million | $4.45 million | $4.45 million | To be decided |
| U.S. Department of Agriculture Policy and programs | | | | Starts LISA program | | Supports sustainable agriculture | |
| Funding recommended | None | None | None | None | None | None | $4.45 million |
| Number of full proposals reviewed | | | | 371 | 318 | 161 | |
| Number of LISA projects funded (%) | | | | 49 (13) | 56 (18) | 45 (28) | |

the Food Security Act (P.L. 99-198), which authorized what is now called the LISA research and education program.

The LISA program might not be recognized by the name of the authorizing section, "Agricultural Productivity Research," Subtitle C of Title XIV on research and education (U.S. Government Printing Office, 1985). Many members of Congress and the agricultural community were, and still are, more comfortable with the term *productivity* than with the term *LISA*.

Subtitle C discussed farmers' need for sound information on alternative production systems that would not only enhance productivity but also reduce soil erosion, conserve energy, and protect the environment. It referred indirectly to the negative side effects of farming practices that rely heavily on purchased chemicals and other inputs. It called for an inventory of studies and told USDA to conduct research projects on alternative farming systems. Farmers, it said, should be involved in those studies, to be sure that the results would be useful to them. The research should also be open to people from all interested universities and private organizations.

The LISA program remained on the drawing board through 1986 and 1987. That was because Congress did not appropriate funds to support a new program, and USDA chose not to redirect existing research funds to get one started. In response to the 1985 act, however, USDA formed a task force on alternative farming systems and began to inventory existing research.

The launching of the LISA program finally came in January 1988, after Congress passed an agriculture appropriations bill for 1988 with funds earmarked for such a program.

The House Appropriations Committee report introduced the words *low-input farming*. It said, "The term 'low-input' is used to describe the implementation of alternative systems to lower costs of production" (U.S. Congress, House, 1988). The report then discussed the low-input research that was under way and the fact that the National Agricultural Library had established an Alternative Farming Systems Information Center. The committee recommended $2.6 million for low-input farming research and education, including $1 million for research, $1.5 million for extension education, and $100,000 for the National Agricultural Library.

The Senate committee was more generous. It recommended $9 million, of which $2.1 million was to be used to fund a companion program known as ATTRA that supported appropriate technology transfer for rural areas. The Senate language used *reduced input* instead of *low-input*. It said, "A growing number of farmers are now looking for reliable information on reduced input farming systems. These farmers are interested in learning how alternative farming methods can be used to reduce production costs and control soil erosion and the pollution of underground water supplies caused, in part, by heavy fertilization, pesticide use, and monocultural cropping systems" (U.S. Congress, Senate, 1988).

When the House and Senate appropriations committees ironed out their differences, the result was a $3.9 million budget for a LISA competitive grants program, including $100,000 for the National Agricultural Library's information center. Instead of sending separate funds to USDA research and extension agencies, the conferees agreed that the program should be administered through USDA's Cooperative State Research Service (CSRS), with clear instructions that CSRS involve extension.

## LISA PROGRAM ENACTMENT

Immediately after the bill was signed into law in late December 1987, the process of designing the LISA program began in earnest under the direction of Paul O'Connell, CSRS deputy administrator. Many of the participants at the workshop on which this volume is based and others associated with universities and private organizations played key roles in that process. The USDA's Alternative Farming Systems task force, which was elevated to a USDA subcommittee, also helped. It drafted a departmental memorandum that defined and established support for research and education on alternative farming systems. The draft became an official policy statement that was issued by then Secretary Richard Lyng in January 1988 (U.S. Department of Agriculture, 1988).

The decision was made to invite, review, and approve LISA project proposals mainly at the regional level. To do that, host institutions were selected in each of the four U.S. regions: the University of Vermont, Burlington (Northeast); the University of Georgia, Athens (South); the University of Nebraska, Lincoln (North Central); and the University of California, Oakland (West).

Coordinators were named to guide the program in each region. The current coordinators are Neil Pelsue, Jr. (Northeast), Charles Laughlin (South), Steven Waller (North Central), and David Schlegel (West).

Patrick Madden, who was then on leave from Pennsylvania State University, worked closely with Paul O'Connell (USDA) to develop program guidelines as well as criteria and procedures for inviting, reviewing, and approving LISA project proposals. Technical committees were established in each region, along with a smaller administrative council to act as a regional board of directors and decision-making body.

Each region was given the same amount of money, after setting aside limited funding to support conferences and other activities at the national level. An example is the Farm Decision Support System, a computer-assisted system to help farmers choose their best farming options. This is described by its designer, John Ikerd of the University of Missouri, Columbia, in Appendix A of this volume.

The LISA program was well under way by the middle of the summer of

1988. The early start encouraged the Congress to increase the program's budget to $4.45 million in 1989 and to keep the funding at that level in 1990. Meanwhile, the administration, which did not recommend funding for LISA in 1988, 1989, or 1990, has included a request for $4.45 million in its 1991 budget proposals, which, as of this writing, are being debated in the Congress. In addition, Charles Hess, Assistant Secretary for Science and Education (see the chapter "The U.S. Department of Agriculture Commitment to Sustainable Agriculture," this volume), points out that if Congress appropriates the $100 million requested by the administration for the USDA's proposed Initiative for Research on Agriculture, Food, and the Enviornment, $1 million of the new funding would be added to the funding available for sustainable agriculture research.

## LISA RESEARCH AND EDUCATION

Ongoing food and agriculture research and LISA research are not the same things, but they need each other. LISA research is applied research on alternative low-input sustainable agriculture practices and their feasibility. LISA projects deal with, or are at least oriented to the concept of, a total farm system. In contrast, much of the ongoing production-related research is focused on single practices or relationships involved in farming—for example, on the effects of different levels of a single input or technology on the yield of a particular crop. Often, the latter research can make an important contribution to the development and testing of knowledge about sustainable agriculture practices. The results of LISA research, in turn, can increase the payoff from ongoing research by identifying the component problems that it should address.

The following is a breakdown of the LISA program's record (Madden et al., 1990):

• In the first year (1988), a total of 371 project proposals were received. Of those, 49 (13 percent) were selected to receive funding. Other competitive grant research programs administered by USDA now fund an average of 20 percent of the proposals received. The 20 percent figure is also fairly typical for other federal and nonprofit research organizations.

• In an effort to reduce the disappointment level associated with this very low acceptance rate, the Western region LISA program called for abbreviated preproposals in 1989. Of a total of 143 preproposals submitted, 32 were selected for development of full proposals. Of 30 proposals received and reviewed, 11 were funded.

• For the United States as a whole, 318 full proposals were evaluated in 1989; 56 (18 percent) were approved, including 27 renewal projects that were first funded in 1988 for which continued support was requested and 29 new projects.

• In 1990, the Southern and North Central regions called for abbreviated preproposals. A total of 438 were submitted; 91 of these preproposals were selected for development into full proposals, and 86 proposals were submitted. Twenty-one of these were funded: 12 in the Southern region (including 11 new projects) and 9 in the North Central region (5 new projects).

• In the Northeast region, 67 proposals were reviewed and 16 were funded, including 7 new projects.

• The Western region did not invite new proposals in 1990. They had agreed in 1989 to continue supporting previously approved projects, if their progress was satisfactory, leaving insufficient funds for new projects.

• Therefore, 28 percent of the full proposals reviewed in 1990 were funded.

Without a higher level of funding, the LISA program will be able to support either a limited number of proposals for 1 or 2 years or an even smaller number of proposals for several years. Neither choice is satisfactory. The need to provide support for more than 1 or 2 years is especially critical for projects dealing with the feasibility of expanded rotations and other practice changes that extend over many years. The need to assist more of the prospective LISA research teams that are interested in and able to shed light on low-input sustainable practices is just as compelling. Moreover, a relatively high proportion of LISA proposals is not likely to find support elsewhere.

## QUESTIONS AND ANSWERS ABOUT THE LISA PROGRAM

The following are some of the questions people ask about the LISA program and the current answers:

*Does every state have a LISA project?*
Almost. Only four states cannot claim a principal coordinator or major participant in a LISA-supported project.

*To what extent are farmers really participating in the program?*
About 1,183 farmers are now involved as participants in LISA-funded projects. Of these, 521 have generated ideas for projects, and 155 have provided land for experimentation and other inputs (Table 2-2). Of a total of 74 people who are members of the regional technical committess, 19 are farmers. Eight farmers are serving on the regional administrative councils, which have a total of 39 members.

*How important is the participation of private organizations?*
Nineteen of the projects now supported have principal participants who are with private organizations. In the 2-year period from 1988 to 1989, some 20 institutes, associations, and other private organizations have added

**TABLE 2-2** Participation of Farmers in LISA Projects in Each Region and the United States

| Farmers' Roles | Number of Farmers Serving This Role (by region) | | | | |
|---|---|---|---|---|---|
| | Northeast | Central | North Southern | Western | United States* |
| Provide land for | | | | | |
|   Replicated experiments | 50 | 42 | 33 | 30 | 155 |
|   Unreplicated studies | 98 | 33 | 45 | 151 | 327 |
|   Demonstration plots | 44 | 44 | 38 | 27 | 153 |
| Generate ideas | | | | | |
|   for project | 118 | 129 | 67 | 202 | 521 |
| Help to manage project | 67 | 97 | 50 | 45 | 259 |
| Evaluate project | 138 | 110 | 77 | 167 | 497 |
| Provide information on | | | | | |
|   Yields | 188 | 169 | 49 | 358 | 767 |
|   Costs and returns | 133 | 158 | 49 | 253 | 596 |
|   Labor requirements | 85 | 74 | 41 | 320 | 523 |
| Presenter at conference | 27 | 60 | 38 | 96 | 221 |
| Total, one or more roles | 217 | 223 | 86 | 652 | 1,183 |

*Includes one national project with five farmers (the Farm Decision Support System).

SOURCE: Survey of LISA project coordinators, conducted by Madden Associates Inc., Glendale, Calif., May 1990.

over $900,000 in matching funds to the $750,000 in funding provided to them by the program.

*How much support are LISA projects actually receiving from the program?*

The average amount of support per project per year is $40,000. The project awarded the largest support so far received $220,000 for 2 years. At the other extreme, a project to develop educational displays requested and received $4,000.

*The LISA program has a matching fund requirement. How much have participating organizations contributed to LISA projects?*

In 1988, they added $3.6 million. Matching funds were $3 million in 1989 and $5.1 million in 1990. Each year private organizations contribute over $500,000 in matching resources as participants in LISA projects.

*What proportion of LISA projects deals primarily with organic farming?*

So far, only about 4 percent.

## THE FUTURE

The future status of the LISA program rests to a great extent on the outcome of the 1990 farm bill debate, which was under way at the time of this writing, as well as on annual appropriations. The current authorization ends with expiration of the Food Security Act of 1985. Different reauthorization language has been suggested by various interest groups and USDA.

Several versions now label the proposed legislation, "Sustainable Agriculture Research and Education," in part because the term *low-input* continues to cause confusion and misunderstanding. Most of the proposed language reaffirms the purpose of the current LISA program and its structure and procedures. Some extend the program's scope—for example, to support research on the social and economic implications of widespread adoption of LISA, in addition to providing farmers with practical information on sustainable agriculture practices. Several versions would authorize an increased funding level from $40 million to $50 million.

The year 1990 is surely critical to the life of LISA. From all indications, however, sustainable agriculture research and education could be well on the way to becoming mainstream agriculture research and education.

**AUTHOR'S NOTE**: When this chapter was written, the 1990 farm bill debate had just begun, and fiscal 1991 federal appropriations had not been passed. During 1990 and early 1991, more LISA projects were funded, and more farmers became involved in the program. Readers who are interested in updating the information in this chapter may obtain the following reports from the LISA Program, Cooperative State Research Service, U.S. Department of Agriculture, Room 342, Aerospace Building, 14th and Independence Avenue, SW, Washington, DC 20250-2200 (202/401-4640):

• USDA. 1991. 1991 Annual progress report. Sustainable Agriculture Research and Education Program, Cooperative State Research Service, USDA. 20 pp.

• Madden Associates, Inc. 1990. Farmers participating in LISA projects, directory 1988 to 1990. Low-Input Sustainable Agriculture Research and Education Program, Cooperative State Research Service, USDA. 50 pp.

## REFERENCES

Madden, J. P., J. A. De Shazer, F. R. Magdoff, N. Pelsue, Jr., C. W. Laughlin, and D. E. Schlegel. 1990. Low-Input Sustainable Agriculture Research and Education Projects Funded in 1988 and 1989. LISA 88-90. Washington, D.C.: Cooperative State Research Service, U.S. Department of Agriculture.

U.S. Congress, House of Representatives. 1988. House Agricultural Appropriations Committee Report. Washington, D.C.

U.S. Congress, Senate. 1988. Senate Agricultural Appropriations Committee Report. Washington, D.C.

U.S. Department of Agriculture. 1988. Office of the Secretary. Alternative Farming Systems. Secretary's Memorandum 9600-1. January 19. Washington, D.C.: U.S. Department of Agriculture.

U.S. Government Printing Office. 1985. Food Security Act of 1985. P.L. 99-198. Washington, D.C.: U.S. Government Printing Office.

# 3

# Challenges and Rewards of Sustainable Agriculture Research and Education

*R. James Cook*

For the past 50 or more years, the emphasis in agricultural research and education has been on improving crop plants and livestock as biological resources. This emphasis must continue. There is another equally great biological resource, however, that has the potential to increase both the productivity and sustainability of agriculture that remains largely untapped. This biological resource is external (or sometimes internal) to, while interacting with, crops and livestock, and it is represented by a nearly limitless variety of biological systems, interactions, and natural cycles on farms. The farming systems identified in the report *Alternative Agriculture* (National Research Council, 1989a) have as their common thread the goal to make greater use of these biological resources.

John Pesek, who was the chairman of the committee that produced the report *Alternative Agriculture*, made the central theme of the report very clear in remarks in Washington, D.C., on the day the report was released and in the following remarks he made to the International Fertilizer Industry Association (Pesek, 1989, p. 6).

> Alternative farming practices . . . are far more than conventional agriculture with lowered inputs of fertilizer and pesticides. They are an array of options that emphasize management and take advantage of biological relationships that occur naturally on the farm. The objective is to enhance and sustain rather than to reduce and simplify these relationships—to make them relevant to the production system rather than irrelevant; not to mask them with excesses. No present technologies have been ruled out—the intensity and, in some cases, the frequency of their use is moderated.

This idea is not new. Charles Benbrook, executive director of the Board on Agriculture, has brought together some quotes from the writings of Aldo Leopold, after whom the recently opened and dedicated Leopold Center for Sustainable Agriculture is named at Iowa State University in Ames. To set the stage for why better use should be made of natural biological interactions and cycles on the farm, one of these quotes is appropriate here (Benbrook, 1990, p. 23):

> . . . Few educated people realize that the marvelous advances in technique made during recent decades are improvements in the pump, rather than the well. Acre for acre, they have barely sufficed to offset the sinking level of fertility. In all of these cleavages, we see repeated the same basic paradoxes: man the conqueror versus man the biotic citizen; science the sharpener versus science the searchlight on his universe; land the slave and servant versus land the collective organism.

Almost 70 years ago, Carl Hartley of the Bureau of Plant Industry, U.S. Department of Agriculture (USDA), recognized the power of what he termed the *biological factor* in producing pine seedlings in nurseries for reforestation (Hartley, 1921). The nursery soils were treated with live steam to kill the pathogens responsible for seedling blights and damping off, but following steam treatment, any inadvertent recontamination of the soil with these pathogens resulted in even more seedling blight than occurred in the natural soil.

He attempted to reproduce the biological factor by adding common soil microorganisms back to the steamed soil. This was the first attempt at biological control of plant diseases by the deliberate release of microorganisms into soil. The effectiveness of his treatments approached but could not duplicate that of the natural biological factor of soil that was suppressive to the pathogens of pine seedlings. Thirty years later, in the 1950s, Kenneth Baker, who was with the California Agricultural Experiment Station, developed a method whereby steam and air could be mixed to selectively treat soils at temperatures just high enough to kill pathogens but not so high as to eliminate the biological factor (Baker, 1957, 1962). This relatively simple technology, plus the use of pathogen-free seeds planted into these soils, was the basis for the *U.C. System for Producing Healthy Container-Grown Plants* (Baker, 1957) and revolutionized the ornamental and bedding plant industries in the United States and other countries.

Hartley's biological factor is just as operative above ground as it is in soil, and a "revolution" is just as possible with field-scale as with container-grown plants. It stands to reason that taking greater advantage of the enormous potential of the biological interactions and natural cycles working in consortium with crops, especially the important food and agronomic crops, is the next big step in making agriculture both economically more viable and ecologically more sustainable.

## CONCEPTS AND DEFINITIONS

Before discussing the challenges and rewards of sustainable agriculture research and education, some basic concepts must be introduced. These definitions refer to the plant side of agriculture, but they could possibly be adapted to livestock systems as well.

### The Elements of Crop Production Systems

Any crop production system can be subdivided, on the basis of component elements, into (1) inputs, (2) biological processes, and (3) depletions or net losses (Figure 3-1A).

The biological processes include photosynthesis, genetics of the crop in terms of its adaptation to the soils and climate and resistance to pests and diseases, biological nitrogen fixation, nitrogen cycling in the soil, phosphorus uptake by mycorrhizal fungi associated with roots, plant defense by plant-associated microorganisms and natural enemies of insect pests, and soil sanitation by the natural soil microbiota.

The inputs include the fertilizers, water, where irrigation is practiced, pesticides, labor, and energy.

The depletions or net losses are largely earth resources and include the organic matter and mineral nutrient contents of the soil, water reserves and water quality, soil lost through erosion, and fossil fuels.

The relative contributions of these three component elements to crop production on any given farm vary with the farming system. Some systems attempt to reduce inputs and make greater use of biological processes; others use more inputs and depend less on biological processes. These components refer only to those elements that are involved directly in crop production and do not include broader considerations such as food safety.

Crop plants growing wild in their native ecosystems depend only on the biological processes (Figure 3-1B). There are no external inputs and basically no net depletions; more likely, there are net gains in earth resources through the soil-building processes of undisturbed ecosystems. Crop plants in the wild present a valuable source of germ plasm and biological control agents for transfer to agricultural systems, and the natural ecosystems present a wealth of clues for assisting in the recognition of potentially useful biological interactions, but these natural systems, by themselves, cannot support the human race.

Sustainable agriculture is a goal aimed at not only allowing no net depletions or net losses in earth resources but, ultimately, at rebuilding or restoring the productive capacity of agricultural soils as well (Figure 3-1C). Sustainable agriculture must meet many other criteria, and these are identified in the many definitions provided by Charles M. Benbrook in the

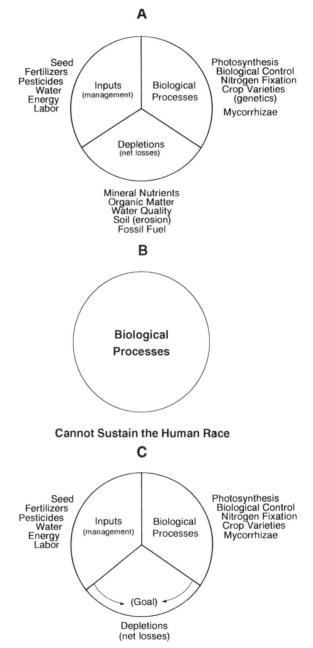

FIGURE 3-1 Conceptual illustrations of the major elements that are directly involved in or affected by (A) crop production, (B) crops in the wild (natural ecosystems), and (C) sustainable agriculture. Source: R. J. Cook and R. J. Veseth. 1991. Wheat Health Management. St. Paul, Minn.: APS Press.

Introduction to this volume, but fundamentally, so long as agriculture is responsible for a steady depletion of earth resources, it cannot claim to be sustainable. It may move to new unexploited areas and tap new earth resources, but in the long run, it will not be sustainable. Farming systems aimed at minimizing or eliminating the net-depletions element will also reduce many of the external costs of agriculture to society such as the cost of soil and other pollutants in lakes and rivers.

## TOWARD SUSTAINABLE SYSTEMS OF WHEAT PRODUCTION IN THE PACIFIC NORTHWEST

The challenges and rewards of research and education toward making agriculture both more profitable and more sustainable can be illustrated by the efforts aimed at identifying and removing the barriers to full and efficient production of wheat in the Pacific Northwest. The work shows:

• the remarkable significance of microbial interactions in the rhizosphere—negative and positive—to the ability of a crop to produce fully and efficiently,
• how science can be held up or even misdirected for decades by misinterpretations and misdiagnoses, and
• the wealth of clues to biological control that can be forthcoming through both empirical and experimental studies of natural processes.

Much of this work was done under the auspices of the tri-state research and education program known as STEEP—Solutions to Environmental and Economic Problems (see the paper by R. I. Papendick, this volume). Some of the work reported here was done after the report *Alternative Agriculture* (National Research Council, 1989a) went to press, but it could easily have been included in that report.

The Pacific Northwest produces soft white winter wheat on mostly deep silt loam soils with wet mild winters and dry warm (or occasionally hot) summers with long days. These conditions are ideal for winter wheat. Semi-dwarf wheat was introduced into the region starting in 1961 with the release by O. A. Vogel of the variety Gaines (Vogel et al., 1956). This variety and its successors were the first wheat varieties with the ability, when adequately fertilized with nitrogen, to take full advantage of the ideal temperatures, long spring and summer days, and natural water supplies in deep soils in the Pacific Northwest. These new wheat varieties also responded to irrigation, and soon after Gaines was available, a farmer near Ellensburg, Washington, produced a record-setting 209 bushels per acre (bu/acre) with irrigation. Vogel already knew in the 1950s that his then late-generation, semi-dwarf lines were capable of producing 130 bu/acre (dryland) at Pullman, Washington. However, while the occasional commercial field produced a yield of 100 or more bu/acre, the majority of fields planted to these new wheat varieties in the high-production areas known as the Palouse yielded only 60 to 80 bu/acre, despite the inputs.

On the other hand, farmers with yields nearly double those obtained with earlier varieties of standard height were delighted with their higher yields—the highest in the United States and two to three times the U.S. average for wheat. Moreover, with foreign markets starting to open up, especially in Asia, where people were switching from rice to wheat as staples in their diets, wheat was planted "fence row to fence row." Meanwhile, Vogel continued to point out that his wheat varieties were not performing up to their proven high potential.

Moreover, if a formula developed for wheat more than 30 years ago is correct—that wheat in this region can produce an average of 6 to 7 bu/acre per acre-inch of available water after the first 4 inches (Leggett, 1959)—then Vogel was right: fields in eastern Washington and adjacent northern Idaho with 18 to 20 inches or more of annual rainfall had enough water and probably enough nitrogen for 110 to 130 bu/acre but were producing only 70 to 80 bu/acre. A vast body of experimental evidence from field trials confirms this basic relationship between available water and average yield potential for soft white winter wheat in the Pacific Northwest (Figure 3-2).

A fact not commonly recognized is that while the attainable yield of a crop increases in proportion to the increasingly more favorable growing conditions, the actual yield responds relatively less because these conditions also favor more damage from pests, diseases, lodging, nutrient shortages, or other hazards (Figure 3-3). However, the potential of the crop to respond to elimination of such constraints is also greater in proportion to the attainable yield. This is not to suggest that crops that are grown under less than ideal conditions are not subject to the effects of pests and diseases. On the contrary, the effects are basically the same; only the margin of response to elimination of these constraints is potentially less.

The ideal growing conditions and the high yield potential for wheat in the Pacific Northwest were the right combination for the work reported below.

## The Soil Fumigation Effect

In 1974, William Haglund of Washington State University and R. J. Cook began studies using soil fumigation as a research tool to reveal experimentally the full production capability of the semi-dwarf wheat varieties (Figure 3-4). Yields were higher by 20, 30, and even as much as 50 bu/ acre in response to the treatments, but with the same available water and nitrogen in the soil (Cook and Haglund, 1982; Cook et al., 1987) (Figure 3-5). Those investigators routinely documented 110 to 120 bu/acre and as much as 150 bu/acre of dryland in the fumigated plots.

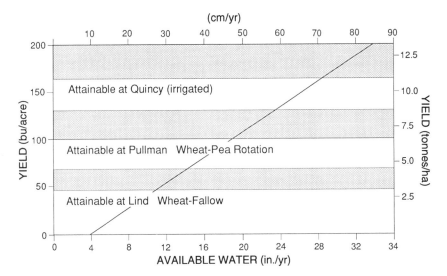

FIGURE 3-2  Relationship between available water and average yields of soft white winter wheat in Washington State.  The horizontal bars indicate the range of attainable yield and available water at Lind for wheat after fallow, Pullman for wheat after peas, and Quincy for irrigated wheat.  The relationship is based on an average of 7 bu/acre per inch of available water (500 kg/ha per 2.5 cm of available water) beyond the first 4 inches and has been verified repeatedly by the best yields for the respective areas.  Source:  Modified from G. E. Leggett. 1959. Relationship Between Wheat Yield, Available Moisture, and Available Nitrogen in Eastern Washington Drylands. Washington Agricultural Experiment Station Bulletin 609. Pullman, Wash.: Washing-ton Agricultural Experiment Station (in R. J. Cook. 1986. Wheat Management Systems in the Pacific Northwest. Plant Disease 70:894–898).

The often spectacular soil fumigation effect has been demonstrated many times over the past 40 to 50 years (Cook, 1984), but routinely, it has been dismissed by scientists as a response of the crop to nitrogen or other mineral nutrients released from the killed microorganisms.  Researchers have left this phenomenon on the "back burner" as something mysterious and too complicated to sort out.  The simple explanations that root health has been improved, that the plants are growing more nearly to their potential, and that root diseases are indeed this important (Cook, 1984; Wilhelm and Paulus, 1980) have not been generally accepted until very recently.

Soil scientists had already shown more than 30 years ago that the fumigation effect cannot be explained by changes in soil chemistry (Aldrich and Martin, 1952).  An initial release of about 10 to 15 pounds (lbs) of ammonium-nitrogen (N) per acre in the top 6 inches of soil was measured, which was minor compared with the 120 lbs of inorganic-N made available by fertilization on top of the 50 to 60 lbs released by natural mineralization (Cook and Haglund, 1982). Cook and Haglund (1982) also showed, as would be expected, that nitrification of the ammonium to nitrate was delayed in the fumigated soils.

The breakthrough came in the mid-1970s, when it was discovered (Cook, 1984; Cook and Haglund, 1982; Cook et al., 1987) that two slightly different and relatively mild fumigants gave identical releases of ammonium-N, and both delayed nitrification, but only one gave the increased growth response. These results showed that the increased growth response of a crop to soil fumigation cannot be explained on the basis of changes in the availability of nitrogen. The one fumigant giving the increased growth response also killed the spores of *Pythium* spp. in soil (Cook and Haglund, 1982; Cook et al., 1987). This was the first direct evidence of the widespread importance of this soilborne pathogen.

Cook et al. (1980) obtained proof of the importance of *Pythium* damage to roots with a new soil fungicide, metalaxyl, which is specific for *Pythium* and closely related fungi and which duplicated the soil fumigation effect (Figure 3-6).

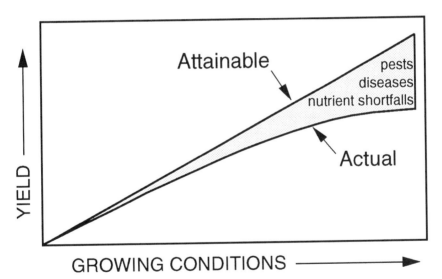

FIGURE 3-3 Diagrammatic illustration of increases in crop yield with increases in conditions favorable to the yield of that crop. While the attainable yield (the yield possible) increases in direct proportion to the more favorable conditions, the actual yield also increases, but less responsively because of specific diseases, pests, and other production hazards also favored by the more ideal growing conditions. Source: R. J. Cook and R. J. Veseth. 1991. Wheat Health Management. St. Paul, Minn.: APS Press.

FIGURE 3-4  Soil fumigation in progress in the fall just before winter wheat was planted in a field near Walla Walla, Wash.  The field was in a 2-year rotation of dry peas (the previous year) alternated with winter wheat.  Soil fumigation was used as a research tool to determine the full production capability (attainable yield) of the wheat with all root diseases controlled. Source:  R. J. Cook and W. A. Haglund. 1982. Pythium Root Rot: A Barrier to Yield of Pacific Northwest Wheat. Washington State College of Agriculture Research Bulletin No. XB0193. Pullman, Wash.: Washington State College of Agriculture.

This capacity of wheat in commercial fields to yield 20, 30, and even 50 bu/acre more with no increase in nitrogen supply leads to the inescapable conclusion that wheat in the nonfumigated plots, which were typical of the rest of the farmer's fields, must be leaving as unused that amount of nitrogen that is necessary to produce these higher yields of wheat.

Cook and colleagues were fortunate in their trials, in that phosphorus was not limiting in the soils and the dense fibrous root system of healthy wheat is very efficient in exploring the soil.  With other crops in other soils, yields have sometimes been lower in response to soil fumigation because the treatment eliminates mycorrhizae—the fungus-root associations involved in phosphorus uptake by plants.  On the other hand, an experiment that shows yield depressions when mycorrhizae are eliminated should lead to the question: How much could yields be enhanced if these beneficial fungi could be enhanced?

Thus far, three root diseases have been diagnosed:  pythium root rot, which is caused by several *Pythium* spp. (Chamswarng and Cook, 1985;

Ingram and Cook, 1990); rhizoctonia root rot, which is caused by at least two *Rhizoctonia* spp. (Ogoshi et al., 1990; Weller et al., 1986); and take-all, which is caused by *Gaeumannomyces graminis* var. *tritici* (Cook and Weller, 1987). *Pythium* spp. strip away the root hairs and fine lateral rootlets of wheat (Figure 3-7), *Rhizoctonia* spp. girdle and sever both the lateral and main roots, and take-all develops as lesions on all roots and progresses into the tiller bases. Of these three, pythium root rot is the most subtle and

FIGURE 3-5 Increased growth response of winter wheat cultivar Stephens following soil fumigation in a commercial field near Pullman, Wash. Wheat in the foreground (natural soil) averaged about 95 bu/acre, compared with wheat in the fumigated soil in the background that averaged 120 bu/acre. Sources: R. J. Cook and W. A. Haglund. 1982. Pythium Root Rot: A Barrier to Yield of Pacific Northwest Wheat. Washington State College of Agriculture Research Bulletin No. XB0193. Pullman, Wash.: Washington State College of Agriculture. R. J. Cook, W. Sitton, and W. A. Haglund. 1987. Increased growth and yield responses of wheat to reduction in the *Pythium* populations by soil treatments. Phytopathology 77:1192–1198.

FIGURE 3-6    Increased growth response of winter wheat cultivar Stephens to a seed treatment with the fungicide metalaxyl (specific for *Pythium* spp. in the trials). This response is typical of the responses of several trials done in the Palouse area in 1979 to 1981, and shows the uniformly debilitating effect of *Pythium* spp. on the growth of wheat. Source:  R. J. Cook and W. A. Haglund. 1982. Pythium Root Rot: A Barrier to Yield of Pacific Northwest Wheat. Washington State College of Agriculture Research Bulletin No. XB0193. Pullman, Wash.: Washington State College of Agriculture.

was the most difficult to diagnose, but overall, it may be the most important.

Plants affected by any combination of these diseases look underfertilized (see Figure 3-6) because of the reduced capacity of their root systems to take up nutrients, and often, farmers who notice the uniformly yellowed and stunted appearance of the crop apply more nitrogen, which may or may not help the crop. The mixture of pathogens is dynamic, in that control of only one pathogen can open the way for more damage by the other pathogens and, hence, no overall change in the performance of the crop. With little or no benefit of standard seed treatments and the total impracticability and unacceptability of soil fumigation for wheat, researchers have been forced to look for alternatives that can be used to control these diseases.

## The Crop Rotation Effect

The average yield response of wheat to soil fumigation in areas with at least 18 to 20 inches of rainfall or rainfall plus irrigation has been 70, 22, and 7 percent, respectively, in fields that are cropped every year (monoculture), every other year (2-year rotations), and every third year (3-year rotations) to wheat (Figure 3-8). As the length of the rotation was increased, up to a maximum of 2 years between wheat crops, the yields in nonfumigated plots were proportionally higher and the yields in fumigated plots were proportionally less (Cook, 1990). In other words, crop rotation is nearly as effective as soil fumigation as a means of achieving the high yields of semi-dwarf wheats in the high-production systems of the Pacific Northwest.

According to the evidence, the fumigation effect and the rotation effect are the same: both provide a means of eliminating root diseases as production constraints, one that is chemical, which takes about 2 days, and one that is biological, which takes about 2 years. Soil fumigation has become a substitute for crop rotation with many high-value vegetable and fruit crops throughout the United States (Wilhelm and Paulus, 1980), but it is not a viable alternative for agronomic crops.

Research by Crookston and associates (1988, 1991) at the University of Minnesota similarly shows that yields of corn are 10 to 15 percent higher after soybeans than after corn, and that yields of soybeans are 8 to 17 percent higher after corn than after soybeans (Figure 3-9). Even the yields for these crops grown in alternate years (2-year rotations) are less than yields for the same crops in the first year after a break from corn or soybeans for 3 or more years.

Like the results described above for wheat in the Pacific Northwest, the work of Crookston and associates (1988, 1991) indicates that yields of corn and soybeans are potentially highest with 3-year (or longer) rotations, such as might be achieved with small-grain crops, in addition to corn and soybeans. And like the results described above for wheat in the Pacific Northwest, these higher yields cannot be attributed to greater nitrogen availability. This is elegant work carried out over several years and is in direct contradiction to earlier claims that corn can be grown continuously without sacrificing yield potential.

It should be pointed out, however, that since the rotation effect in the Pacific Northwest is basically a process of soil sanitation achieved with the microbial activity of soil—a biological factor—it is probable that the lower the soil temperature or the drier the soil the slower the process. It may be that 2-year rotations in the central or southern states are the equivalent of

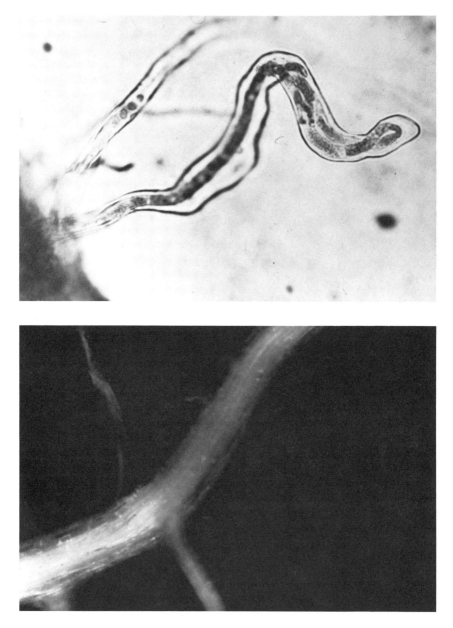

FIGURE 3-7  Typical damage of *Pythium* spp. to the fine rootlets and main roots of wheat, starting with (top) invasion and destruction of root hairs, followed by (bottom) a complete stripping away of the root hairs and rootlets and discoloration of the cortical tissues of the main roots.

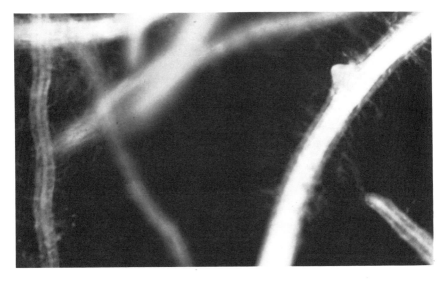

FIGURE 3-7 *(Continued)* These white healthy roots came from fumigated soil. Source: R. J. Cook, J. W. Sitton, and W. A. Haglund. 1987. Increased Growth and Yield Responses of Wheat to Reduction in the *Pythium* Populations by Soil Treatments. Phytopathology 77:1192–1198.

3-year rotations in northern states, such as Washington and Minnesota. More work on this is needed.

Crookston (1984) has not yet diagnosed the factor(s) that is responsible for what he calls the "monoculture effect" controlled by the "rotation effect." He has ruled out common factors such as corn rootworm, nematodes, and brown stem rot of soybeans. The results in the case of corn point clearly to the presence of a negative factor that persists in the soil when corn is grown continuously and not a growth-promoting influence on corn left in the soil by growing soybean (Crookston et al., 1988). Some root diseases can be very difficult to diagnose, especially those such as the ubiquitous pythium root rot where the root hairs and fine lateral roots destroyed by the disease remain in the soil by most standard methods of root recovery. It is well established that the continual presence of the roots of one crop selects for microorganisms that can break down the roots of that crop; Cook et al. (1987) allowed time for the antagonistic microorganisms to displace or destroy the root pathogens specialized for one crop while growing other unrelated crops.

It is not the presence of different crops that leads to the rotation effect, but rather, it is not growing the same crop year after year in the same field that allows for expression of the rotation effect.

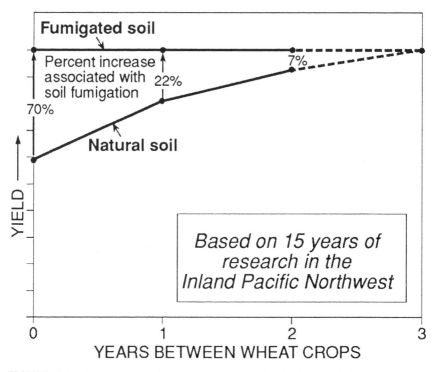

FIGURE 3-8  Response of winter wheat to soil fumigation as influenced by the length of the crop rotation in eastern Washington and northern Idaho. Source: From R. J. Cook and R. J. Veseth. 1991. Wheat Health Management. St. Paul, Minn.: APS Press.

The idea that root diseases can be controlled by crop rotation goes back many years. In 1909, H. L. Bolley (see Stack and McMullen, 1988, p. 8) at what was then the North Dakota Agricultural College (now North Dakota State University) pointed this out to farmers in a poster that read:

> The reason for crop rotation is not particularly to prevent loss of fertility. It is a sanitary measure. Proper rotation frees th  soil from specific crop diseases. No matter how fertile the land, you cannot raise heavy seed if the mother seeds carry fungus diseases internally. Flax does this, wheat does, oats and barley do. Nor can you raise heavy seed wheat if soil is wheat sick. Our old wheat lands are not "worn out"— they are full of diseased wheat roots and stubble. ROTATE.

The challenge in educating farmers on the value of the rotation effect comes when half or all of their farm is wheat base. Moreover, farmers

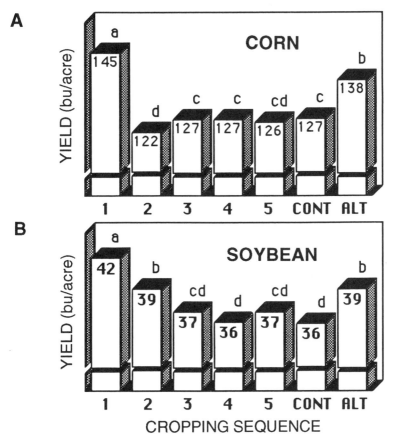

FIGURE 3-9 (A) Grain yield of corn grown under monoculture (CONT) or alternated (ALT) with soybean. (B) Grain yield of soybean grown in monoculture (CONT) or alternated (ALT) with corn. Bars with the same letter are not significantly different at $p = 0.05$. Source: R. K. Crookston, J. E. Kurle, R. J. Copeland, J. H. Ford, and W. E. Lueschen. 1991. Rotational cropping sequence affects corn and soybean yield. Agronomy Journal 83:108–113.

averaging 70 bu/acre in each of 3 years produce a total of 210 bu in a 3-year cycle, compared with those with 3-year rotations who produce an average of only 100 to 110 bu in a 3-year cycle. With a government-guaranteed target price and subsidy per bushel of wheat up to the so-called proven yield on the farm, they are assured of an acceptable net return with continuous wheat, even with root diseases.

Yet, a yield of 110 bu/acre every third year has the lowest cash cost per bushel and the maximum return per acre of wheat. Moreover, the net return from the respective rotation crops, namely, spring barley and then either

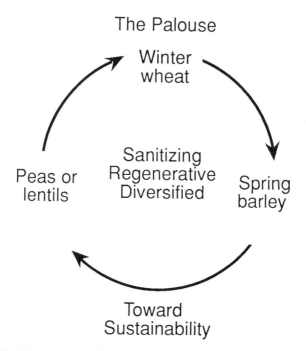

FIGURE 3-10  Diagrammatic illustration of the winter wheat, spring barley, pea, or lentil rotation used as a 3-year rotation in eastern Washington and northern Idaho and some projected benefits.  Source: Modified from R. J. Cook. 1989. Biological control and holistic plant-health care in agriculture. American Journal of Alternative Agriculture 3:51–62.

lentils or peas, needs to be competitive only with the net profit of 50 to 60 bu of wheat per acre per year while the field is not planted to wheat.

In addition, fields in a 3-year rotation typically can be planted directly with no prior tillage without the yield penalty when the no-till method is used in combination with continuous crops of wheat (see next section), and farmers have no reason to burn the stubble, as many do when they grow wheat year after year in the same field. Equally important, a field in winter wheat only once in 3 years is injected with a heavy dose of anhydrous ammonia only once rather than three times in each 3-year cycle, which should delay soil acidification. Wheat grown in the 3-year rotation meets the definition of both maximum economic yield and sustainable agriculture (Figure 3-10).

Wheat growers attempting to produce maximum yields with intensive wheat cropping commonly claim that their crops respond to more nitrogen or phosphorus than is recommended from soil tests.  With the absorptive capacity of the roots greatly reduced because of disease, it seems likely that the surplus fertilizer and better placement compensates, in part, for the lack of roots.

The reward comes in watching the average yields in these high-production areas increase by 20 to 30 bu/acre without changes in the rates of nitrogen and with the same amount of rainfall when the farmer switches to a 3-year rotation. The reward also comes when a grower reports that he is putting more wheat in the bin with only one-third of his farm in wheat than he used to put in the bin with half his farm in wheat.

Despite its simplicity and the overwhelming amount of evidence in favor of the importance of crop rotation for root disease control, like the soil fumigation effect, the notion is still not widely accepted among plant and soil scientists. It is still relatively uncommon in research reports on results of fertilizer trials, water use efficiency, or similar agronomic factors to state the crop history of the field in which the work is done, apparently because the previous crops (other than a nitrogen-fixing legume) are not considered important to the yield potential of the next crop in any given field. Individual state recommendations on intensive crop management or maximum economic yield (MEY) focus intently on planting, fertilizing, and then caring for the crop with little or no reference to what was or should be grown in the field the previous year or years.

A true MEY system applied to any given crop in any given field needs to be started 2 and even 3 years before the field is planted to that crop rather than the year the field is planted without regard to the cropping history.

One of the greatest challenges in education is how to effect a paradigm shift within the plant and soil sciences toward a greater appreciation for the significance of root health in matters now attributed to soil fertility.

### The Crop Residue Effect

The high cost to progress in agricultural research of not recognizing the importance of crop rotation to root health is perhaps best illustrated by the misinterpretations and even the misdiagnoses of the crop residue effect on wheat in the western states starting in the late 1940s and early 1950s.

Stubble-mulch farming was introduced in these areas in the wake of the Dust Bowl days to slow soil erosion. Almost immediately, researchers observed that in the wetter areas or years, yields of wheat were lower when the crop residue was left on the soil surface, which is typical of mulch tillage, than when it was buried, which is typical of clean tillage. Zingg and Whitfield (1957) reported that in 108 separate comparisons at eight locations in the Great Plains and Pacific Northwest, the wetter the climate of the area, the greater the shortfalls in yields with mulch tillage relative to those with clean tillage (Figure 3-11). Total yields were higher in the

50

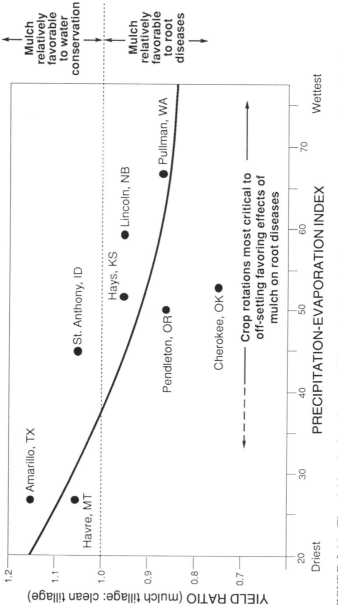

FIGURE 3-11 The yield ratio for wheat at different locations in the western states grown with mulch tillage (crop residue left maximally on the soil surface) versus clean tillage (crop residues buried). The wetter the area, the lower the yield with mulch tillage compared with that with clean tillage. Source: A. W. Zingg and C. J. Whitfield. 1957. Stubble-Mulch Farming in the Western States. Pp. 1–56 in U.S. Department of Agriculture Technical Bulletin 1166. Washington, D.C.: U.S. Department of Agriculture. Explanations from R. J. Cook and R. J. Veseth. 1991. Wheat Health Management. St. Paul, Minn.: APS Press.

wetter areas than they were in the drier areas, regardless of tillage, but the performance of wheat with mulch tillage dropped below that with clean tillage proportionally more with increasing rainfall.

Wheat grown with mulch tillage in the problem areas looks undernourished, and therefore, early work focused on nutrient deficiencies that were thought to be caused by a relatively greater nutrient tie-up if the crop residue was left on the soil surface than if it was buried. This hypothesis was ruled out (reviewed in Cook [1990]), and attention was directed to the crop residue itself, or its decomposition products, as possibly being phytotoxic or allelopathic to the wheat (reviewed in Elliott et al. [1978]). Work on the suspected phytotoxins in wheat straw continued over nearly four decades in several U.S. states as well as in Australia and England (reviewed in Cook [1990] and Rovira et al. [1990]).

In virtually all of the past studies on wheat crop residue effects and wheat health and yield, the sites have been either wheat crops year after year or wheat-fallow-wheat. How else can one plant wheat into wheat stubble unless the site is cropped to wheat after wheat? Even with an intervening fallow, the field is often allowed to green-up early in the fallow period because of the grass weeds and volunteer wheat that can serve as a green bridge between crops of wheat—in a sense, another wheat crop.

Moore and Cook (1984) confirmed, for conditions in eastern Washington, that the problem was microbiological in origin by showing that it could be eliminated with soil fumigation (Figure 3-12). Indeed, when the soil and residue were fumigated, yields were even higher if residues were left on the soil surface than if they were buried (Moore and Cook, 1984), presumably because the surface residues served as a barrier to the evaporative loss of water from the soil so that more water was available to the wheat. On the other hand, burning of wheat stubble in subhumid areas such as the Palouse region not only is detrimental to soils in the long term in lost organic matter but it also costs yield potential in the short term because the bare soils lose valuable water and, hence, yield potential.

It has also been shown (R. J. Cook, unpublished data) that fresh wheat straw layered as a mulch on a site cropped to lentils the previous year had no negative effect on the growth and yield of wheat. On the contrary, yields were greater by 12 bu/acre in response to the mulch. Yields of winter wheat are consistently higher with reduced tillage than with conventional tillage if the field is in a 3-year crop rotation with wheat no more than every third year.

Take-all, pythium root rot, and rhizoctonia root rot each have been shown experimentally to be more severe in fields with reduced tillage than in fields with no tillage (Cook et al., 1980; Moore and Cook, 1984; Rovira, 1986; Weller et al., 1986). The surface residues serve the pathogens as a secondary source of energy (with the living roots of wheat being their

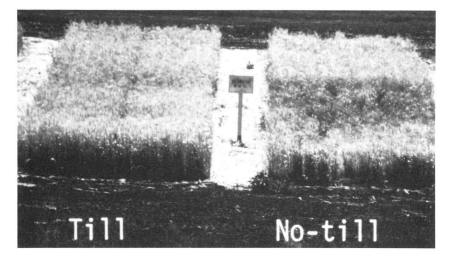

FIGURE 3-12 Spring wheat planted at Pullman, Wash., into soil cropped to winter wheat the 2 previous years (monoculture) with (left) or without (right) tillage and with (top) or without (bottom) soil fumigation. These two drill strips were typical of all four replicates in the experiment and has been repeated in the Palouse region many times.

primary source of energy), but more importantly, the surface residues are thought to keep the top few inches of soil, where these fungi reside, more ideally moist for their activities as root pathogens.

Across the United States, grass weeds, leaf diseases such as tan spot, and several insect pests of wheat, most notably the Hessian fly, are also favored by leaving the crop residue on the soil surface and can be controlled by crop rotation. However, these problems can be seen above ground and are relatively easy to diagnose, whereas the root diseases, because they are out of sight, are not so obvious, and have therefore gone mostly undiagnosed or misdiagnosed.

Because studies of root diseases did not rule out allelopathic chemicals as still another factor, tests were conducted to determine whether the micro-organisms responsible for the effect and eliminated by soil fumigation were in the soil or in the straw. These experiments, which were conducted in the field at two locations, showed that the etiologic agents are in the soil (probably in the old root tissues) and not in the straw, where they would need to be if phytotoxic decomposition products from the straw were important (Cook and Haglund, in press; Figure 3-13).

Crookston and Kurle (1989) have also tested the hypothesis that the monoculture effect with corn results from an allelopathic effect of the corn residue. They placed corn residue on plots planted to soybeans or corn, but the only effect they could demonstrate was that of crop rotation.

This focus on the phytotoxin hypothesis and failure to recognize the importance of the rotation effect to root health have held up progress toward solving this problem for at least three decades—from about 1950 until work aimed in the right direction got under way in the 1980s. A proper diagnosis is usually the first big step toward solving any problem of this kind. Any progress toward improving the root health of crops is progress toward the more efficient use of fertilizers and higher net returns, and if this progress also allows farmers to use less tillage without the penalty of reduced yield, then farming systems have moved toward greater sustainability.

These findings and experiences with the crop residue effect support an

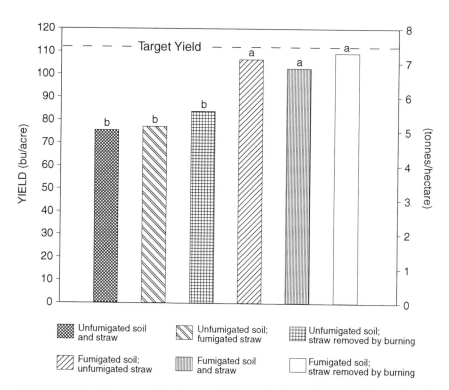

FIGURE 3-13 Influence of fumigation (with methyl bromide gas) of the soil only, straw only (as a mulch on the soil), both soil and straw, or neither (all natural) on yields of winter wheat after winter wheat (no crop rotation) in eastern Washington. The only significant yield response occurred when the soil was fumigated, regardless of the nature of the straw on the soil surface. Bars with the same letter are not significantly different at $p = 0.05$.

important message of the *Alternative Agriculture* report (National Research Council, 1989a) that correct diagnoses are beginning to be made of the biological and ecological problems responsible for the recent disappointing performance of many potentially high-yielding cropping systems—systems that have worked well for the short term or in some areas but that are increasingly unreliable.

## The Herbicide Effect

In the Pacific Northwest, farmers who grow continuous small grains with no tillage use herbicides as a preplant treatment to eliminate the weeds before planting into the wheat or barley stubble. Of all the possible weeds, the most common are volunteer plants of wheat and barley, especially in the combine row, where the lighter-weight seeds of these crops are most concentrated. The common practice has been to allow the fields to green up with volunteer plants of wheat and barley, grass weeds, and other weeds and then spray them with a nonselective herbicide (e.g., glyphosate [Roundup]) just before planting—usually only 1 or 2 days before planting or even on the day of planting.

Wheat or barley planted in this way developed unusually severe root disease, especially rhizoctonia root rot. Volunteer wheat and barley serve as a "green bridge" for many pests and pathogens between the time of harvest of one crop and the time of planting of the next crop of wheat or barley. In the case of continuous no-till wheat or barley in the Pacific Northwest, the volunteer plants are a reservoir of root pathogen inocula. Upon application, the herbicide weakens these growing plants and their roots systems, in effect shutting down the plant defense mechanisms that normally retard disease development and, hence, the populations of these pathogens. The pathogens, which are already on and within the roots of these host plants at the time of the herbicide treatment, build up to unusually high populations as their hosts begin to die from the herbicide, and this accelerated buildup, when coincidental with the sowing of the next crop, is perfectly timed for a major epidemic of root disease on the next crop (A. G. Ogg, R. W. Smiley, and R. J. Cook, unpublished data).

Volunteer plants and weeds that are killed outright by tillage are subject almost immediately to takeover by the common saprophytic soil microflora, while those plants that are allowed to die slowly, which is typical of the herbicide effect, provide a competitive advantage to the root pathogens that are already in the roots and that have the ability to act as both parasites and saprophytes. Some farmers went broke trying to grow continuous cereals with no tillage and spraying the volunteer plants just before planting. Farmers noticed that the problem was worse when herbicide had been used, even if the field was then also tilled, than when the field was tilled only

**TABLE 3-1** Yields of Spring Grains in Eastern Washington Directly Drilled into Standing Stubble of Cereal Grains

| Soil Type | Spring Wheat (bu/acre) | Spring Barley (lbs/acre) |
|---|---|---|
| Fumigated soil | 63a | |
| Natural soil; volunteer control by glyphosate | | |
| 2 Weeks before planting | 55ab | 1,838a |
| 2 Days before planting | 43b | 1,573b |

NOTE: The spring wheat was directly drilled into standing stubble and volunteer plants of winter wheat and included soil fumigation (methyl bromide) to produce a "pathogen-free" check. The spring barley was directly drilled into stubble and volunteer plants of spring barley grown the previous year on the same site. Volunteer cereal plants and weed plants in nonfumigated plots were eliminated 2 weeks or 2 days before planting. The volunteer plants serve as a "green bridge" for root pathogens. Values followed by the same letter are not significantly different at $p = 0.05$ according to the Duncan's multiple range test.

with no use of an herbicide. Early theories suggested that the herbicide was residual, and the term *Round-up injury* was proposed, but all the evidence now points to root disease and not herbicide injury.

The proper diagnosis also gave rise to a method of control, at least for no-tilled spring grains, for which this problem has been greatest. By applying the herbicide (e.g., glyphosate) 10 days to 2 weeks before planting, there is enough time for the pathogens to reach their peak levels and then decline to safer levels because of the normally rapid succession of non-patho-genic microorganisms that move into and rot these tissues (Table 3-1).

The difference between planting immediately and planting 10 days to 2 weeks after herbicide application is the amount of time for the biological factor in soil to have its effect. In this case, the biological factor is the competing soil microorganisms that displace the root pathogens and take over or destroy their energy source.

## IMPROVEMENT OF ROOT HEALTH WITH BENEFICIAL MICROORGANISMS IN THE RHIZOSPHERE

Crop rotation is not a permanent solution in all areas of the United States to the problem of root diseases on important food crops such as wheat. In many areas, alternative crops are not available. Moreover, the demand for food crops such as wheat can only be expected to increase, and as the area of land and intensity of cropping increase, so do the pressures from plant pests and diseases increase.

A significant advancement for the biological control of plant diseases is the discovery that on every plant or within every population of plants, there reside microorganisms with the ability to defend those plants. These beneficial microorganisms occur at a very low frequency, but several independent studies show that following several successive outbreaks of the disease, their numbers increase proportionally until an equilibrium is reached or future disease outbreaks are suppressed (Cook and Baker, 1983).

These beneficial plant-associated microorganisms are another example of the biological resource—a biological factor—external (or internal) to the plant.

The importance of this biological factor was highlighted in a report from the National Research Council (1989b) supported by the National Science Foundation and the Competitive Research Grants Office of USDA that points out the enormous potential of this resource and calls for more basic research on the ecology of the microorganisms. These microorganisms hold tremendous potential for increasing plant productivity in ways consistent with making agriculture more sustainable—and possibly also for the development of new products of biotechnology for use in crop protection.

Studies have been conducted in Washington State since 1968 on a phenomenon known as take-all decline, whereby take-all increases in textbook fashion for the first 3, 4, or even up to 7 or 8 years of consecutive wheat crops but then declines while yields recover (Figure 3-14). This pattern has been documented virtually everywhere in the world where soils are favorable to take-all and where wheat has been grown over many years without crop rotation (Shipton, 1975). It has been shown both in Australia and in the United States that there are qualitative and quantitative shifts in the makeup of root-associated bacteria that produce antibiotics inhibitory to the take-all fungus (reviewed in Thomashow and Weller [1990]). These bacteria increase in numbers in the lesions caused by the take-all fungus, inhibit continued progress of the disease, and then cohabit root tissues with the pathogen during its survival or saprophytic phase in soil between the harvest of one crop and the sowing of the next one (reviewed in Cook and Weller [1987]).

Unfortunately, but not surprisingly, a shift in the microbial population toward suppression of one disease may not provide suppression of other diseases in the same ecosystem. Thus, when wheat was grown continuously for 20 years under irrigation in central Washington, the yields declined for the first 7 years because of increasing take-all, recovered over the next 8 years because of take-all decline, and then declined again because of increasing rhizoctonia root rot (Figure 3-14).

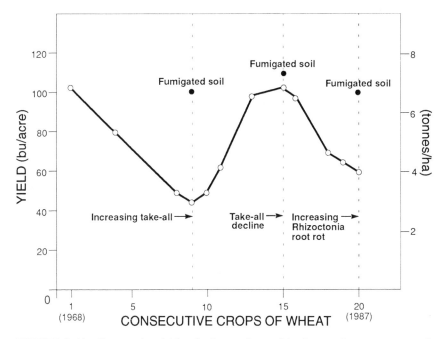

FIGURE 3-14  Changes in yields of winter wheat with changes in occurrence and severity of root diseases.  Source:  R. J. Cook. 1989. Biological control and holistic plant-health care in agriculture. American Journal of Alternative Agriculture 3:51–62.

For the past 10 years, researchers have tested the possibility that strains of suppressive bacteria from the rhizosphere of wheat can protect wheat in fields where take-all or other root diseases are yield-limiting factors, if the numbers of these bacteria can be increased in the rhizosphere in advance of infection.  Strains were isolated from the rhizosphere of wheat growing in soils from fields where take-all had declined.  These strains have been applied singly and as mixtures to the seed with significant biological control (Figure 3-15).

Bacteria that are applied as a living seed treatment are carried passively downward into the soil with the elongating root, where they multiply to the limits of the energy available from the root in competition with the native rhizosphere microorganisms (Howie et al., 1987).  In some 10 trials carried out over several locations in Washington in fields where take-all was the dominant yield-limiting factor, an average 10 percent greater yield was achieved with one mixture of strains and 15 percent greater yield was achieved with a different single strain (Tables 3-2 and 3-3).  If there were a wheat variety that averaged a 10 to 15 percent greater yield than

FIGURE 3-15  Response of spring wheat at Pullman, Wash., to biological control of take-all by seed bacterization with natural strains of fluorescent *Pseudomonas* spp. The plots had three rows each, as follows:  left, no disease and no seed bacterization;   center, pathogen introduced the same as done with plot on left, but seed treated with two bacteria active against the pathogen;  right, severe disease (pathogen introduced) and no bacterization.  Source:  D. M. Weller and R. J. Cook. 1983. Suppression of take-all of wheat by seed treatments with fluorescent pseudomonads. Phytopathology 73:463–469.

**TABLE 3-2**  Wheat Yields in Washington State in Response to Seed Treatment with Indicated Strains of Fluorescent *Pseudomonas* spp.

| Location, Year | Untreated (check) (bu/acre) | Yield (bu/acre)(percent change) Strain 30-84 | | Strain 2-79 + 13-79 | |
|---|---|---|---|---|---|
| Wilbur, 1982 | 90.1 | | | 109.8 | (22) |
| Mt. Vernon, 1982 | 64.2 | | | 70.4 | (9.7) |
| Ephrata, 1983 | 103.8 | | | 105.0 | (1.3) |
| Ephrata, 1984 | 103.0 | | | 105.0 | (1.9) |
| Mt. Vernon, 1984 | 55.2 | | | 60.8 | (10.1) |
| Ephrata, 1985 | 110.2 | 105.8 | (−4) | 110.2 | (0) |
| Mt. Vernon, 1985 | 27.4 | 40.6 | (48) | 36.4 | (32) |
| Mt. Vernon, 1986 | 81.9 | 95.5 | (18) | 86.1 | (5.1) |
| Yakima, 1986 | 90.6 | 87.5 | (−2.7) | 89.0 | (−1.1) |
| Mt. Vernon, 1987 | 50.8 | 60.0 | (18.5) | 65.5 | (25) |
| | | 15.6 | | 10.6 | |

NOTE:  Fields were cropped repeatedly to wheat (no crop rotation) with take-all as the yield-limiting factor.  Each value is the average of at least four replicate treatments in a commercial field.  The plots were planted as drill strips by the farm owner or with the owner's drill.

SOURCE:  R. J. Cook and D. M. Weller (unpublished data).

the standard, it would be considered resistant and would be released, assuming it met all other standards as well. These bacteria, together with the wheat roots, present the equivalent of resistance to this disease, and they can be used in combination with virtually any existing wheat variety.

While single strains have been effective in some trials, the best overall and most consistent performance has come from mixtures of strains (Cook et al., 1988; Pierson and Weller, 1990; Weller and Cook, 1983). Like artley's (1921) classic work, mixtures are best, but even these have not performed up to the standard of the naturally suppressive soils.

Using the tools of recombinant DNA technology, Thomashow and Weller (1988) and Thomashow et al. (1990) showed for the first time that antibiotic production occurs in the rhizosphere and can protect roots against infections. The antibiotic in this case is phenazine-1-carboxylate (Brisbane et al., 1987; Gurusiddaiah et al., 1986), although other mechanisms of inhibition are also operative. To test the role of an antibiotic such as phenazine, producer strains were made deficient in their ability to make this antibiotic by inactivating a specific gene in the organism needed for antibiotic biosynthesis (Thomashow and Weller, 1988). This inactivation was accomplished by Tn5 mutagenesis, whereby a transposable element of foreign DNA is inserted randomly into the organism's chromosome until

**TABLE 3-3** Yields of Stephens Soft White Winter Wheat at Mt. Vernon, Washington, in 1987

| Seed Treatment | Yield (bu/acre) | Percent Change from Untreated Check |
|---|---|---|
| Check (no treatment) | 50.7c | |
| Methylcellulose check | 51.6c | 2 |
| Triadimefon (alone) | 64.8a | 28 |
| + metalaxyl + captan | 64.9a | 28 |
| + strain 30-84 | 64.5a | 27 |
| Strain 30-84 (alone) | 69.0ab | 18 |
| Strain 2-79 (alone) | 55.5bc | 9 |
| + 13-79 + R4a-80 | 63.4a | 25 |

NOTE: Seed was treated as indicated in the table and planted in a field cropped to wheat for the fourth consecutive year. Each value is an average of six replicates as drill strips planted with the farmer's drill in a field naturally infested with *Gaeumannomyces graminis* var. *tritici*. Values with the same letters are not significantly different at $p = 0.05$.

SOURCE: Data from R. J. Cook, D. M. Weller, and E. N. Bassett. 1988. Effect of bacterial seed treatments on growth and yield of recropped wheat in western Washington, 1987. P. 53 in Biological and Cultural Tests, Vol. 3. St. Paul, Minn.: APS Press.

strains appear that are no longer able to make the antibiotic. By using a technique known as complementation, DNA from a library of the wild-type DNA is used to restore the ability of the mutant strain to make the antibiotic.

Antibiotic-negative mutants generated by Tn5 mutagenesis were greatly reduced in their ability to protect wheat against take-all, but mutant derivatives that were restored in their ability to produce the phenazines were coordinately restored in their ability to protect wheat roots (Thomashow and Weller, 1988) (Figure 3-16). The genes for antibiotic production have now been cloned and have been expressed in other bacteria, including strains of bacteria inhibitory to take-all by other mechanisms (D. Essar, L. S. Thomashow and L. S. Pierson, USDA, Agricultural Research Service, Pullman, Washington, personal communication, 1990). It is expected not only that wild-type strains will be mixed but also that genetic traits within the same strain will be mixed to improve the effectiveness of this biological control.

The antibiotic is delivered in molecular quantities only where and when needed, in the same way that natural antibiotics produced by plants in their own defense are produced in molecular quantities only at the sites of infection and usually not throughout the entire plant. Energy is not spent to any significant extent until or unless the plant is threatened by infection.

It should not be surprising that plants would support populations of microorganisms on and within their leaves, stems, roots, and other parts—a biological factor—with the ability to provide the first line of defense against infections and insect attack. The challenge is in how to identify and understand these associations, but the reward is a new dimension to plant improvement—a genetic system external to but complementary of the plant's genetic system. These beneficial microorganisms also become a source of genes for future plant improvement and pest control with transgenic plants. However, it seems doubtful that transfer of a single gene or a few genes from one of these beneficial bacteria to the plant can duplicate the multiple mechanisms and benefits of the organisms when they are present and are associated with the plant.

Greater knowledge of these biological interactions and cycles could help reverse a situation of too little progress on biological control. Too few programs have approached the development of microbial biocontrols by starting with nature's own best systems as a source of candidate organisms.

In concluding this section, the following question must be asked: How many other biological interactions and processes of this kind are working or could be working for the benefit of crop production but are still waiting to be discovered?

FIGURE 3-16 Suppression of take-all by phenazine-producing and nonproducing bacterial strains. Seedlings were from experiments in which (A) nontreated seeds or (B) seeds treated with antibiotic-negative Tn5 mutant, (C) mutant strain restored for antibiotic-producing ability or (D) wild-type antibiotic producer was applied. Source: L. S. Thomashow and D. M. Weller. 1988. Role of phenazine antibiotic from *Pseudomonas fluorescens* in biological control of *Gaeumannomyces graminis* var. *tritici*. Journal of Bacteriology 170:3499–3508.

## ALTERNATIVE AGRICULTURE

The report *Alternative Agriculture* (National Research Council, 1989a) calls attention to the evidence that agriculture stands to gain in both economic terms and ecological sustainability by taking greater advantage of the biological interactions and natural cycles that are already at work or available to work on the farm. When pesticides upset or mask this natural biological resource, it may be advantageous to use them in a more sophisticated way or find alternatives that more nearly optimize the benefits of these biological processes external to but associated with the crop. This was the approach used by Baker (1957) in the 1950s to get around the problem of the excess heat treatment of soil that destroyed Hartley's (1921) biological factor that was suppressive to soilborne pathogens. Baker did not eliminate steam treatment. He turned, instead, to milder treatments to eliminate the pathogens selectively while leaving a complex microbiota as a source of biological buffering against reinvasion of the soil by pathogens. It is not proposed that a herbicide such as glyphosate be eliminated as a tool for the management of volunteer plant cereals and weeds, but it is proposed that it be applied to permit time for soil microorganisms to compete with and reduce the energy supply of the pathogens.

Alternative agriculture is a process or strategy used to guide decisions with the goal of making the farming enterprise more sustainable both economically and ecologically. It is not a distinct set of farming practices, methods, or systems (Figure 3-17). Moreover, there is no intrinsically

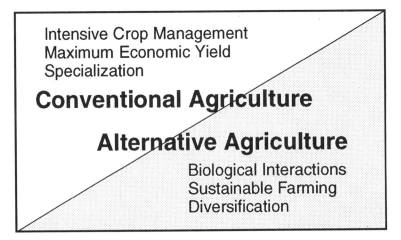

FIGURE 3-17 Diagrammatic illustration of some features that distinguish conventional and alternative agriculture, but it should be pointed out that these farming systems also tend to overlap in many ways.

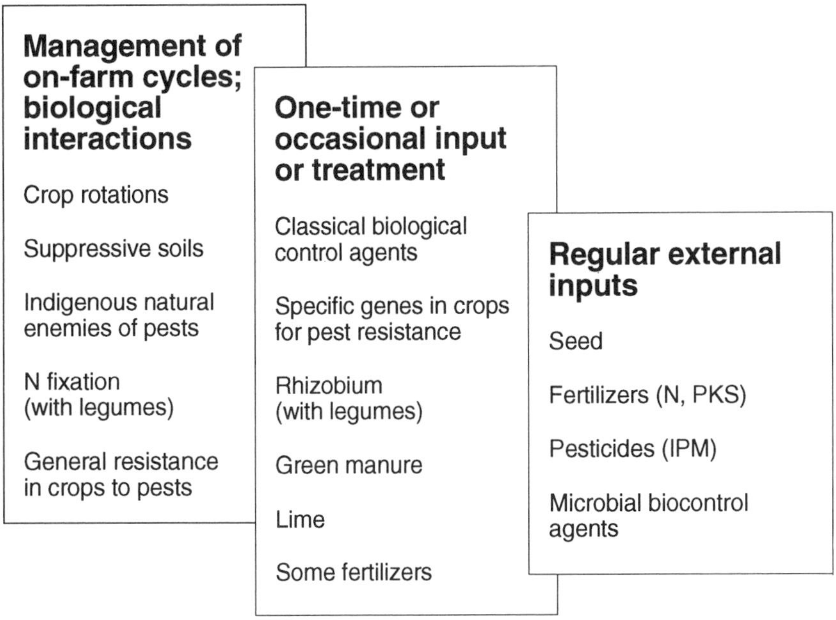

**Management of on-farm cycles; biological interactions**

Crop rotations

Suppressive soils

Indigenous natural enemies of pests

N fixation (with legumes)

General resistance in crops to pests

**One-time or occasional input or treatment**

Classical biological control agents

Specific genes in crops for pest resistance

Rhizobium (with legumes)

Green manure

Lime

Some fertilizers

**Regular external inputs**

Seed

Fertilizers (N, PKS)

Pesticides (IPM)

Microbial biocontrol agents

FIGURE 3-18 Diagram of the levels of practices, inputs, and mechanisms that could be considered in designing sustainable farming systems. N, nitrogen; IPM, integrated pest management; PKS, phosphorus, potassium, sulfur. Source: Modified from R. J. Cook and R. J. Veseth. 1991. Wheat Health Management. St. Paul, Minn.: APS Press.

correct way to proceed since different soils, climates, and markets require different practices, methods, or cropping systems. Nevertheless, the same general ecological principles can be used to guide the process, whether in a given field, on the farm, within a specific region, or across the United States.

The basic ecological principles and benchmark indicators of sustainability have been covered adequately in other chapters of this volume. Instead, a pragmatic guide to the process of designing sustainable farming systems is proposed in Figure 3-18. This guide examines the alternatives starting with

• making maximum use of biological interactions and cycles already working on the farm;

• making a one-time or occasional input that is long-lasting or perhaps even self-maintaining; and

• making regular inputs in the form of purchased fertilizers, pesticides, plant-asociated microorganisms, and seeds.

## Making Maximum Use of Biological Interactions and Cycles on the Farm

Among the on-farm practices or methods are the use of crop rotations and disease-suppressive soils together with less tillage, the optimization of biological controls of insects by using indigenous natural enemies within and across agroecosystems, more innovative use of nitrogen-fixing crops, and maximization of general mechanisms of resistance to pests in plants. Examples of these are provided below.

### Crop Rotations and Tillage

Crop rotations and tillage need little further discussion, but the problem of poor performance of crops when they are grown without adequate crop rotation is a general phenomenon that involves much more than small grains, corn, and soybeans. The following terms have emerged to describe the kinds of problems that are encountered and, some presume, to reflect the diagnosis: soil sickness, allelopathy, autotoxicity, the interference effect of corn crops year after year, tired and worn out soil, phytotoxic crop residue, monoculture injury, monoculture effect, and replant problem. Most of these terms are euphemisms for root diseases.

### Suppressive Soils

Suppressive soils are soils in which, because of their unique microbiological properties, pathogens, nematodes, or insect pests will not establish; they establish but do not reach economic population thresholds; or they establish and produce disease or cause damage for awhile but then decline while yields recover (Baker and Cook, 1974). As described above for take-all decline, these suppressive soils hold a wealth of clues to biological control (Cook and Weller, 1987). In some cases, it may be possible to recover a single strain of microorganism, genetically alter it to make it more competitive in the rhizosphere, and in this way duplicate the effect. Other cases are too complicated to manage in any way other than empirically by cultural practices or with mixtures of beneficial microorganisms. However, like so many related areas, there are too few investigators of these problems and the progress is too slow.

Boswell (1965) pointed this out 25 years ago. Boswell was then with the Crops Research Division of the Agricultural Research Service, USDA, and a member of the Board on Agriculture of the National Research Council. Working through the National Research Council, he helped obtain funding for the first international symposium on the ecology of soilborne plant pathogens, which was held at the University of California, Berkeley, in 1963 and published as a proceedings (Baker and Snyder, 1965). In the forward to that volume, which he termed "A Landmark in Biology," Boswell (1965, p. 3) wrote:

> [I]f the search for disease resistance is not successful, and no industrial chemical or physical treatment is available for economically controlling a soil-borne pathogen, the disease generally goes uncontrolled. Why is the microecological approach so rarely tried on a substantial basis? Is it too slow, too expensive, or just too hard? In view of the stakes to be won, and of some of man's efforts today, none of those terms seem to be applicable.

No area of research offers more to improving both the productivity and the ecological sustainability of cropping systems than does the area of rhizosphere microbiology (Rovira et al., 1990). Understanding and managing the interface between roots and soils are the key not only to root health but also to the greater use of nitrogen fixation, more efficient phosphorus uptake by mycorrhizae, and many other agriculturally important biological processes. Thus far, however, the field suffers from benign neglect.

*Biological Control of Insects with Indigenous Natural Enemies*

Biological control of insects with indigenous natural enemies is another largely unexplored approach to making greater use of the biological resource external to a crop. One such approach involves the use of predatory insects that have the ability to maintain their populations on alternate or secondary food sources during periods when the population of the target insect pest is low. The potential for this approach has been shown by Coll and Bottrell (1991) for biological control of the European corn borer. The peak density of two insect predators (*Orius insidiosis* and *Coleomeqilla maculata*) on corn in western Maryland occurred in response to corn pollen, which these beneficial insects use as a secondary food source, but coincided with the buildup of the corn borer, which these predators then turned to as their primary food source. Buildup of the predators can occur simultaneously with or even prior to buildup of the pest, rather than in response to the pest and often too late for maximum biological control. D. Bottrell (International Rice Research Institute, personal communication, 1990) has suggested that a mixture of corn hybrids, each with the ability to produce pollen at a slightly different time, could help to maintain and stabilize the populations of these pollen-feeding predatory insects even more, provided they are not destroyed by insecticides.

*Biological Nitrogen Fixation Using a Legume*

Biological nitrogen fixation using a legume in the crop rotation can provide most or, in many cases, all of the nitrogen needs of corn or small grains, depending on the legume and how it is managed (National Research Council, 1989a). The land-grant universities and USDA have all but ceased agronomic research on legumes and their management as alternative sources of nitrogen, but this trend may now be reversed. Nitrogen fertilization accounts for the largest single direct cost to producing crops such as corn and small grains and is often more than the cost of all other off-farm purchases combined (National Research Council, 1989a). Nitrogen losses from the rooting zone also account for much of the present concerns for groundwater quality. Clearly, the greatest impact on reducing the cost of inputs, rebuilding soils, and protecting groundwater will come from bringing more legumes back into the rotations.

Like other challenges and rewards in research and education on matters pertaining to changing agricultural practices, illustrated, for example, by the accounts of recent work on the crop rotation and crop residue effects presented above, the placement of legumes back into the cropping systems may require that researchers and growers dispense with misconceptions, preconceived ideas, and conventional wisdom and try new ideas or reexamine old ones.

## Maximizing the General Resistance in Crops to Disease

The general resistance in crops to disease can be maximized by cultural practices that minimize the disease-favoring physiological stresses so common in intensively managed crops. The diseases, in these cases, are caused by the so-called weak parasites or opportunistic pathogens that take advantage of crops under physiological stresses, such as those caused by temperature extremes, salinity, drought, and nutritional deficiencies.

Returning to the case of semi-dwarf wheats in the Pacific Northwest, while farmers in areas with higher amounts of rainfall were pleased with their relatively high yields, farmers in areas with low amounts of rainfall where wheat is grown on summer fallow were not satisfied with their relatively low yields and expected that they, too, should reap the high yields with these "miracle" wheats. By 1964, only 3 years after the Gaines variety of wheat was introduced in the Pacific Northwest, fusarium foot rot had become a major yield-limiting factor in the low-rainfall area (Cook, 1980). This disease is similar to fusarium stalk rot in sorghum and maize and is favored by plant water stress.

Research over about a 10-year period showed that the general resistance of wheat strains to this disease was masked by the water stress caused by too much top growth in response to the high rates of nitrogen (Cook, 1980; Papendick and Cook, 1974). In a nutshell, farmers were fertilizing for 80- and even 100-bu/acre yields in fields that had enough water for only 50 to

60 bu/acre. When nitrogen rates were cut back more in line with the available water supplies, the disease ceased to be a problem, except during unusually dry years.

This experience and relatively simple control of a disease illustrates an important principle of sustainable agriculture: that the limits of the cropping systems should be known and respected (Cook, 1989).

## Making an Occasional Release, Introduction, or Treatment

The on-farm biological processes and cycles can make major contributions to the productivity of the cropping systems but are not adequate in and of themselves as a means to reach and sustain the full production capability of the crops. Ordish and Dufour (1969, p. 31) put it this way:

> [F]arming is a most unnatural activity. Man has imposed on the environment a system of survival of what he wants to use over the Darwinian system of the fittest to survive. Consequently the farmer is engaged in a constant struggle with nature.

Meeting the challenge of this struggle requires inputs as a supplement to the background biological control and natural sources of soil fertility. It can be an even greater challenge to design or come up with inputs that are required only once or occasionally and that become somewhat self-maintaining. Yet, some of the greatest success stories in agriculture have involved this approach and have been based on public-supported research.

### Classical Biological Control of Insects

Classical biological control of insects, whereby exotic natural enemies of either a native or introduced insect pest are introduced into habitats or environments where the pest is a problem, has one of the best records of success for investment of all approaches to pest control. According to Ehler (1990), more than 500 releases of predators or parasites have been made worldwide for the control of nearly 300 different species of insect pests over the past 100 years, of which 40 percent are now providing substantial control and 15 percent are providing complete control of insect pests. This is a remarkable record of accomplishment; yet this approach, like other approaches to the biological control of pests and diseases, continues to be inadequately funded and remains largely the work of a few people.

### The Deployment of Genes for Disease and Pest Resistance

The deployment of genes for disease and pest resistance in new varieties also has a remarkable record of success. Stem rust of wheat in the Great

Plains has been limited to only a few local epidemics since 1954, the year of the last major epidemic of stem rust in North America. This has been accomplished by the strategic deployment of genes for resistance in new varieties released in response to changing patterns of virulence in the pathogen (Roelfs, 1988). The southern corn leaf blight epidemic of the 1970s is lesson enough of what can happen without proper ongoing attention to disease resistance in major crops.

On the other hand, only a few of the many diseases and insect pests of crops are amenable to control by resistant varieties, and all known sources of resistance to some diseases are now in use and have not kept up with the variable populations of pathogens or new biotypes of insect pests. Increasingly, unconventional genetic sources and unconventional methods of breeding will need to be examined to stay ahead of some pathogens or pests and to control others for which there has been no source of resistance to date.

## Making Regular Inputs

The on-farm processes or cycles and the on-time or occasional inputs or treatments are rarely, if ever, adequate by themselves for full and efficient crop production, and therefore, this unnatural activity called farming also depends on regular inputs. These are the familiar off-farm sources of fertilizers, pesticides, seed, and possibly in the future, plant-associated microorganisms.

However, these inputs should be used as a supplement to and not a replacement for making maximum use of the on-farm renewable resources. Soil fumigation is necessary for some vegetable and fruit crops (Wilhelm and Paulus, 1980), but to what extent has it become a substitute for crop rotation? Insecticides are essential for the control of some insect pests, but to what extent have they become substitutes for equally effective cultural practices, taking advantage of indigenous natural enemies, or growing the right variety?

One new approach that uses existing technology is to grow varieties as mixtures (Wolfe and Barrett, 1980). The evidence is clear that mixtures of varieties having common maturity characteristics and end-use properties but differing in their traits for resistance or tolerance to environmental or biological stresses can produce both higher average yields and more stable yields over a wider range of conditions (Figure 3-19). Yet, changes from the traditional thinking that pure seed lines are best come slowly. More research data on the value of diverse populations versus single genotypes of crop plants can always be used, but the greatest need at this stage may be in education.

Microbial biocontrol agents typically do not persist in the soil or on plants when they are introduced into these habitats, and therefore, they must be reintroduced on a regular basis. For 70 years, since the pioneering work of Hartley (1921), plant pathologists have been attempting to obtain biological control of soilborne plant pathogens by introducing microorgan-

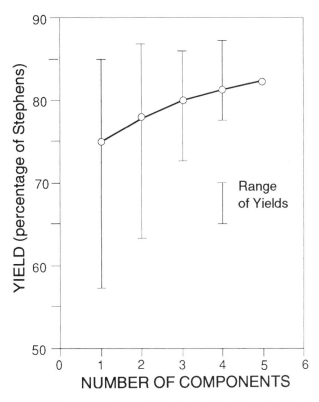

FIGURE 3-19 Yield as a percentage of standard soft white wheat in Oregon when varieties of similar end-use characteristics were mixed. Unpublished data of C. C. Mundt, L. S. Brophy, and M. R. Finckh.

isms into soil, with there being fewer than half a dozen successes to date, because the agents do not survive in high enough populations long enough or in the right places to do the job. Looking to the future, this situation may change with new approaches. Rather than isolating microorganisms from soil and reintroducing them in mass numbers back into soil, they are now isolated from plants, screened for their ability to inhibit the target pathogen, and then reintroduced back onto the plant (Cook and Baker, 1983). By making selections of candidate strains from plants growing where disease once occurred but has declined, as described above for bacteria that suppress take-all, the chances of finding effective strains are increased.

The International Rice Research Institute is now taking this approach to the control of rhizoctonia sheath blight in Asia (Mew and Rosales, 1986), which is now among the three most important diseases of rice in that part of the world.  There is no source of genetic resistance to this disease, and chemical control is either too inconsistent or too expensive.  In March 1990, there was an historical planning workshop where rice pathologists from six Asian countries and the International Rice Research Institute decided to concentrate their collective efforts for the next 3 years on screening and testing some 10,000 candidate strains of plant-associated microorganisms representing 15 to 20 environments and rice-growing ecosystems in Asia in an effort to evaluate the potential of this approach to biological control of rhizoctonia sheath blight and other rice diseases in Asia.

This is indeed an exciting approach.  Just as public efforts in plant breeding gave rise, eventually, to some highly successful private seed companies, so will public efforts in this new dimension to plant improvement eventually give rise to successful private companies with biocontrol products. For now, however, progress seems to depend largely on public-supported research and education.

### Rhizobium *Inoculations*

*Rhizobium* inoculations have been carried out by farmers since the turn of the century, yet knowledge of the ecology and practical management of introduced *Rhizobium* species and strains has not increased appreciably in the past 30 years.  Virtually all of the modern-day attention on *Rhizobium* species is directed at studies of the molecular biology of the root-bacterium association and of nitrogen fixation (Djordjevic et al., 1987; Long, 1984), but there has been virtually no progress on how to establish a preferred strain in competition with less efficient but more aggressive native strains already in the soil.

It is hoped that some of the current molecular biology research results will point the way to future productive research in soil ecology, because unless strains engineered for more efficient nitrogen fixation can be established and maintained in natural soils, continuation of progress will largely be on the present plateau in which only limited benefits are derived from nitrogen fixation.  The problem is very similar to that faced by plant pathologists who try to introduce root-associated microorganisms into fields with the seed to control root diseases: how to manage microorganisms in the rhizosphere of crop plants.

### THE EXPANDING AGENDA FOR AGRICULTURAL RESEARCH AND EDUCATION

The discussion here is not meant to suggest that research and education should shift the emphasis away from production as a means of making agriculture more sustainable.  A temporary shift away from strictly production-type research may be needed in some cases, but in the long term, both can and must be done. The agenda is not changing; it is expanding.

U.S. agriculture has long given emphasis to ecologically sound farming practices and methods, going back to the establishment of the Soil Conservation Service and Soil Conservation Districts; but political, economic, cultural, and technical forces have worked at cross-purposes to this goal. The problems of the 1980s have compelled U.S. agriculture researchers and growers to think more deeply about the early successes with specialization and the high-yield cropping systems introduced in the 1960s and 1970s and whether they can be sustained. Many have concluded that ecology in all of its aspects must now take its rightful place on a par equal with both increasing production and increasing production efficiency.

A point often overlooked is that farms with highly productive soils benefit more from new technology than do those with severely eroded soils. The introduction of high-yielding semi-dwarf wheats together with new fertilization practices was of benefit on farms with eroded soils but was an even greater boon to farms with deep rich soils. Likewise, the farms that benefited most from hybrid corn were those with soils that could support the full production capability of these new plant types. The actual decline in productivity of a soil is usually so gradual as to go unrecognized in the context of year-to-year variations in yields, and the impact of the loss of each inch of topsoil is also relatively small early in the erosion process. The impact can then increase dramatically, however, with a further loss of topsoil, particularly in shallow soils or over subsoils with low productivities.

It is not good enough to simply decree that 50, 60, 75 and eventually, 100 percent of the acreage of major agronomic crops grown in the United States will be grown with conservation tillage by the year 2000, 2010, or some other target date. Farmers need solutions to the problems—mainly biological problems—that they face with these high-risk management systems. These problems, as pointed out in *Alternative Agriculture* (National Research Council, 1989a), are amenable to biological solutions.

When considering the major approaches used for pest and disease control—physical, biological, and chemical methods (Figure 3-20)—the emphasis of the past 30 years has been disproportionately weighted toward physical and chemical methods. Tillage, open-field burning, as is often done with wheat stubble in the Pacific Northwest, fumigation, and pesticides can produce spectacular results that farmers can see. In contrast, biological methods are slower acting and less spectacular, and the effects often go unnoticed. Nevertheless, those involved in research and education have a responsibility to help elevate the status and credibility of this approach to the level it so richly deserves. The ability to make agriculture both economically more productive and ecologically more sustainable depends on it.

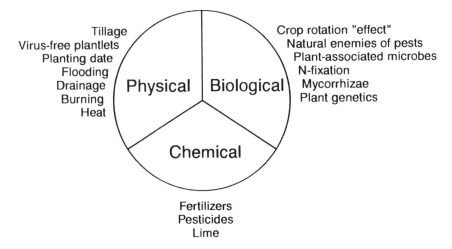

FIGURE 3-20 Examples of the physical, biological, and chemical elements of crop health management. Source: Modified from R. J. Cook and R. J. Veseth. 1991. Wheat Health Management. St. Paul, Minn.: APS Press.

The integrated pest management (IPM) efforts aimed at specific disease and pest problems must continue, but it is no longer good enough to fragment research and education efforts into IPM for insects, IPM for diseases, IPM for weeds, and IPM for nematodes, leaving the real integration to farmers. The agenda must be expanded to include more holistic approaches, perhaps developed around the principles of plant health management (Cook and Veseth, 1991).

Pests, diseases, and the abiotic constraints to plant health must be dealt with in ways that are both ecologically sustainable and economically profitable, but doing so will also become increasingly more technical. Some people question whether educational institutions are providing the right kind of training for practitioners of plant health and suggest that there is a need for a doctor of plant health comparable to the doctor of veterinary medicine or general practitioner in human medicine (Browning, 1983). Making greater use of biological controls and biological nitrogen fixation may come in the form of new products, but it will just as likely come in the form of new practices where the only commodity for sale is information. Agriculture would benefit enormously by highly trained private as well as public-supported practitioners of plant health.

The ideas presented here are not new, and the notion that agriculture can become both more productive and more sustainable through biology and ecology goes back a long way. Boswell (1965), in his reference to what he called the "microecological approach," predicted that it could lead to "more enduring and wiser farming practices." Nevertheless, this is an idea whose time has come. There are both the need and the tools, as never before, to do the research and carry out the education. The challenge is great, but so will be the rewards.

## REFERENCES

Aldrich, D. G., and J. P. Martin. 1952. Effect of soil fumigation on some chemical properties of soils. Soil Science 73:149–159.

Baker, K. F., ed. 1957. The U.C. System for Producing Healthy Container-Grown Plants. California Agricultural Experiment Station Manual 23. Berkeley: University of California.

Baker, K. F. 1962. Principles of heat treatment of soil and planting material. Journal of the Australian Institute of Agricultural Science 28:118–126.

Baker, K. F., and R. J. Cook. 1974. Biological Control of Plant Pathogens. San Francisco: W. H. Freeman.

Baker, K. F., and W. C. Snyder, eds. 1965. Ecology of Soil-Borne Plant Pathogens. Prelude to Biological Control. Berkeley: University of California Press.

Benbrook, C. M. 1990. Why a Leopold center for sustainable agriculture? Paper presented at the dedication ceremony, Aldo Leopold Center for Sustainable Agriculture, Ames, Iowa, February 6, 1990.

Boswell, V. R. 1965. A landmark in biology. P. 3 in Ecology of Soil-Borne Plant Pathogens. Prelude to Biological Control, K. F. Baker and W. C. Snyder, eds. Berkeley: University of California Press.

Brisbane, P. G., L. L. Janik, M. E. Tate, and R. F. O. Warren. 1987. Revised structure for phenazine antibiotic from *Pseudomonas fluorescens* 2-79 (NRRL B-15132). Antimicrobial Agents and Chemotherapy 31:1967–1972.

Browning, J. A. 1983. Goal of plant health in the age of plants: A national plant health system. Pp. 45–57 in Challenging Problems in Plant Health, T. Kommedahl and P. H. Williams, eds. St. Paul, Minn.: APS Press.

Chamswarng, C., and R. J. Cook. 1985. Identification and comparative pathogenicity of *Pythium* species from wheat roots and wheat-field soils in the Pacific Northwest. Phytopathology 75:821–827.

Coll, M., and D. G. Bottrell. 1991. Microhabitat and resource selection of the European corn borer (Lepidoptera: Pyralidae) and its natural enemies in field corn. Environmental Entomology 20:526–533.

Cook, R. J. 1980. Fusarium foot rot of wheat and its control in the Pacific Northwest. Plant Disease 64:1061–1066.

Cook, R. J. 1984. Root health: Importance and relations to farming systems. In Organic Farming: Current Technology and its Role in a Sustainable Agriculture, D. F. Bezdicek and J. F. Power, eds. American Society of Agronomy Special Publications, Madison, Wis.

Cook, R. J. 1986. Wheat management systems in the Pacific Northwest. Plant Disease 70:894–898.

Cook, R. J. 1989. Biological control and holistic plant-health care in agriculture. American Journal of Alternative Agriculture 3:51–62.

Cook, R. J. 1990. Diseases caused by root-infecting pathogens in dryland agriculture. In Advances in Soil Science, Vol. 13, B. A. Stewart, ed. New York: Springer-Verlag.

Cook, R. J., and K. F. Baker. 1983. The Nature and Practice of Biological Control of Plant Pathogens. St. Paul, Minn.: APS Press.

Cook, R. J., and W. A. Haglund. 1982. Pythium root rot: A barrier to yield of Pacific Northwest wheat. Washington State University College of Agriculture Research Bulletin No. XB0913. Pullman, Wash.: Agricultural Research Center, Washington State University.

Cook, R. J., and W. A. Haglund. In press. Wheat yield depression associated with conservation tillage caused by root pathogens in soil not phytotoxins from the straw. Soil Biology and Biochemistry.

Cook, R. J., and R. J. Veseth. 1991. Wheat Health Management. St. Paul, Minn.: APS Press.

Cook, R. J., and D. M. Weller. 1987. Management of take-all in consecutive crops of wheat or barley. Pp. 41–76 in Nonconventional Methods of Disease Control, I. Chet, ed. New York: John Wiley & Sons.

Cook, R. J., J. W. Sitton, and J. T. Waldher. 1980. Evidence for *Pythium* as a path-ogen of direct-drilled wheat in the Pacific Northwest. Plant Disease 64:102–103.

Cook, R. J., J. W. Sitton, and W. A. Haglund. 1987. Increased growth and yield responses of wheat to reduction in the *Pythium* populations by soil treatments. Phytopathology 77:1192–1198.

Cook, R. J., D. M. Weller, and E. N. Bassett. 1988. P. 53 in Effect of bacterial seed treatments on growth and yield of recropped wheat in western Washington, 1987. Biological and Cultural Tests, Vol 3. St. Paul, Minn.: APS Press.

Crookston, R. K. 1984. The rotation effect. What causes it to boost yields? Crops and Soils Magazine, March, 12-14.

Crookston, R. K., and J. E. Kurle. 1989. Corn residue effect on the yield of corn and soybean grown in rotation. Agronomy Journal 81:229–232.

Crookston, R. K., J. E. Kurle, and W. E. Lueschen. 1988. Relative ability of soybean, fallow, and triacontanol to alleviate yield reductions associated with growing corn continuously. Crop Science 28:145–147.

Crookston, R. K., J. E. Kurle, P. J. Copeland, J. H. Ford, and W. E. Lueschen. 1991. Rotational cropping sequence affects yield of corn and soybean. Agronomy Journal 83:108–113.

Djordjevic, M. A., D. W. Gabriel, and B. G. Rolfe. 1987. *Rhizobium*—the refined parasite of legumes. Annual Review of Phytopathology 25:145–168.

Ehler, L. E. 1990. History and future of classical biological control. P. 93 in Biological Control of Insects: Directions for the Future. AAAS Annual Meeting Abstracts. Washington, D.C.: American Association for the Advancement of Science.

Elliott, L. F., T. M. McCalla, and A. Waiss, Jr. 1978. Phytotoxicity associated with residue management. Pp. 131–146 in Crop Residue Management Systems, W. R. Oschwald, ed. American Society of Agronomy Special Publication No. 31. Madison, Wis.: American Society of Agronomy.

Gurusiddaiah, S., D. M. Weller, A. Sarkar, and R. J. Cook. 1986. Characterization of an antibiotic produced by a strain of *Pseudomonas fluorescens* inhibitory to

*Gaeumannomyces graminis* var. *tritici* and *Pythium* spp. Antimicrobial Agents and Chemotherapy 29:488–495.

Hartley, C. 1921. Damping-off in forest nurseries. Pp. 1–99 in U.S. Department of Agriculture Bulletin 934. Washington, D.C.: U.S. Department of Agriculture.

Howie, W. J., R. J. Cook, and D. M. Weller. 1987. The effect of soil matric potential and cell motility on wheat root colonization by fluorescent pseudomonads suppressive to take-all. Phytopathology 77:286–292.

Ingram, D. M., and R. J. Cook. 1990. Pathogenicity of four *Pythium* species to wheat, barley, peas, and lentils. Plant Pathology 39:110–117.

Leggett, G. E. 1959. Relationship between wheat yield, available moisture, and available nitrogen in eastern Washington dryland areas. Washington Agricultural Experiment Station Bulletin No. 609. Pullman, Wash.: Washington Agricultural Experiment Station.

Long, R. R. 1984. Genetics of *Rhizobium* nodulation. Pp. 265–306 in Plant-Microbe Interactions, T. Kosuge and E. W. Nester, eds. New York: Macmillan.

Mew, T. W., and A. M. Rosales. 1986. Bacterization of rice plants for control of sheath blight caused by *Rhizoctonia solani*. Phytopathology 76:1260–1264.

Moore, K. J., and R. J. Cook. 1984. Increased take-all of wheat with direct-drilling in the Pacific Northwest. Phytopathology 74:1044–1049.

National Research Council. 1989a. Alternative Agriculture. Washington, D.C.: National Academy Press.

National Research Council. 1989b. The Ecology of Plant-Associated Microorganisms. Washington, D.C.: National Academy Press.

Ogoshi, A., R. J. Cook, and E. N. Bassett. 1990. *Rhizoctonia* species and anastomosis groups causing root rot of wheat and barley in the Pacific Northwest. Phytopathology 80:784–788.

Ordish, G., and D. Dufour. 1969. Economic bases for protection against plant diseases. Annual Review of Phytopathology 7:31–50.

Papendick, R. I., and R. J. Cook. 1974. Plant water stress and development of Fusarium foot rot in wheat subjected to different cultural practices. Phytopathology 64:358–363.

Pesek, J. 1989. Comments on alternative agriculture. Paper presented to International Fertilizer Industry Association, Cancun, Mexico, November 29, 1989.

Pierson, E. A., and D. M. Weller. 1990. Recent work on control of take-all of wheat by fluorescent pseudomonads. P. 3 in Proceedings of the Second International Workshop on Plant Growth-Promoting Rhizobacteria, October 14-19, 1990, Interlaken, Switzerland, G. Défago, ed. Zurich: Institut für Pflanzenwissenschaften Bereich Phytomedizin, ETH-Zentrum.

Roelfs, A. P. 1988. Genetic control of phenotypes in wheat stem rust. Annual Review of Phytopathology 26:351–367.

Rovira, A. D. 1986. Influence of crop rotation and tillage on Rhizoctonia bare patch of wheat. Phytopathology 76:669–673.

Rovira, A. D., L. F. Elliott, and R. J. Cook. 1990. The impact of cropping systems on rhizosphere organisms affecting plant health. Pp. 389–435 in The Rhizosphere, J. M. Lynch, ed. New York: John Wiley & Sons.

Shipton, P. J. 1975. Take-all decline during cereal monoculture. Pp. 137–144 in

Biology and Control of Soil-Borne Plant Pathogens, G. W. Bruehl, ed. St. Paul, Minn.: APS Press.

Stack, R. W., and M. McMullen. 1988. Root and crown rots of small grains. NDSU Extension Service PP-785 (Rev.). Fargo, N.D.: North Dakota State University.

Thomashow, L. S., and D. M. Weller. 1988. Role of a phenazine antibiotic from *Pseudomonas fluorescens* in biological control of *Gaeumannomyces graminis* var. *tritici*. Journal of Bacteriology 170:3499–3508.

Thomashow, L. S., and D. M. Weller. 1990. Application of fluorescent pseudomonads to control root diseases of wheat and some mechanisms of disease suppression. Pp. 109–122 in Biological Control of Soil-Borne Plant Pathogens, D. Hornby, ed. Slough, England: C.A.B. International.

Thomashow, L. S., D. M. Weller, R. F. Bonsall, and L. S. Pierson III. 1990. Production of the antibiotic phenazine-1-carboxylic acid by fluorescent *Pseudomonas* species in the rhizosphere of wheat. Applied and Environmental Microbiology 56:908–912.

Vogel, O. A., J. C. Craddock, Jr., C. E. Muir, E. E. Everson, and C. R. Rohde. 1956. Semidwarf growth habit in winter wheat improvement for the Pacific Northwest. Agronomy Journal 48:76–78.

Weller, D. M., and R. J. Cook. 1983. Suppression of take-all of wheat by seed treatments with fluorescent pseudomonads. Phytopathology 73:463–469.

Weller, D. M., R. J. Cook, G. MacNish, E. N. Bassett, R. L. Powelson, and R. R. Petersen. 1986. Rhizoctonia bare patch of small grains favored by reduced tillage in the Pacific Northwest. Plant Disease 70:70–73.

Wilhelm, S., and A. O. Paulus. 1980. How soil fumigation benefits the California strawberry industry. Plant Disease 64:264–270.

Wolfe, M. S., and J. A. Barrett. 1980. Can we lead the pathogen astray? Plant Disease 64:148–155.

Zingg, A. W., and C. J. Whitfield. 1957. Stubble-mulch farming in the western states. Pp. 1–56 in U.S. Department of Agriculture Technical Bulletin No. 1166. Washington, D.C.: U.S. Department of Agriculture.

# 4

# Overview of Current Sustainable Agriculture Research

*John C. Gardner, Vernon L. Anderson, Blaine G. Schatz,
Patrick M. Carr, and Steven J. Guldan*

To most Americans, "nature" occupies the country's national parks
and wildlife preserves. The vast stretches of the United States that are
dedicated to agriculture, however, make farmers and their land the most
important components of the contemporary environment. Agriculture is
among the most intimate experience that people have with nature, since
it is a two-way interaction. It is perhaps this link, the direct impact of
people on nature through agriculture, that is at the heart of the current
interest in public agricultural policy. The rural population is the most
rapidly shrinking and economically stressed sector of U.S. society, and
is the sector that has most closely interacted with the nation's most
important natural resource base—the land. Alternatives in agriculture
must be sought to ensure the permanence of soil and water, an economy
that rewards stewardship and maintains rural communities, and an overall social understanding of both the true costs of production and the
risks associated with further neglect in developing a global agricultural
policy.

Sustainable agriculture research has been widely discussed recently. It is
a time of rapid change as farmers, members of industry, researchers, and
educators adapt to new ideas. The objectives of this overview are to identify the course that research has taken to date and discuss, largely by
example, the process that will be needed to maintain the rapid pace of
discovery that has occurred over the past few years.

## THE CHALLENGES

Driven largely by urban environmental and rural economic and social concerns, the long-envied U.S. crop and livestock production process is now being questioned. Are too many chemicals being used? Is the water safe to drink? How much more topsoil can the United States afford to lose? Are foods contaminated with pesticides? Can farmers afford the rising costs of energy and purchased inputs? Can farmers farm without these inputs? While each question may be a legitimate concern, they all deal with symptoms. To date, most of the concern of both the urban public and the agricultural research community has been to describe and quantify the symptoms rather than to uncover the causes and discover the solutions to urgent problems.

It seems that the United States is on the threshold of a new vision for agriculture and the important role it plays in society. Appropriately, agriculture's reexamination has begun with the production process itself. History has repeatedly suggested that economic and social policy will only succeed if it is based on the sound ecological use of natural resources. The use of ecology and its principles, however, is largely unapplied in agricultural settings and seems to gain attention only after events that expose unsuccessful agricultural production practices. Hanson (1939) addressed the Entomology Society of America about ecological thinking after the Dust Bowl and grasshopper problems of the 1930s. Jackson and Piper (1989), Paul and Robertson (1989), and Elliott and Cole (1989) are echoing the plea in the midst of today's soil, water, and food safety concerns. An ecological approach in the study and development of crop production practices, however, is fundamentally incompatible with the typical agricultural research paradigm of reductionistic science. The reductionistic approach assumes that answers to problems are always at the next lower level of system organization; thus, agronomists become physiologists and physiologists become biochemists. This has largely led to the virtual abandonment of adequately funded and staffed applied interdisciplinary systems research programs (Buttel, 1985), including those with an ecological orientation.

Despite the apprehension, use of an ecological perspective could become the fundamental foundation for most sustainable agriculture research. At present, it holds the most potential to guide the search for ways to avoid or ameliorate the negative side effects and to find environmentally harmless approaches to the solutions of U.S. agricultural production problems. It may also help to reveal where potentially dramatic improvements can be made in agricultural productivity and resource-use efficiency. Even the most simple ecosystem models (Figure 4-1) graphically reveal the importance of cycling and the interdependency of all components within the ecosystem. Previously, those working in the agricultural sciences saw

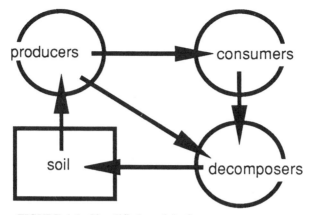

FIGURE 4-1  Simplified model of an ecosystem.

only the differences between natural and agroecosystems (Figure 4-2).  Now they are beginning to see the value in discovering the similarities (Lowrance et al., 1984).  Most of the negative symptoms of modern agriculture are a direct result of either bypassing or ambitiously attempting to remove vital ecosystem components.  Under such circumstances, what remains is an ecosystem with a limited ability to cycle nutrients and organic matter and one that becomes increasingly expensive to maintain.

It is within this "new" vision of agricultural production that clues to future improvement may be hidden.  Altieri (1983) and others have suggested the ecological approach as the scientific basis of an alternative agri-

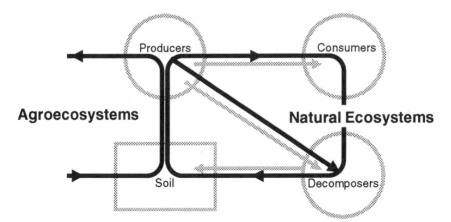

FIGURE 4-2  Differences in primary flow of nutrients and biomass between agricultural and natural ecosystems.

culture system that includes not only soils, plants, and animals but also the people who direct it. The observational and managerial skills of farmers are a key component. Such an agriculture system is now being widely tested through the low-input sustainable agriculture (LISA) research and education program of the U.S. Department of Agriculture.

## LISA: SUCCESSES AND LIMITATIONS

The LISA program has perhaps been the most visible and controversial of any agricultural research and education program recently introduced. It was first greeted with skepticism by both the agricultural chemical industry and many land-grant institutions. The popular agriculture trade journals frequently referred to it as a "smokescreen for organic agriculture" (Progress, 1988) and denounced the involvement of farmers and nonprofit institutions as "unscientific." In retrospect, the involvement of farmers and others outside of the traditional agricultural research circles has offered perhaps the most creative and practical advice yet (Kirschenmann, 1988). LISA has bridged many gaps between farmers, nonprofit organizations, and land-grant institutions, as evidenced by the projects supported thus far and reviewed in this volume (Madden et al., 1990).

LISA has also undoubtedly contributed to agriculture's self-evaluation in recent months. Articles in fertilizer trade journals today refer to the challenge of LISA concepts, the importance of the environment, and the need for knowledge of soil tests and how to apply their results for reasonable application rates and safety. *Successful Farming* now has an environmental column alongside market reports and herbicide recommendations. In addition, most farm magazines now carry regular sections on how farmers can cope with problems and remind them to savor the joys of modern rural life. A resurgence of agrarianism and LISA have occurred simultaneously, neither of which is coincidental.

One of the most surprising successes of LISA has been the ability to attract matching funds and encourage other institutions to invest likewise in alternative agricultural research and education programs. Ironically, the very mechanism of steering the public research agenda with outside grant money, as suggested by Hightower (1973) in reference to agricultural business and industry, is occurring in the direction of alternative agriculture with LISA. To many field-oriented and applied research programs across the country, LISA has been the first opportunity in some time to compete legitimately for a new source of outside funds. Long-forgotten extension programs involving on-farm research and demonstration are also being revitalized.

While LISA's success in broadening the research team and topics is admirable, it is not without its limitations. As the program matures, it must retain its original and most valuable quality: that of expanding the

TABLE 4-1 Major focus of Low-Input Sustainable Agriculture Projects Funded in 1988 and 1989

| Project Topic | Percentage of All Projects |
| --- | --- |
| Crops | 43.8 |
| Economics or sociology | 12.3 |
| Fertility or tillage | 16.3 |
| Insect management | 5.0 |
| Livestock | 13.8 |
| Pathogen management | 2.5 |
| Weed management | 6.3 |

NOTE: Of the projects funded through LISA in 1988 and 1989, about 25 percent were for demonstration or education and 75 percent were for research. The research projects are displayed by major focus. Studies dealing primarily with economics or sociology make up the remaining 12.3 percent not shown.

boundaries of agricultural research possibilities. Examination of the current research funded through the LISA program by major topic reveals that more than two-thirds of all projects have remained within the comfortable confines of traditional agroecosystems (Table 4-1). Even livestock have largely been ignored because of the scarcity of livestock-based project proposals submitted to the LISA program. Soil microbial aspects beyond a disease control emphasis are incorporated in very few of the projects, and only a handful of the projects funded in 1988 and 1989 examined the whole picture of soils, crops, and livestock.

Much of the LISA research has been the testing of alternative treatments in traditional settings. Beyond the topics of study themselves, the methods need to advance with the discoveries. What may have been novel 3 years ago may now appear simplistic. Old tendencies must be resisted; otherwise, demonstration projects will begin to lose their depth and research projects will lose their breadth. The creation of new methods and innovative inquiry must be seen to be as equally as important as exploration of the performance of an alternative crop or tillage method. Without the development of these new methods, researchers will succumb to the same old boundaries that limited them before. The inquiries must selectively challenge long-held agricultural beliefs that do not fit new ecological, economic, or social realities.

## ECOLOGICAL, ECONOMIC, AND SOCIAL CHANGES
## ON THE GREAT PLAINS

It is difficult to anticipate, much less identify, changes that occur slowly over several generations. Yet, it is anticipation that is a principal feature of sustainable agriculture. There are many important relationships that must be anticipated: agriculture and the environment, urban and rural society, agriculture and the economy. The ability to perceive changes can be viewed through two examples that greatly affected the land and its people in the Great Plains.

### Fallow on the Great Plains

Much of the Great Plains of the United States was transformed from native, perennial prairie grasses to domesticated, annual wheat over a century ago. The effect of this change on the nitrogen and carbon content of the soils is well documented (Haas and Evans, 1957; Hobbs and Brown, 1957). When those reports were published over 30 years ago, there had

Fallowed, or idle, land annually occupies up to one-quarter of all land in the wheat-growing regions of the Great Plains. Traditionally this land is tilled several times each summer to control weeds and mineralize nutrients. Over time, such practices have also increased susceptibility to soil erosion and hastened loss of soil organic matter.

Virgo black medic (*Medicago lupulina*) is grown as a "living" mulch fallow substitute crop after grain sorghum in the Central Great Plains. The optimum living mulch would be one that fixes nitrogen symbiotically, is capable of competing with weeds, and requires little soil moisture.

been a decline of approximately 46 percent in the organic carbon content and 42 percent in the organic nitrogen content of Great Plains soils. With the aid of fallow periods to conserve moisture and allow mineralization of nutrients, the prairie pioneers essentially "mined" the soil.

Faced with this reality, agricultural scientists first recognized the importance of nitrogen and spent considerable time and effort in the study of how best to replace it with nitrogen fertilizer. Over the long term, however, the loss in carbon content may be an equal, if not more serious, problem than the lack of soil nitrogen. As revealed in a recent review on soil tilth (Karlen et al., 1990, p. 158), which was defined as "the physical condition of a soil described by its bulk density, porosity, structure, roughness, and aggregate characteristics as related to water, nutrient, heat, and air transport; stimulation of microbial and microfauna populations and processes; and impedance of seeding emergence and root penetration," a clear knowledge of its importance was recognized among the scientific community 50 years ago. Such a complex and seemingly unquantifiable property was soon rejected in the reductionistic thinking of the past few decades, however. The past soils textbooks that taught most of today's scientists largely rejected the term *tilth*, using the concept in reference to tillage alone and dedicating less than three pages to its discussion (Brady, 1974).

Today, the soils of the Great Plains are thus dramatically different than they were during the early part of the twentieth century. With the loss of tilth and organic matter from both tillage and erosion since then, productiv-

"Dead" and "living" mulch substitutes for traditional tilled fallow. Although both alternatives help control soil erosion, the living system, which includes a cover crop, also has the potential for nitrogen fixation, temporary grazing of ruminants, and improvement of soil structure and organic matter content. The living system may also rely less on herbicides to maintain the fallow period.

ity has decreased (Bauer and Black, 1981). Acidification due to nitrogen fertilization is also a problem in the southern Great Plains. The ecology of the soil itself has changed, and there has been less nutrient cycling and greater susceptibility to damage from natural forces caused by changes in soil properties. Yet, tilled fallow remains a management practice for crops that annually cover nearly one-quarter of the land area encompassing the central and northern Great Plains.

Alternative tillage systems offer one possible means of regaining organic matter (Bauer and Black, 1981). Another possibility exists in alternative cropping systems that include the use of legumes for both nitrogen and carbon content improvement, but the scientific literature has discouraged such study. Most work on legumes in rotations in the Great Plains took place in the early 1900s. Summaries of these studies clearly reported that legumes were of no benefit to succeeding wheat crops: "The results of 20 years of experiments with green manure crops show nothing to recommend them" (Sarvis and Thysell, 1936). Similar conclusions were offered 20 years later, but qualifiers began to creep into the summaries: "The crop weather data . . . suggest that with present cultural techniques green manures should not be used" (Army and Hide, 1959). Army and Hide acknowledged that the research reported was set up with ideas based on the traditional approaches used in humid regions and implied that these approaches would not work in drier regions.

Duley and Coyle (1955) speculated that the value of using green manures in the Great Plains may not be realized until after the land had been farmed for a longer time and until further experiments determined the most effective culture methods. Brown (1964) recognized the changes that had occurred in Great Plains soils because of erosion and cropping over time. He also emphasized that the full benefits of cropping systems that included perennial grasses and legumes would not be detected from research plots located on favorable, relatively protected sites atypical from the norm of the Great Plains.

Researchers and farmers must learn to see and adapt to new realities such as the ecological changes that have occurred in the soils of the Great Plains over time. Carbon and nitrogen are needed in Great Plains soils, and legumes used as green manures could help increase the levels of both elements. Rather than rely on old data gathered under a different set of circumstances, researchers and farmers must look to alternative legumes and/or management strategies that are suited to contemporary conditions (Sims, 1989). The analysis must be multifaceted, recognizing all the factors that legumes contribute over time. Reduction of the inquiry to only a water use consumption comparison or some other single factor may mask the future potential of such systems (MacRae and Mehuys, 1989). Had the southern Australians approached the use of green manure legumes as being applicable only to humid regions, they would have missed the development of one of the most innovative approaches of raising wheat and ruminant animals in a dryland region in the world (Puckridge and French, 1983).

## The Changing Roles of Livestock

To the typical American, who is at least several generations off the farm, the mental image of a farm always includes livestock. Today this image is mostly nostalgic because farms are typically specialized into either crops or livestock. For example, in North Dakota, farm numbers have dropped at a rate of nearly 1,000 per year since the early 1960s. The number of farms with cattle has dropped at twice this rate (North Dakota Agricultural Statistics Service, 1988). The fewer and larger farms left are mostly crop-only operations. Economic and social changes over the past century have had a profound effect on the presence of livestock in agroecosystems.

The economies-of-scale and transportation were probably the first reasons for concentrating the livestock industry. Meat-packing plants built alongside the rail centers at Kansas City and Chicago gave easy access for the incoming cattle and the outgoing carcasses. The livestock industry is still largely directed by these strong economic forces. The poultry industry is concentrated in the southeast United States, and feeding of beef cattle has dominated in southwest Kansas and the southern High Plains. The largest packing plants are now located near the feedlots, to allow shipping of boxed beef rather than live cattle. The economy of packaging and delivering the highest-quality, most-uniform, and inexpensive meat products to consumers has thus changed the distribution of livestock across the country.

Social factors have further separated livestock from many farms. With the advent of larger tractors and implements concurrent with larger farms and fields, the mechanization of the entire farm has been appealing. Depending on the type of operation, livestock may require care throughout the year, competing with crop management and leisure time. Although cattle have romantically remained a part of the image of western agriculture, increasingly they do not fit the mold of a contemporary farm. There are also other social factors that have discouraged meat consumption. Health concern about animal fats and animal rights and welfare issues, whether perceived or real, have had an impact on the livestock industry.

Although agroecosystems are possible without livestock, domesticated animals have long been recognized as possible consumers for the agroecosystem (Joandet and Cartwright, 1975). Ruminants have the most potential to diversify and revitalize the agroecosystem because they use forage-based rations. Much of the livestock production, however, has concentrated on the use of high-quality feed grains. Switching from grain-based to forage-based rations would have widespread implications for both the livestock industry and the farm (Wedin et al., 1975). Even the recent attention of organic farming methods has been accompanied by criticism of the perception of requiring too much livestock (Bender, 1988). The enhanced use of livestock to enrich the agroecosystem thus seems confined by rigid economic and social boundaries.

## SCIENTIFIC LIMITATIONS

These examples indicate that new methods and thinking will be as important as new plants, animals, or tools in the development of alternative agriculture systems. Although reductionistic science has its strengths, it also has its weaknesses. The scientific community must internalize the criticism that it is not only new ecological, economic, and social situations that must be understood. It must also struggle with the scientific boundaries that limit attempts to meet these new challenges (MacRae et al., 1989).

One of the boundaries is how to perceive and deal with new technologies as they are applied to exercise control over agroecosystems. Many tools for this control have been added in the past century: first, there were mechanical tools, such as tractors, plows, and other inventions; then, there were chemical fertilizers and pesticides. Now, with the current understanding of molecular genetics, biotechnology could spawn a new era of biological tools. Yet, each technology has been slowly understood in how it relates to already existing technologies and the agroecosystem as a whole. Each new technology has been seen as a substitute for the previous one (Figure 4-3): herbicides for tillage, predators for insecticides. In practice, however, the relationships among these technologies in the field seldom

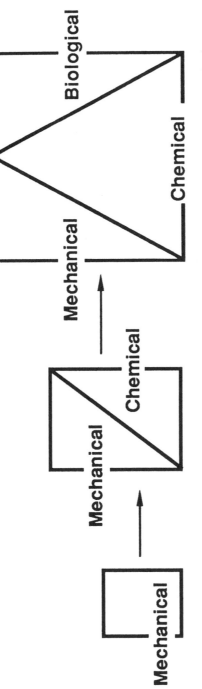

FIGURE 4-3  Evolution and current perception of the relationship among agricultural technologies.  Linear substitution of one technology for another is currently assumed in many research designs.

fit the one-for-one substitution vision. Like the resurgence of the term *soil tilth*, there must also be a broader vision of the agroecosystem's complexity. Most likely, there will be a move beyond studying the substitution of one technology for another and, instead, the interrelationships of technologies will be studied, given that there are likely other influential factors that have yet to be discovered (Figure 4-4).

A logical place to begin practicing such a vision is on a farm itself. Several groups of scientists have begun such studies. The U.S. Department of Agriculture Soil Tilth Laboratory is carefully studying the farm of Dick Thompson, which was featured as a case study in *Alternative Agriculture* (National Research Council, 1989), and his neighbor. The Northern Plains Sustainable Agriculture Society, along with North Dakota State University, is likewise carefully studying nine farms with different management styles across North Dakota. These and other similar ecological studies could identify key aspects of how current agricultural practices actually interrelate with the environment.

Once the agroecosystem is refined on the ecological level, new economic and social policies must be built upon it. Certainly, a longer-term economic vision is needed for agriculture. Too often, good agronomic ideas are aban-

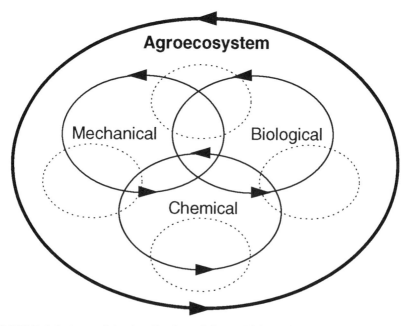

FIGURE 4-4  A possible visualization of the multidimensional relationship among current and yet undiscovered agricultural technologies and the ecosystem.

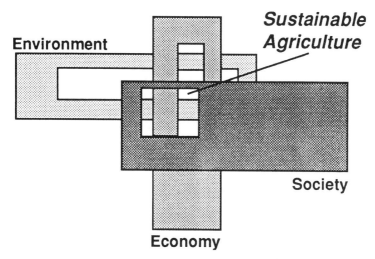

FIGURE 4-5 Three of the important boundaries that will shape future agricultural policy emphasizing the few possibilities that may exist that meet universal approval.

doned when the local lender analyzes the plan on an annual ledger sheet. Policies that will reduce the economic risk of ecologically sound, alternative agricultural production systems must be developed. Although many production systems may pass the rigors of environmental protection, economic security, and social acceptability individually, few will satisfy all the facets necessary for a successful and sustainable agriculture (Figure 4-5).

## CONCLUSION

While the current emphasis on sustainable agricultural research is on alternative practices, many of these are being tested and studied under rather confining ecological, economic, social, and scientific boundaries that must be tested on an equal basis. Both the treatments and methodologies of research must be expanded to continue the advance toward sustainable agriculture.

As suggested previously by many other investigators, the ecosystem model has been suggested as a reference point to help guide and reveal important missing components in sustainable agriculture research. If such research is to be carried out by individual, discipline-based scientists, ready and willing access to the other disciplines is necessary. Use of a farm itself along with full participation of the people who manage it may serve as an excellent starting point to focus on the whole system, rather than the parts of the system, that makes up agriculture. The involvement of farmers also provides early tests of the appropriateness and practicality of applied research. Many other, more basic areas of research may also be suggested with the aid of farmers and on-farm research.

Using the metaphor that agriculture is a conversation between humans and nature, perhaps there has been too much talking and not enough listening. People's observatory skills must be sharpened. People often learn from the most simple and ordinary of experiences. As written in 1854 by Henry David Thoreau in *Walden*, after his experience at Walden Pond:

> It is remarkable how easily and insensibly we fall into a particular route, and make a beaten track for ourselves. I had not lived there a week before my feet wore a path from my door to the pond-side; and though it is five or six years since I trod it, it is still quite distinct. It is true, I fear others may have fallen into it, and so helped to keep it open. The surface of the earth is soft and impressible by the feet of men; and so with the paths which the mind travels.

While LISA and other sustainable agriculture research has proved controversial and challenging, researchers and farmers must not be blinded by its early success, nor satisfied with its current vision. Many challenges lie ahead in the further advancement of agriculture.

## REFERENCES

Altieri, M. A. 1983. Agroecology. The Scientific Basis of Alternative Agriculture. Berkeley, Calif.: Division of Biological Control, University of California.

Army, T. J., and J. C. Hide. 1959. Effects of green manure crops on dryland wheat production in the Great Plains area of Montana. Agronomy Journal 51:196–198.

Bauer, A., and A. L. Black. 1981. Soil carbon, nitrogen, and bulk density comparisons in two cropland tillage systems after 25 years and in virgin grassland. Soil Science Society of America Journal 45:1166–1170.

Bender, J. 1988. Does organic farming require too much livestock? American Journal of Alternative Agriculture 3:2.

Brady, N. C. 1974. The Nature and Properties of Soils, 8th ed. New York: Macmillan.

Brown, P. L. 1964. Legumes and Grasses in Dryland Cropping Systems in the Northern and Central Great Plains. Miscellaneous Publication No. 952. Washington, D.C.: U.S. Government Printing Office.

Buttel, F. H. 1985. The land-grant system: A sociological perspective on value conflicts and ethical issues. Agriculture and Human Values, Spring.

Duley, F. L., and J. J. Coyle. 1955. Farming where rainfall is 8–20 inches a year. Pp. 407–415 in U.S. Department of Agriculture Yearbook, Water. Washington, D.C.: U.S. Department of Agriculture.

Elliott, E. T., and C. V. Cole. 1989. A perspective of agroecosystem science. Ecology 70:1597–1602.

Haas, H. J., and C. E. Evans. 1957. Nitrogen and Carbon Changes in Great Plains Soils as Influenced by Soil Treatments. USDA Technical Bulletin 1164. Washington, D.C.: U.S. Department of Agriculture.

Hanson, H. C. 1939. Ecology in agriculture. Ecology 2:111–117.

Hightower, J. 1973. Hard Tomatoes, Hard Times: A Report of the Agribusiness Accountability Project on the Failure of America's Land Grant College Complex. Cambridge, Mass.: Schenkman.

Hobbs, J. A., and P. L. Brown. 1957. Nitrogen changes in cultivated dryland soils. Agronomy Journal 49:257–260.

Jackson, W., and J. Piper. 1989. The necessary marriage between ecology and agriculture. Ecology 70:1591–1593.

Joandet, G. E., and T. C. Cartwright. 1975. Modeling beef production systems. Animal Science 41:1238–1246.

Karlen, D. L., D. C. Erbach, T. C. Kaspar, T. S. Colvin, E. C. Berry, and D. R. Timmons. 1990. Soil tilth: A review of past perceptions and future needs. Soil Science Society of America Journal 54:153–161.

Kirschenmann, F. 1988. Switching to a Sustainable System. Windsor, N.D.: Northern Plains Sustainable Agricultural Society.

Lowrance, R., B. R. Stinner, and G. J. House. 1984. Agricultural Ecosystems, Unifying Concepts. New York: John Wiley & Sons.

MacRae, R. J., and G. R. Mehuys. 1989. The effect of green manuring on the physical properties of temperate-area soils. Advances in Soil Science 3:71–94.

MacRae, R. J., S. B. Hill, J. Henning, and G. R. Mehuys. 1989. Agricultural science and sustainable agriculture: A review of the existing scientific barriers to sustainable food production and potential solutions. Biology, Agriculture, and Horticulture 6:173–219.

Madden, J. P., J. A. De Shazer, F. R. Magdoff, N. Pelsue, Jr., C. W. Laughlin, and D. E. Schlegel. 1990. LISA 88-89. Low-Input Sustainable Agriculture Research and Education Projects Funded in 1988 and 1989. Washington, D.C.: Cooperative State Research Service, U.S. Department of Agriculture.

National Research Council. 1989. Alternative Agriculture. Washington, D.C.: National Academy Press.

North Dakota Agricultural Statistics Service. 1988. North Dakota agricultural statistics. No. 57. Fargo, N.D.: North Dakota Agricultural Statistics Service.

Paul, E. A., and G. P. Robertson. 1989. Ecology and agricultural sciences: A false dichotomy? Ecology 70:1594–1597.

Progress. 1988. The smokescreen of "sustainable" agriculture. Progress 19:20.

Puckridge, D. W., and R. J. French. 1983. The annual legume pasture in cereal-ley farming systems of Southern Australia: A review. Agriculture, Ecosystem, and the Environment 9:229–267.

Sarvis, J. T., and J. C. Thysell. 1936. P. 71 in Crop rotation and tillage experiments at the Northern Great Plains Field Station Mandan, ND. USDA Technical Bulletin No. 536. Washington, D.C.: U.S. Department of Agriculture.

Sims, J. 1989. CREST farming: A strategy of dryland farming in the Northern Great Plains inter-mountain region. American Journal of Alternative Agriculture 4:85–90.

Wedin, W. F., H. J. Hodgson, and N. L. Jacobson. 1975. Utilizing plant and animal resources in producing human food. Animal Science 41:667–686.

# 5

# Economic Considerations in Sustainable Agriculture for Midwestern Farmers

*Michael Duffy*

Since World War II, chemical pesticide and fertilizer use has increased steadily. In 1988, 96 percent of the acreage planted to corn and soybeans in the United States was treated with a herbicide. Insecticides were used on 35 percent of the corn and 8 percent of the soybeans (U.S. Department of Agriculture, 1989). Similarly in 1988, 97 percent of the acreage planted to corn was fertilized. The average application was 137 pounds of nitrogen, 63 pounds of phosphate, and 85 pounds of potash per acre. Labor use has declined sharply since 1950, while chemical use has increased (Figure 5-1).

The land resources that are used have not changed significantly (U.S. Department of Agriculture, 1986). With changing technology, labor productivity has increased much more rapidly than has crop yield per acre or output/input ratios (Figure 5-2).

Recognition of these costs and unintended environmental costs have given rise to low-input sustainable agriculture (LISA) research. The situation today can be characterized as one in which chemical techniques dominate agricultural production. This domination is so great that many farmers do not count the contribution of internal resources. For example, 52 percent of the farmers in an Iowa survey indicated they ignored the nutrient content of animal manure when deciding how much chemical fertilizer they should apply to their fields (Padgitt, 1987). Several recent surveys in Iowa have confirmed that up to one-fourth of the farmers never take soil tests (Lasley and Kettner, 1989; Padgitt, 1985, 1987).

The Iowa Rural Life Poll reported that (1) 78 percent of the farmers agreed or strongly agreed with the statement that modern farming relies too heavily on insecticides and herbicides, and (2) 76 percent responded to the

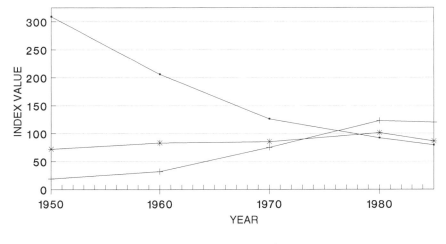

FIGURE 5-1 U.S. farm input use (1977 = 100). Source: U.S. Department of Agriculture.

statement that modern farming relies too heavily on chemical fertilizers (Lasley and Kettner, 1989). Another survey of the Iowa Farm Business Association members showed that over two-thirds of the respondents recognized that pesticides and fertilizers were a source of groundwater contamination and that pesticides threaten their health (Duffy, 1989b).

The 1980s was a time of financial upheaval in agriculture. At the same

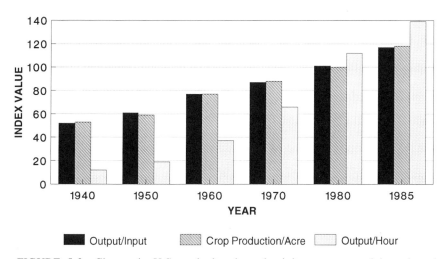

FIGURE 5-2 Change in U.S. agricultural productivity as measured by selected indices (1977 = 100). Source: U.S. Department of Agriculture.

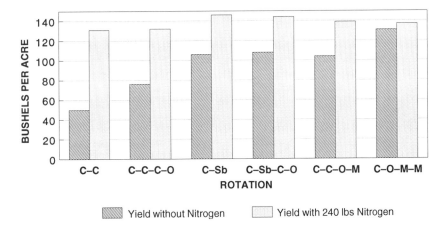

FIGURE 5-3 Average corn yield by rotation and using no nitrogen or 240 lbs of nitrogen on just corn acres, 1984 to 1989. C, corn; Sb, soybeans; O, oats, M, hay. Source: Kanawha Research Farm, Iowa State University, Ames.

time that farmers had the highest output per unit of input, tens of thousands of farmers had to endure severe financial stress. Research efforts to decrease the costs of production per unit of output has been furthered by the LISA program.

## THREE LOW-INPUT FARMING SYSTEMS

Crop rotations are an integral part of most sustainable agriculture systems (with perennial crops and permanent pasture being major exceptions). The lengths of and crops in the rotation have several impacts, which can include nitrogen fixations and reductions in pest populations. Figure 5-3 shows corn yield responses to nitrogen and rotation. Note in Figure 5-3 that corn every other year had almost identical average yields regardless of the crop followed or the level of nitrogen available. In the designation of crop rotation, C represents corn, O is oats, Sb is soybeans, and M is alfalfa and grass meadow. For example, C-Sb-C-O is a 4-year rotation of corn-soybean-corn-oats where corn is grown every other year. Because of the nitrogen fixation by legumes, the corn in the C-O-M-M rotation produced almost identical yields with and without commercial nitrogen fertilizer.

This is an example of the many studies that are examining the impact of rotations and various amounts of fertilizer use, all of which show essentially the same results. One major aspect of sustainable agriculture research is to understand and evaluate rotation benefits. It is obvious from the data that there are more than fertility benefits from a crop rotation.

Manure usage is another integral part of many sustainable agriculture systems. Although animals are not absolutely necessary on every farm, sustainable agriculture recognizes their benefits in terms of added income, more equal distribution of labor demands throughout the year, useful by-products for crop production (notably manure), and productive use of crop residue (such as corn stalks) as feedstuffs.

Studies of three systems of low-input farming in Iowa and Pennsylvania are described below.

### Chemical and Organic Production System Demonstration Project in Northeastern Iowa

This study is a comparison between chemical and organic production systems. The chemical system uses current chemical pest management and fertilizer techniques by which the farm manager decides each year which material and amount of material should be used. Two chemical-based rotations are examined: continuous corn and corn-soybeans. The organic system uses no pesticides or commercial fertilizer. However, in 3 of 12 years, an emergency herbicide treatment was applied. This system follows a 3-year corn-oat-meadow (C-O-M) rotation. Beef feedlot manure is applied at the rate of 20 tons/acre before the corn is planted.

This study is being conducted on 1-acre plots at the Iowa State Northeast Research Center in Floyd County, Iowa. Each crop in each rotation is grown once a year. The study results presented here are for 1978 through 1989 (Duffy and Chase, 1989a).

Figure 5-4 shows the average expenses of the system for machinery,

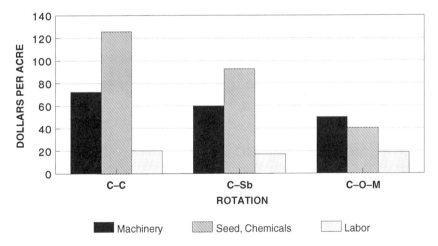

FIGURE 5-4 Average expenses by input category and rotation, 1978 to 1989. C, corn; Sb, soybeans; O, oats; M, hay. Source: Nashua Chemical/Organic Demonstration Project, Iowa State University, Ames, Outlying Research Centers reports.

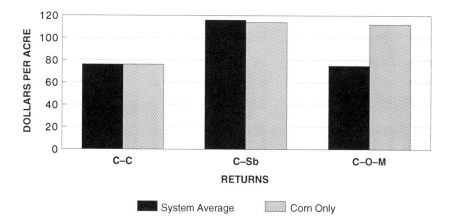

FIGURE 5-5 Average return to land, labor, and management for three alternative rotations and production systems. C, corn; Sb, soybeans; O, oats; M, hay. Source: Nashua Chemical/Organic Demonstration Project, Iowa State University, Ames, Outlying Research Centers reports.

input, and labor. Machinery expenses are estimated based on Iowa State University Extension Service data for every operation for each crop. The input costs are for the amount used and are estimated by using unpublished average price lists. Manure is charged at spreading costs. Labor is for the fieldwork time only, which is charged at $6/hour.

Figure 5-5 shows the preliminary findings of the average returns to land and management both for the rotation system and corn alone. No land or overhead charges were subtracted because these were constant across all systems. Yearly average prices were used in the calculations. The average corn yields were 138, 119, and 98 bushels/acre for corn after soybeans, corn after corn, and corn after meadow, respectively. The average returns with corn in the rotation were $120, $78, and $117 from corn after soybeans, corn after corn, and corn after meadow, respectively (Figure 5-5). However, when the returns for the entire rotation system were calculated, continuous corn earned about the same as the C-O-M rotation, but the C-Sb rotation earned more than the continuous corn or C-O-M rotation did.

Another observation from this project was the relative comparison between C-C and C-O-M. Without government program benefits or premium prices for organic produce, there was essentially no difference in the returns from these two systems. Not all of the benefits from rotation were reflected in annual net returns, however. Reduced use of pesticides and chemical fertilizers and lower erosion rates can yield long-term benefits in terms of water quality, the environment, and human health.

The nonmonetary benefits between the systems have not been examined thus far. One area of consideration in sustainable agriculture is energy use. It is an issue in both the source and amount used. Commercial fertilizers (especially nitrogen) and pesticides also require energy for their production and use. Figure 5-6 presents one view of energy use in this demonstration project. The energy produced is measured by its value as animal feed. The energy consumed is for machinery operation on the farm plus the energy required for use in the production of inputs.

Very few differences were found in the total energy value of feed produced by the three systems. Energy consumption, however, varied significantly. The greatest factor in energy use was fertilizer. Over three-fourths of the energy used for C-C and C-Sb rotations was fertilizer. It takes approximately 1 gallon of a diesel fuel equivalent of energy to produce 4 pounds of nitrogen.

Three general conclusions can be drawn from this particular study. The C-Sb rotation produced the highest average returns, and the C-O-M rotation was a viable alternative. The C-C and C-Sb rotation systems, however, were more vulnerable to external shocks, especially in energy prices.

## Rodale Conversion Project in Kutztown, Pennsylvania

This project is operated and supported by the Rodale Research Center. Its original purpose was to estimate the impact of the starting crop of a rotation when converting to an organic system. The Rodale Research Center uses the term *low input* to describe their study; pesticides or commercial

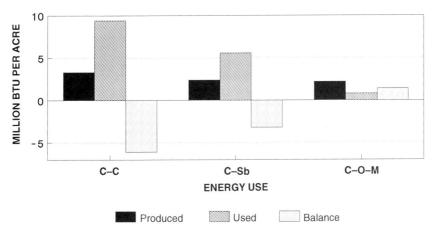

FIGURE 5-6 Energy balance from chemical organic demonstration project, 1978 to 1989. Values are average British thermal unit (Btu) equivalents (in millions). C, corn; Sb, soybeans; O, oats; M, hay. Source: Nashua Research Farm, Iowa State University, Ames.

fertilizers were not used in two of three systems that were evaluated. This study was similar to the Iowa chemical and organic production system demonstration project discussed above.

The conversion project examined three alternative production systems with three alternative starting crops. Each system and starting crop was replicated eight times on 20-by-300-foot plots (Duffy et al., 1989).

The first system used no chemicals or commercial fertilizers. Animal manure was used to supplement soil fertility. The rotation was small grain-hay-corn-soybeans-corn silage. The three starting crops were small grain, corn, and corn silage.

The second system in the project did not use chemicals, commercial fertilizers, or animal manure. The rotation was small grain-corn-small grain-corn-soybean. A legume was planted with the small grain and plowed under to help augment soil fertility needs. The three starting crops were small grain, soybeans, and corn.

The third system followed a conventional chemical and fertilizer program and used the standard recommendations of The Pennsylvania State University (University Park). This system followed a corn-corn-soybean-corn-soybean rotation.

The choice of the starting crop had a major effect on returns over the first rotation cycle. Row crops require more pest management and soil nutrients. Without the rotational benefits for soil fertility and pest management provided by previous legume crops, returns were greatly reduced when the conversion rotation was started with corn.

This finding has implications for farmers. It means that if they are going to use a rotation-based system, then pest management and fertility needs must be augmented in the initial years of the conversion.

The second major conclusion in the Rodale study was that returns for the organic system with manure and the conventional system were not significantly different. Both systems, however, produced significantly higher returns than those of the organic cash grain system without manure.

### Farming Systems Project in Central Iowa

A third farming system project is the Iowa State Farming Systems Project (Duffy, 1988, 1989a; Honeyman et al., 1989). This 5-year project started in 1987 on the Allee Research Farm near Newell, Iowa, in Buena Vista County. The three alternative systems examined in this study are based on the level of management used. The first system is low management with very little field information to determine pest management or fertility needs. Pesticides and fertilizers are applied on a routine basis. There are two low-management rotations: continuous corn and corn-soybeans.

The second system is a high-management system that uses pest scouting, soil tests, ridge-till, banded herbicide applications, and manure application. This system also has two rotations: continuous corn and corn-soybeans.

The third system is also high management, but it has the added goal of low chemical usage. Chemicals are used only in emergency situations. This system follows an oat-meadow-corn silage-rye/soybean-corn rotation. In the fourth year, rye and soybeans are double-cropped by planting rye the previous fall and harvesting it as hay the following spring.

Each crop and system is replicated four times on 1.2-acre plots. The choice of materials, the timing of operations, and other management details are determined by a steering committee and by the farmer.

Three years of this 5-year project have been completed. Although definitive conclusions cannot be drawn, some tentative findings are emerging. The most important finding thus far is the importance of farm manager performance in farming systems projects. Experiment stations, on-farm experiments, and other single-operator projects typically hold this key factor constant. No matter how many replications are included in the experimental design, there is only one manager. Timeliness, attention to detail, carefulness, and attitude are a few of the essential managerial attributes. Although they are hard to quantify, these skills are extremely important in determining the success or failure of an alternative production system.

Another important finding is the success with which additional management information can replace the need for capital. Figure 5-7 shows the decrease in variable costs as more management is added.

Figure 5-8 presents the 3-year average return to land and management. As in previously reported studies, no land or overhead charges are included. The high-management, low-chemical system is not described here because of technical difficulties. High management significantly increased returns, especially when a crop rotation (even 2 years) was used.

## Conclusion

The three studies described here show the potential for alternative agriculture production practices. As knowledge increases and available tools increase, production practices and profitability will improve.

Results of these studies suggest several areas where further economic consideration must be given.

## SUGGESTIONS FOR FURTHER RESEARCH

### Profitability

Individual farm profitability is of paramount importance. Understanding of how alternative systems compare must be increased. Similarly, there

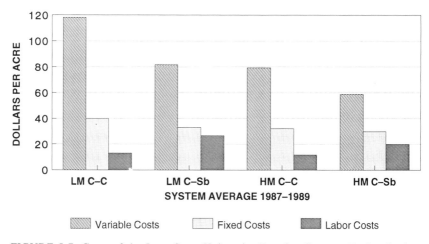

FIGURE 5-7  Costs of the Iowa State University Farming Systems Project by input category, rotation, and management intensity.  LM, low management; HM, high management; C-C, continuous corn; C-Sb, corn-soybean rotation.  Source: Farming Systems Project, Iowa State University, Ames, Outlying Research Centers Report 88-31.

must be a better appreciation of how pieces or parts of different systems can be combined.  Trade-offs exist between chemical, cultural, mechanical, and biological techniques.  What are these trade-offs for the individual farmer?  There is no one best system for all farms.  All farmers are unique, and so is the land that they farm.  Many of the innovations in agriculture came about because of the need to overcome natural boundaries.  As a consequence, much of the current technological research must be devoted to correcting mistakes from past innovations.  Rather than trying to overcome natural limitations, sustainable agriculture uses the land  and other natural resources and management to determine the best systems.

## Societal Costs and Benefits

A second area for further consideration is the efficiency of resource use from a societal perspective.  As noted above, agriculture production practices can produce unintended social benefits and costs.  For sustainable agriculture to be understood, it is critical that these nonmarket impacts be recognized and that an attempt be made to place a value on them.

Soil erosion provides the best example of an external cost.  Farmers suffer the loss of future productivity because of soil erosion; however, it also creates on-farm and downstream costs. Organic matter and topsoil are lost.  Roadway ditches must be cleaned of the topsoil washed from fields.

Silt accumulates in reservoirs, and recreation areas deteriorate. Fish and other wildlife are impaired or eliminated. Water quality deterioration, increased municipal water treatment costs, and other environmental problems are other examples of unintended external costs.

Another often overlooked societal aspect of farm resource use is the quality of rural life. Changes in farm production practices and farming systems have led to a decline in the farm population. These demographic changes affect the health and viability of rural communities. In 1990 there are more part-time farmers, more megafarmers, and fewer middle-sized family farms than in previous decades. Better understanding is needed of how current practices affect food safety, rural communities, environmental quality, and resource use.

### Farm Family Resources

A third area for economic consideration in sustainable agriculture involves the allocation of resources for farm families. It is essential that the appropriate balance be achieved for sustainable agriculture.

Labor is a major area for consideration in resource allocation. Too much work for laborers can decrease work quality, while too little work for laborers can affect profitability. To understand the labor constraint, labor availability, the effects on timeliness, labor quality, and the trade-offs between capital and labor must be evaluated.

Another farm family resource issue is capital availability—in particular,

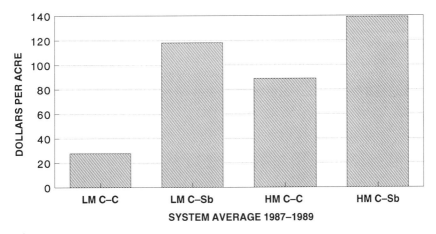

FIGURE 5-8 Return to land and management, Iowa State University Farming Systems Project. LM, low management; HM, high management; C-C, continuous corn; C-Sb, corn-soybean rotation. Source: Farming Systems Project, Iowa State University, Ames, Outlying Research Centers Report 88-31.

how much the production system has an impact on capital requirements. The farming system must be a compatible match with the farm family's goals and resources.

A final farm family resource issue is managerial skills. Management is crucial in determining the success or failure of a farm. As complexities and options in agriculture increase, farms must consider hiring outside experts, such as those used for pest control scouting.

## Government Policies

Government programs and policies are a fourth general area for economic consideration in sustainable agriculture. Although most agriculture policy attention is focused on the 1990 farm bill, many other government policies and decisions, including environmental and health regulations, have an impact on agriculture. Less obvious, but also important, are the impacts of government monetary and fiscal policies. U.S. agriculture is inextricably intertwined with the national and world economies. Inflation, tax policies, trade barriers, and the value of the U.S. dollar all have an impact on agriculture and influence the profitability of agriculture production practices. A complete discussion of government influence is beyond the scope of this chapter.

The conservation compliance provision and the conservation reserve program of the Food Security Act of 1985 are targeted toward protecting U.S. natural resources. The commodity programs, on the other hand, favor the production of certain crops such as corn and wheat. This leads to higher input use and penalizes farmers for adopting sustainable agriculture rotations.

Regardless of the program, farmers respond to what they perceive to be in their best interest. Some programs provide unintended incentives or disincentives. For example, the current corn price support program rewards past corn production and encourages its continuation (Duffy and Chase, 1989b). These features mean that the more corn there is in the rotation, the higher the reward. However, the more corn there is in a rotation, the more dependent the farmer is on chemical pesticides and fertilizers.

## Risk Management

Risk is one of the most important considerations in sustainable agriculture. Risk can be defined and quantified in many different ways. It occurs anytime there is a less than certain outcome. Many kinds of risk are associated with farming, including price and production risks, worker safety risks, genetic risks, consumption risks, and transportation risks.

Farmers must be able to assess the risks of alternative production practices accurately if they are to make informed choices. Everything involves trade-offs, and there is no such thing as a riskless agriculture production system. The goal in sustainable agriculture research and education efforts such as LISA is to understand and reduce the severity of these trade-offs.

## Macrolevel Impacts

The macrolevel effects—those beyond the farm—associated with the widespread adoption of sustainable farming systems should also be considered. In the United States there are regional and distributional questions to be answered. Areas with marginal levels of output could be forced out of production. There are international considerations in sustainable agriculture as well. The United States depends on a positive agriculture trade balance to help with the nation's overall balance of trade. U.S. agriculture is tied to the rest of the world through trade and competition. Alternatives must be thoroughly evaluated.

## STEPS TOWARD SUSTAINABLE AGRICULTURE

Thus far, this chapter has provided examples of sustainable agriculture research and has discussed the many areas in which more and better information is needed. This section examines some currently available techniques that can move farmers toward sustainable agriculture.

First, however, it is interesting to note two findings regarding farmers and sustainable agriculture. In the Iowa Rural Life Poll, farmers identified the extent to which they used 11 different practices to reduce pesticide or fertilizer use. The practices were soil testing, crop rotation, manure application, mechanical cultivations, planting of legumes, self-scouting, professional scouting, pheromone traps, degree days, tillage, and nonconventional products. Most farmers are currently doing some of the sustainable agriculture practices themselves (Lasley et al., 1990).

Another study was conducted by the U.S. General Accounting Office (1990). In that study, farmers were asked to identify what they perceived to be the barriers to the adoption of sustainable agriculture practices. The top five reasons mentioned by over three-fourths of the farmers contacted were greater management requirements, fear of lower yields, concerns over weed pressure, possibility of lower profits, and the need to maintain base acres.

Most farmers are already using some sustainable agriculture practices, indicating that they are thinking about these agricultural issues. The following six management techniques can help farmers begin to move toward a more sustainable agriculture system immediately.

Step 1 is recognizing fertilizer and yield benefits from rotations and manures. Many studies have estimated the available nitrogen provided by rotations and manure applications, as well as the impact on crop yields.

Step 2 is performing accurate soil tests and using the results to improve fertility management. A proper representative sample is absolutely crucial. Too often soil samples are not representative. A good soil sample and test from a reputable laboratory shows many things, including the available phosphorus (P) and potassium (K) and the need for lime. Plants need adequate amounts of P and K for efficient production. Plants can utilize P and K from the soil, manure, or other sources. Most Iowa corn farmers follow a P and K application schedule where nutrients are applied in amounts equal to those that the crops remove.

While this seems sustainable on the surface, it ignores the P and K that is already available in the soil. Several studies have shown that beyond certain soil test levels, crops do not respond to added P or K (Webb, 1988). Soil acidity affects many aspects of soil microbiology, soil chemistry, and crop physiology. Maintenance of a proper soil pH can enhance the efficiencies of fertilizers and chemicals.

Step 3 is evaluating tillage trips and methods. Elimination of unproductive trips can improve profitability and enhance sustainability. Farmers must have a tillage plan for each crop and field.

Step 4 is to evaluate alternative production systems. Farmers must continually be receptive to new ideas and techniques. They should look for pieces or parts of systems that can work for their farms. Different economic conditions, different soils, and different managerial skills all indicate the need to continue to search for alternatives.

Step 5 is careful evaluation of chemical applications and application techniques. Farmers need to know the chemicals they are using and the trade-offs of using various chemicals. Price is one discriminating factor. Relative efficacy and ability to control particular pest species are also important. The relative toxicities to humans, animals, and beneficial species, as well as persistence in the environment, also vary.

Application techniques vary in their costs and efficacies. Banded herbicide applications and the use of strictly mechanical controls (such as cultivation) have been shown to be profitable alternatives to the broadcasting of herbicides in many instances (Iowa State Extension Service, 1987, 1988).

Understanding of pest population dynamics and the available alternative techniques must be improved. Pest population monitoring and other integrated pest management techniques have been proven to be effective tools.

Step 6 is the adoption of farming practices based on the available resources. The inherent productivity of the land is often omitted from determining the land's highest and best use. For example, spending $20 an acre for weed control costs $0.40 a bushel for a 100-bushel yield and only $0.27 a bushel for a 150-bushel yield. It is essential to stay within the internal resources of the farm and the farmer. Every manager has strengths and weaknesses. The farming system should accentuate the positive.

There are other examples of currently available practices that support sustainable agriculture. Sustainable agriculture looks at not only better use of existing technologies but also development of new and better technologies—better in terms of profit, social acceptability, and environmental harmlessness.

## REFERENCES

Duffy, M. 1988. ISU Farming Systems Project 1987 Start-Up Year: Overview. Ames, Iowa: Department of Economics, Iowa State University.

Duffy, M. 1989a. ISU Farming Systems Project Observations on 1988 Crop Year. Ames, Iowa: Department of Economics, Iowa State University.

Duffy, M. 1989b. Farmers' Attitudes and Opinions Concerning Records and Sustainable Agriculture, Selected Survey Results, Iowa Farm Business Association, 1989. Unpublished paper presented at the Iowa Farm Business Association Executive Workshop.

Duffy, M., and C. Chase. 1989a. Costs and Returns Comparison for Chemical Versus Organic Rotations in Northeast Iowa, 1978–1988. Presented at the Annual Meeting of the Northeast Iowa Growers Association.

Duffy, M., and C. Chase. 1989b. Impacts of the 1985 Food Security Act on Crop Rotations and Fertilizer Use. Staff Paper No. 213. Ames, Iowa: Department of Economics, Iowa State University.

Duffy, M., R. Ginder, and S. Nicholson. 1989. An Economic Analysis of the Rodale Conversion Project: Overview. Staff Paper No. 212. Ames, Iowa: Department of Economics, Iowa State University.

Honeyman, M., M. Duffy, E. Dilworth, D. Grundman, and D. Shannon. 1989. ISU Farming Systems Project. Report No. ORC88-31. Ames, Iowa: College of Agriculture, Iowa State University.

Iowa State Extension Service. 1987. Integrated Farm Management Demonstration Program 1987 Summary Report. Report No. Pm-1305. Ames, Iowa: Iowa State University Extension Service.

Iowa State Extension Service. 1988. Integrated Farm Management Demonstration Program 1988 Progress Report. Report No. Pm-1345. Ames, Iowa: Iowa State University Extension Service.

Lasley, P., and K. Kettner. 1989. Iowa Farm and Rural Life Poll, 1989 Summary. Report No. Pm-1369. Ames, Iowa: Iowa State University Extension Service.

Lasley, P., M. Duffy, K. Kettner, and C. Chase. 1990. Factors affecting farmers' use of practices to reduce commercial fertilizers and pesticides. Journal of Soil and Water Conservation 43(1):132–136.

Padgitt, S. 1985. Farming Operations and Practices in Big Spring Basin. Report No. CRD 229. Ames, Iowa: Iowa State University Extension Service.

Padgitt, S. 1987. Monitoring Audience Response to Demonstration Projects. Baseline Report: Audubon County. Report No. CRD 273. Ames, Iowa: Iowa State University Extension Service.

U.S. Department of Agriculture. 1986. Outlook '87 Charts. Sixty-Third Annual Agents Outlook Conference. Washington, D.C.: Economic Research Service, U.S. Department of Agriculture.

U.S. Department of Agriculture. 1989. Agricultural Resources, Situation and Outlook. Publication No. AR13. Washington, D.C.: Economic Research Service, U.S. Department of Agriculture.

U.S. General Accounting Office. 1990. Alternative Agriculture, Federal Incentives and Farmers Opinions. Publication No. U.S. GAOI/PEMD-90-12. Washington, D.C.: U.S. General Accounting Office.

Webb, J. 1988. Phosphorus and Potassium Fertilization. Report No. ORC87-13. Ames, Iowa: Northeast Research Center, Iowa State University.

# PART TWO
# Research and Education in the Western Region

# 6

# Comparative Study of Organic and Conventional Tomato Production Systems: An Approach to On-Farm Systems Studies

*Carol Shennan, Laurie E. Drinkwater,*
*Ariena H. C. van Bruggen,*
*Deborah K. Letourneau, and Fekede Workneh*

This chapter describes an on-going study of existing organic and conventional tomato production systems in California supported by the low-input sustainable agriculture (LISA) program of the U.S. Department of Agriculture (USDA). The goal of the project is to investigate various soil, plant, and animal processes that function within the agroecosystem as they respond to different amounts and types of inputs. In addition, economic data are being collected to document the costs and trade-offs associated with the various management systems. Such information can then be used to help assess the long-term sustainability of these production systems in terms of productivity, efficiency of resource use, reduced inputs and off-farm impacts, and maintenance of the resource base, notably, the soil. The project was initiated in 1988, and the first-season data were collected in 1989. Because this project represents a relatively unique design for comparing different production systems, the major focus of this chapter will be to discuss and evaluate the approaches taken in developing this on-farm study.

## TERMINOLOGY

For this study, farms are considered organic when the management strategy for at least the past 3 years has emphasized reliance on biological processes. Plant nutrients are supplied primarily through the use of green manures, organic soil amendments, or both, and synthetic fertilizers and pesticides are not used. Farms that use synthetic fertilizers, pesticides, or both and that do not add organic soil amendments (other than crop residues) are considered conventional. Several sites are intermediate between these two extremes and are referred to as transitional. Most transitional farms fit into one of two categories: (1) farms that have grown crops organically for less than 3 years, or (2) organically managed farms that do not have a soil management program that includes the regular use of organic soil amendments or green manures. Although these management designations are convenient, they are still somewhat artificial, since farming practices in reality fall along a continuum rather than into discrete groups. For example, several of the conventional farmers did not spray insecticides in 1989, and in 1990 one conventional farmer used legume cover crops for nitrogen fertility but will still use synthetic fertilizers and insecticides as needed. This problem of farm categorization will be discussed further later in the chapter.

## BACKGROUND RATIONALE

California currently leads all other states in vegetable production (Scheuring, 1983). In the Central Valley alone, the annual tomato acreage (215,000 acres) is valued at $400 million, while other vegetables, predominantly melons, occupy 245,000 acres with a value of $540 million (California Farmer, 1987). Vegetable production systems utilize large quantities of inputs such as pesticides, fertilizers, and irrigation water; therefore, a decrease in inputs in these systems could have a profound effect on California's agroecosystems as a whole. Fresh market tomato production systems have been targeted for the present study, because production from California's Central Valley represents 30 percent of the total U.S. fresh market tomato production, and most importantly, a variety of management systems exist in this region, including long-term organic tomato production.

The widespread adoption of intensive conventional agriculture in California has been accompanied by the appearance of symptoms of poor soil structure (Chancellor, 1977). One symptom is decreased porosity, which can inhibit water infiltration, root penetration, and, thus, plant nutrient and water acquisition (Oades, 1984; University of California, Davis, 1984). To compensate for deteriorated soil structure or poor root development, farmers may increase applications of water, nitrogen, or both (Chancellor, 1977; E. M. Miyao, Yolo County Farm Advisor, personal communication, 1990), which, in turn, raises the potential for leaching of nutrients into groundwater (Freidrich and Zicarrelli, 1987). Furthermore, continuous cropping with vegetables and the decline in soil structure have been accompanied by increased losses caused by root diseases, such as phytophthora root rot of tomatoes (University of California, Oakland, 1985). Moreover, excessive soil nitrogen and water application have been implicated in the increased susceptibility of tomatoes to some pathogens (Ristaino et al., 1988; Schmitthenner and Canaday, 1983; van Bruggen and Brown, in press), further compounding the problem.

The beneficial effects of increased organic matter on soil structure and biology have been well documented (Chaney and Swift, 1984; Oades, 1984; Tate, 1987; Tisdall and Oades, 1982). Suppression of several plant diseases by certain soils has been attributed to high levels of microbial activity (Chen et al., 1988), and there is also evidence that use of green manures can decrease crop disease severity (Cook, 1984). Slow release of nitrogen from organic sources is reputed to lead to lower nitrate concentrations in the soil, which could potentially reduce losses via leaching, in addition to ameliorating some root disease problems.

Taken together, these considerations suggest that management systems in which an effort is made to improve soil organic matter may help alleviate many of these problems. Most of the studies cited, however, refer to climates where organic matter turnover is relatively slow and increased levels can be maintained over time by appropriate management (Johnston, 1986; Reganold, 1988). Little information is currently available for semi-arid irrigated systems in which high temperatures and frequent water applications favor rapid organic matter turnover. Long-term studies of cover-cropped orchard systems in California concluded that it is not possible to increase organic matter in this environment significantly (Proebsting, 1952, 1958). A corollary of this has been the assumption, therefore, that soil structural properties similarly could not be improved by efforts to enhance organic matter in California soils. In contrast, work in progress (Groody, 1990; C. Shennan, C. Griffin, and T. L. Pritchard, unpublished data) suggests that leguminous winter cover crops may improve the structural characteristics of soil and that various combinations of cover crops, tillage, and gypsum applications can improve orchard soils (Moore et al., 1989). Earlier work by Williams and Doneen (1960) and Williams (1966) also demonstrated the beneficial effects of a variety of cover crops on water infiltration in a Central Valley soil. However, it should be noted that, in general, the links between specific management practices, changes in soil properties, and their effects on plant growth have not been well established (Karlen et al., 1990).

In conventional farming systems, levels of damage caused by insects remain high on many crops, even doubling in the past 30 years in some cases, despite continual development of sophisticated crop production technologies (Bottrell, 1980). Pesticides, fertilizers, and cropping patterns can drive pest population dynamics and modify damage to crops in various ways (Altieri and Letourneau, 1982; Bethke et al., 1987; Fery and Cuthbert, 1974). Insecticides can rapidly control pest species or can cause reactive outbreaks (Pedigo, 1989). Vegetational diversity in space and time may enhance the maintenance of natural enemies (Altieri and Letourneau, 1982), and for many crops, increased plant nitrogen content is associated with increased attractiveness to insect herbivores (Leath and Ratcliffe, 1974; Letourneau and Fox, 1989). Each of these factors can be influenced by farming practices that can be expected to differ among organic and conventional systems.

## GOALS OF THE PROJECT

It is clear from the preceding discussion that crop yield, the most commonly measured attribute of agroecosystems, represents the outcome of complex interactions among soil, plant, pest, disease, and environmental and management parameters. It is the goal of this project to determine the impact of different combinations of production practices (ranging from organic and transitional to conventional) on various components of the agroecosystem.

More specifically, management practices are being documented, and inputs, outputs, and a variety of soil, plant, pest, and disease parameters are being quantified for each site. A variety of questions are being addressed. For example, do soils on organic and conventional farms differ with respect to nitrogen availability, structural properties, or microbial activity? If so, are these differences reflected in patterns of plant growth and nutrient acquisition? Are the incidence and severity of root diseases or insect pests affected by these soil, plant, or management characteristics; if so, which ones are the most important? Does the structure of arthropod communities associated with the tomato crop change with different management practices; if so, are these changes reflected in differences in crop damage levels or yield loss because of insect pests? Are soils from organic farms more able to suppress the growth of root pathogens, and does this ability correlate with particular soil characteristics such as microbial activity or nitrate levels? These and other questions will be answered by the use of a hierarchical approach in which data are collected both at the field and individual plant levels.

The project's main focus is to develop an understanding of the biological and ecological characterisitics of the production systems. From a practical point of view, however, it is critical to obtain sufficient information to provide a context for economic assessment of the different strategies used. To this end, enterprise budgets are being derived for each tomato production system, and the extent and nature of any financial gains or losses resulting from decisions to reduce, or cease, application of chemical fertilizers and pesticides will be evaluated. Since the economic component of the study is at a very early stage of development, it will not be discussed further.

## WHY AN ON-FARM STUDY?

A great deal of agricultural research has been based on replicated factorial experiments conducted at experiment stations or in growers' fields. In these experiments the effects of varying one or two factors are monitored while steps are taken to maintain all other factors constant. In this way the potential for the factors under study to affect processes of interest is clearly established. However, since the ecosystem processes of interest can potentially respond to, and interact with, many environmental and management factors simultaneously, it is not feasible to approach the study of integrated systems by use of factoral experiments. Furthermore, since there is little or no information available on organically managed vegetable production systems, it was not clear a priori what factors or management practices should initially be targeted. It seemed logical, therefore, to document the characterisitics and functioning of selected complete management systems before attempting to isolate components of these systems for more detailed examination. Significant relationships identified from the systems study can then be targeted in separate experiments to elucidate the mechanisms that are operating.

Having decided upon a systems comparison, the question remains as to whether it is preferable to simulate the systems of interest in some kind of replicated experiment or to study existing farm operations. A number of farming system comparisons have taken the first approach by creating experimental organic, biological, integrated, or conventional treatments to simulate the various production systems of interest (Culik et al., 1983; Daamen et al., 1989; Doran et al., 1988; Sahs and Lesoing, 1985; Steiner et al., 1986; Vereijken, 1989; Weisskopf et al., 1989; Zeddies et al., 1986; see also R. Janke, J. Mountpleasant, S. Peters, and M. Bohlke, "Long-Term Low-Input Cropping Systems Research," this volume). This approach offers a number of advantages by allowing whole management systems to be studied while at the same time reducing the influence of potentially confounding variables such as soil type, surrounding habitat, and microclimate and by allowing for true replication of treatments. A further advantage of this approach is that the experimenters have full control over all management decisions. What is often the case, however, is that because of resource limitations, there must be a trade-off between the scale of the experiment and the number of replications. The sizes of the experimental plots is generally reduced to less than typical field scale to allow for reasonable replication. Alternatively, if field-size plots are chosen, then there may be little or no replication (Vereijken, 1989; Weisskopf et al., 1989).

Deciding upon the scale of experimental plots is a very important consideration, since scale can have significant impacts on the results that are obtained and their interpretation. This is particularly true for studies of mobile insects (Kareiva, 1983; Letourneau and Fox, 1989) and properties that have distinctly patchy distributions, such as root disease incidence or severity (Madden, 1989). Further disadvantages of the experimental approach include the fact that the data obtained are from only a single location and may not be readily extrapolated to other sites. Also, one set of practices is used to represent a type of production system, whereas in reality, there are usually many variations on a central theme.

The second approach to studying existing farm operations also brings its own set of advantages and disadvantages. Indeed, the two approaches are regarded as complementary since they can provide different kinds of information. First and foremost among the assets of the on-farm approach is that the systems under study are realistic for the present time. They represent combinations of practices and decisions made by farmers who are surviving in a world full of practical and economic constraints. Moreover, when the information is derived from multiple locations and management combinations, the robustness of any relationships that can be identified is increased. Of particular importance for this study is the fact that multiple sites could be selected that have been under organic management for various lengths of time—some for many years.

In other studies a transition period has been observed following a switch to organic production methods, during which time yields may be reduced, nutrient availability may be limited, and pest problems may be increased (see the chapter by R. Janke and colleagues in this volume). Presumably, this period represents the time required for the system to attain some kind of dynamic equilibrium with respect to soil biological changes (Paul, 1984; van der Linden et al., 1987) and, perhaps, insect and weed population dynamics. Thus, the first few years of studying experimental organic systems will be spent describing the transition process. Although this is clearly of great interest, in this study the major focus is the potential for organic practices to affect attributes of the agroecosystem over the long term. By observing existing farms that have operated organically for various lengths of time, the present study provides a mechanism for doing this from the outset.

Finally, working on existing farms provides an avenue for considerable interaction and information exchange between farmers and researchers. In particular, the researchers become much more familiar with the kind of decisions and compromises farmers have to make and the issues they feel are most important. Farmers, in turn, can benefit from opportunities to communicate their ideas and concerns directly to the researchers and have access to research results. Some of the approaches taken in this study to maximize this kind of two-way communication are described later in this chapter.

The disadvantages associated with comparisons of existing production systems include the need to account for potentially confounding variables and the extra work load associated with the more extensive data collection this approach requires. Climate, soil type, surrounding habitat, planting date, crop cultivar, and other management details unrelated to those of interest will vary among locations and therefore must be measured and included in subsequent analyses, when appropriate, to avoid misinterpretation of the data. Provided that sites can be selected such that a similar range of these variables exists within each of the broad management categories (e.g., organic and conventional), then the effects caused by management comparisons of interest, such as organic versus inorganic nitrogen input, can be separated from those caused by extraneous variables such as soil type or planting date.

Because of these considerations and the complexity of farm systems, site selection involves considerable time and effort, as does the collection of information and sampling from multiple locations and coordination of these activities with each of the farmers in turn. Finally, the fact that the investigators are not in control of the management decisions may be an asset or may prove to be problematic. On the one hand, the farmers are most knowledgeable about their fields and production options, and any decisions they make are based on the realities of physical, biological, and economic constraints. On the other hand, the decisions that are made may not be in the best interest of the project. For example, based on market considerations, a grower may decide to change the planting date, not plant the crop of interest, or cease managing a field part way through the season. An ability to compensate farmers financially for modifying their plans to accommodate the research project may help avoid such problems. While these potential problems can increase the risk and difficulty of conducting the research, none of these problems are insurmountable, and the rewards from studying existing farming systems outweigh the disadvantages.

## METHODOLOGICAL CONSIDERATIONS

Selected components of the agroecosystem have been emphasized in previous studies on existing farms, such as specific insect abundance (Altieri and Schmidt, 1986), soil properties (Bolton et al., 1985; Doran et al., 1988; Maidl et al., 1988; Reganold, 1988; Reganold et al., 1987), or economics (Goldstein and Young, 1988), but interdisciplinary studies such as the one being conducted with tomatoes in California have rarely been attempted. Two excellent examples of integrated interdisciplinary studies are the comparison of organic and conventional farms in the Midwest United States by Lockeretz and coworkers (1981) and the study of interactions among soil properties, cultural practices, pathogens, and crop yield in existing Australian wheat farms by Stynes and colleagues (1979, 1981, 1983) and Veitch and Stynes (1979, 1981).

The advantages of conducting *integrated interdisciplinary* studies of farming systems are numerous. A distinction is made between integrated interdisciplinary and *multidisciplinary,* because in many cases the latter results in two or more distinct facets of a system being investigated, with few connections being made between them. Integration of the disciplinary approaches can increase the resource use efficiency of the work, since much of the data collected is often common to many areas of investigation. More importantly, however, interactions among different components of the system can only be studied realistically in an integrated interdisciplinary framework. For example, interactions occur among soil properties, nutrient cycling, disease development, plant growth, susceptibility to insect damage, and economic return. To understand these interactions, it is critical that relevant data be collected in a manner that is useful to soil scientists, horticulturists, pathologists, entomologists, and economists. This requires considerable time in planning and discussion, a willingness to compromise, and respect for each investigator's research goals.

Miller (1983), while acknowledging the necessity for team approaches to studying complex systems, has identified psychosocial and institutional barriers to truly integrated interdisciplinary research in the context of the development of integrated pest management and forest management programs. In his study, he found that the researchers had only rudimentary collaborative skills themselves and very little institutional support or incentive to form effective collaborations. In developing the study of tomatoes in California described here, many of the problems Miller identified were encountered. The need for continual dialogue and coordinated decision making throughout the project must be reemphasized. The outcome of this process is a project that is truly cooperative, representing a synthesis of ideas from researchers with diverse backgrounds.

## Site Selection

The most important consideration in choosing study sites for the project described here was to minimize confounding variables. Clearly, what constitutes a confounding variable depends on the questions being addressed. For example, if the question relates to the role of production practices in affecting soil properties, the actual location of a farm may be of little importance, whereas parent soil type is critically important. However, if interactions between soil properties and plant growth are being examined, climatic variability among locations will influence the analysis and interpretation of the data that are collected. For an entomologist, the major variables of concern are surrounding habitat, microclimate, and field size, whereas for a plant pathologist, the presence of the relevant pathogens in fields of all management types is of primary importance.

While differences in preferred selection criteria exist among the disciplines, it became obvious during the development of the tomato project that a few general criteria formed a reasonable compromise. Primary consideration needed to be given to locating farms such that the overlap among the different management types, in terms of geographic location, local climatic conditions, and range of parent soil types, was maximized.

Initially, two important vegetable-producing regions were considered as study areas: the central coastal valleys of California, which produce mainly cool season vegetables, and the Central Valley, where warm season vegetables are produced. Based on the above criteria, site visits and a questionnaire were used to gather information on organic and conventional vegetable producers in both regions. The coastal region was found to have favorable characteristics for some of the project's objectives, most notably, varying levels of a potentially serious root pathogen of lettuce among different types of management systems. However, several problems existed that made this region less suitable for this kind of interdisciplinary study. Conventional and organic sites were geographically separated and experienced very different local climatic conditions. Furthermore, most organic farms were situated in isolated valleys away from other forms of agriculture. In this case, it would be difficult to separate out the effects caused by isolation and climate differences from those caused by management. Indeed, one viewpoint often stated is that organic farms require this type of isolation from other agricultural fields in order to avoid insect pest problems, although there is little evidence in the literature to support or refute this contention.

In contrast, in the Central Valley there was some overlap in the locations of organic and conventional sites, and while there was a climatic gradient, it was relatively slight. Furthermore, all of the organic farms in this area had agricultural neighbors, and some were surrounded by large agricultural fields. This attribute is essential if findings from the study are to be used to make inferences regarding the impact on farms in intensively cultivated areas where organic or reduced chemical input farming is adopted.

The selection of specific sites required a compromise among members of the different disciplines. All members of the research team ranked the sites in order of preference based on priorities related to their particular research interest. In this way, the sites that were most important for each discipline were sampled by the entire team. Finally, additional sites were selected for study by members of individual disciplines to address specific questions. This approach seemed like a realistic compromise, achieving the advantages of both the integrated interdisciplinary approach on a majority of sites while also providing a degree of autonomy to allow sampling methods and selection criteria for each research component to be more rigorously tested. For example, additional conventional fields were selected to be paired with nearby organic sites for measurements of arthropod community structure but were deemed unsuitable for extensive soil analysis because of large differences in field slopes between the sites.

## Working with Farmers

During the initial contact, the project was described and growers were asked about their willingness to participate in the study. After the final selection of sites, all those who decided to cooperate in the study were mailed a letter that described the measurements that would be made and identified the specific field to be sampled. Most information on farm characteristics and production practices was collected orally by the project team members. In addition, data sheets were provided to growers to facilitate record keeping of irrigation applications, pest control, and tillage operations.

Contact is maintained with the cooperating growers on a regular basis, usually through letters and by telephone, to keep them informed of the study's progress. Some growers who are very interested in the project ask more questions about results and also serve as sounding boards for the research team's ideas. Recently, all cooperating farmers were provided with a copy of the progress report and invited to a meeting to discuss the preliminary results and to solicit their input and comments for the future.

## ANALYTICAL APPROACHES

As discussed previously, two different approaches have generally been used to compare different farming systems. Either existing organic and conventional farms have been compared in the same general area (Lockeretz et al., 1981; Niederbudde and Flessa, 1989; Niederbudde et al., 1989; Reganold, 1988; Reganold et al., 1987; Sengonca and Bruggen, 1989) or different farming systems have been experimentally developed side by side (Daamen et al., 1989; Steiner et al., 1986; Vereijken, 1989; Vereijken and Spiertz, 1988; Weisskopf et al., 1989; Zeddies et al., 1986; see also Janke et al., this volume). With few exceptions (notably, Lockeretz et al. [1981] for comparison of existing farms and Janke et al. [this volume] for experimental systems), the number of replications included in these studies was limited or nonexistent (Daamen et al., 1989; Niederbudde and Flessa, 1989; Niederbudde et al., 1989; Reganold, 1988; Reganold et al., 1987; Sengonca and Bruggen, 1989; Vereijken, 1989; Weisskopf et al., 1989; Zeddies et al., 1986), and the data were subjected to minimal statistical analyses.

For this study, complete data sets were obtained for 8 sites in 1989, and selected measurements were made on 10 additional sites to address specific questions (Table 6-1 provides a breakdown of sites by discipline). The term *site* rather than *farm* is used since the sampling unit was a field, and in one

**Table 6-1**  Breakdown of Data Collection in 1989 by Number of Sites

| Number of Sites | Data Collected |
|---|---|
| ≈60* | Farming practices |
| 18 | General soil characteristics |
| 13 | Detailed soil sampling (20 samples/site); analysis of nitrates, ammonium, nitrogen mineralization, aggregate stability, organic carbon |
| 12 | Disease assessment (corky root, *Phytophthora*), microbial activity |
| 12 | Insect diversity, damage assessment |
| 8 | Complete sampling by members of all disciplines |
| 6 | Water use, soil moisture profiles |
| 16 | Economic analysis |

*The growers at the 60 sites who were interviewed consisted primarily of organic, transitional, and conventional mixed vegetable producers. Selected representative large-scale processing and fresh market tomato producers were also included in the survey.

case two separate fields in different locations belonging to one farm were each used as study sites. Field-level data were collected to describe the sites in terms of basic attributes (e.g., parent soil type and texture and surrounding habitat), management inputs (e.g., irrigation, fertilizer, cover crop biomass and nitrogen content, manure or compost, pesticides, and labor), and incidence of communities of associated pests and beneficial organisms. To compare community structure and pest and pathogen incidence among the different management categories more rigorously, efforts were made initially to pair each organic or transitional site with a conventional site on the basis of geographic proximity and, thus, microclimate similarity and shared source pools of colonizing pests.

Various plant, soil, disease, and pest damage parameters were also measured at 20 locations within each field (individual plant level). These data will be subjected to various multivariate analyses, as described below, to determine which factors contribute most to explanations of any major differences observed among farming systems. This approach has been successful in a variety of other studies, for example, in plant pathology (Ratkowsky and Martin, 1974; Thomas and Hart, 1986; Wallace, 1978), community ecology (Gauch, 1982; Pfender and Wootke, 1988; Widden, 1987), soil science (Nolin et al., 1989; Stynes and Veitch, 1981, 1983; Stynes et al., 1979, 1981; Veitch and Stynes, 1979, 1981), and entomology (Baumgartner et al., 1985).

## Sampling Techniques and Variables Measured

For a general description of the selected sites, three or four composite soil samples (composed of 20 systematic subsamples at depths of 8 inches) were collected from each site before the growing season. The following parameters were measured: percent sand, silt, and clay; pH and cation exchange capacity; and concentrations of nitrogen, phosphorus, potassium, calcium, magnesium, sodium, and boron.

During the tomato growing season (at fruit set), 20 individual soil samples were taken from each site. The sampling area varied from about 0.13 to 0.5 acres, depending on the number of rows planted to the same cultivar (as determined by the individual growers). Stratified random sampling was used to take into account the potential patchiness of the variables measured. Each sampling area was divided into 20 blocks, with one random sample taken from each block. Microbial activity (Schnuerer and Rosswall, 1982), water-stable aggregates (Kemper and Rosenau, 1986), organic carbon content (Nelson and Sommers, 1975), available nitrogen (nitrate and ammonia in potassium chloride extracts), nitrogen mineralization rate (Waring and Bremner, 1964), and *Phytophthora parasitica* populations (using the leaf baiting technique described by Tsao [1983]) were determined for each soil sample. Soil pH (in potassium chloride), electrical conductivity, total nitrogen content, and percent clay were determined on four composite samples from each field. At the green fruit stage, 18 inches of the tomato row (usually, approximately one plant) was uprooted at each of the same 20 locations where soil samples were previously taken. Total nitrogen in the tissue, shoot and fruit dry weight, insect injury (visual scoring of feeding signs of flea beetles, thrips, chewing insects, and leaf miners), and root rot severity scores (corky root and phytophthora root rot) were determined for each plant sample.

Pest and beneficial insects were sampled primarily by using malaise traps (flying insects) and vacuum collectors (insects on the tomato plants) for population and community comparisons. Predation and parasitism rates of pest species were assessed by baiting plants with prey and monitoring the degree of predation and parasitism of the baits.

## Statistical Analysis

Initial statistical analysis of the field level data will consist of the derivation of descriptive statistics for each variable by management type (organic, conventional, and transitional), pesticide use, cultivar, and planting date. All variables will be checked for normality, and nonnormal variables will be transformed. Principal component analysis will be used to provide a first indication of the parameters that explain most of the variability among sampling sites on the different types of farms. To test whether the management groupings used reflect natural groupings based on the measured soil, plant, disease, and insect damage parameters, cluster analysis will be performed. In addition, multivariate analysis of variance (MANOVA) will be used to address the question of whether there are significant differences between farm types or pesticide use based on the measured parameters. MANOVA will also be used to identify and compensate for potential confounding factors such as planting date, soil texture, and cultivar.

The interrelationships between soil parameters will be examined by principal component analysis by using correlation matrices. Correlation rather than covariance matrices will be chosen to avoid bias caused by the widely different scales of the variables measured (Stynes et al., 1979). Plant yield and disease will then be regressed on the major soil components identified in the principal component analysis. In a similar manner, insect damage data will be regressed on the major components obtained from a principal component analysis based on soil and plant nitrogen data. Finally, all plants will be classified into two categories (healthy or diseased) with respect to corky root or *Phytophthora* spp., and then discriminant function analyses will be performed to determine whether sampling locations with healthy or diseased plants differ significantly in regard to various soil and plant characteristics (Afifi and Clark, 1984; Thomas and Hart, 1986). This analysis will also enable identification of the relative contributions of each soil and plant variable to the classification and to predict whether a plant growing in soil with a particular set of characteristics is more likely to become diseased than one growing in soil with a different set of characteristics. Similar discriminant analyses will be carried out to distinguish between sampling locations with and without various arthropods.

## PRELIMINARY RESULTS AND EVALUATION

### Site Selection

When scale-related farm characteristics such as total acreage, field size, crop mix, and marketing strategy were considered, it became clear that two different types of tomato growers exist in this region and cross the management designations of organic, transitional, and conventional as defined previously. Theoretically, any size farm could fall into all three management designations, ranging from small-scale mixed vegetable producers to large-scale tomato growers producing either "green-gas" (fresh market varieties harvested while fruit are still green) or processing (for example, machine-harvested varieties selected for processing into tomato sauce, tomato paste) tomatoes. At present, however, the project has not located any long-term, large-scale organic tomato producers, although some are currently in the transitional stage. The smaller-scale conventional mixed vegetable production systems have much in common with the organic systems, such as scale, crop mix, and reliance on direct marketing, but they are similar to large-scale systems with regard to the methods of soil management and pest control (Figure 6-1).

Sixty mostly small-scale mixed vegetable producers were identified as potential sites in an area from Yuba City to Stockton. Approximately 30 sites from the southern part of the region were eliminated because of highly variable soil types and the considerable distances involved, leaving a pool of about 30 sites from which the 18 sites sampled by members of at least some of the disciplines were chosen (Table 6-1). The sites selected in 1989 represent a majority of the available organic within a 50-mile radius of the University of California at Davis campus, with 9 of a possible 13 organic or transitional sites being sampled. A much greater number of potential large-scale sites existed relative to the number of other types of sites. Therefore, the large-scale sites chosen were selected on the basis of proximity to organic or transitional sites and from discussions with local farm advisers to ensure that they were representative of farms in the region. Some of the more southern sites rejected in 1989 will be included in 1990, however, to increase the representation of large green-gas tomato producers, for reasons discussed below.

There was reasonable success in ensuring that the location of sites in each of the broad management categories overlapped (Figure 6-2) and that similar ranges of soil types, climatic conditions, surrounding vegetation, and field sizes were represented. In this way the potential for these variables to obscure the effects of different management systems was minimized. Other variables such as planting date, sampling date, and cultivar were not adequately controlled and therefore must be accounted for in all analyses (see below). For example, the abundance of green peach aphids early in the season was highly correlated with the transplant date ($r = -0.76$, $p = 0.0063$) rather than management. In contrast, the abundance of other insects (thrips, flea beetles, stink bugs, and various beneficial insects) showed no such relationship.

The six organic and three transitional farms sampled typically produced a diverse mix of both winter and summer vegetables and tree crops, with acreages in tomatoes ranging from less than 1 to 5 acres and total farm size ranging from 10 to 160 acres, except for one transitional farm of 800 acres. Conventional mixed fruit and vegetable producers also grew a mix of vegetables and tree crops on acreages ranging from 1 to 1,500 acres (typically, 200 to 300 acres). In general, 10 to 90 percent of this land was in mixed vegetables (1 to 200 acres), and 1 to 10 acres of the total acreage was in

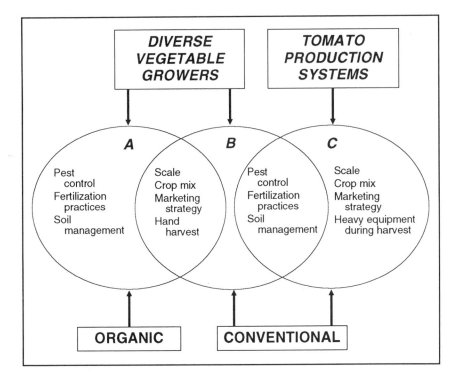

FIGURE 6-1 Diagram illustrating the general relationships between the types of tomato production systems found in California's Central Valley. Type A refers to small-scale organic mixed vegetable producers, type B refers to the smaller conventional mixed vegetable producers, and type C refers to the large-scale processing or fresh market producers (typically, green-gas [see text]). Type B shares characteristics in common with both the small-scale organic growers (type A) and the large-scale conventional growers (type C). At present, some type C growers are in the process of converting part of their land into organic tomato production.

tomatoes. Six of the conventional farming sites sampled in 1989 fell into the category of mixed vegetable producers; the remaining three consisted of one large green-gas tomato field and two processing tomato fields into which fresh market tomatoes were transplanted. Although there are some management differences between fresh market and processing tomatoes, there are many important similarities in terms of fertilizer practices, reliance on chemical pest control, large scale of production (typically 40 to 60 acres per tomato field), use of heavy machinery, and, typically, the presence of more severe soil structural and plant disease problems than those in small-scale systems. For these reasons and the fact that large areas of the

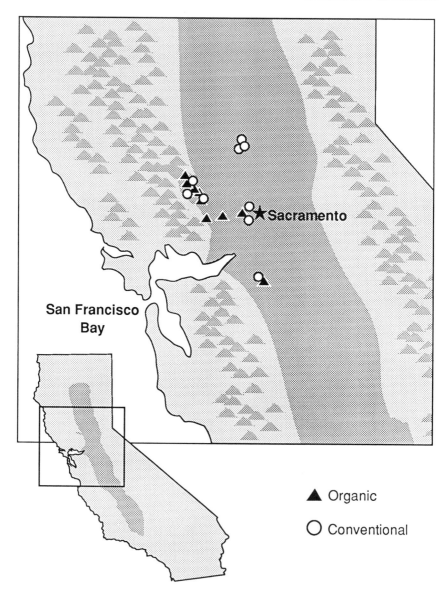

FIGURE 6-2 Map showing the locations of (▲) organic/transitional and (○) conventional sites for the project in 1989.

Central Valley are in these production systems, it was decided that it would be important to include more of the large-scale sites in this study in 1990, even though no long-term organic counterparts in this size category currently exist.

## Management Practices

Forms and quantities of nitrogen input varied considerably among the farms studied, ranging from virtually zero (residue of preceding dry bean crop) to various combinations of legume cover crops, compost, manure, earthworm castings, or recommended rates of chemical fertilizer. Insect pest management ranged from zero intervention to extensive use of organic controls (sulfur, *Bacillus thuringiensis*) or broad-spectrum synthetic insecticides. All farms used some combination of mechanical cultivation and hand hoeing for weed control; in addition, pre- and postemergence herbicides were typically applied on the conventional farms. Irrigation practices also differed between the organic and conventional farms, with about half the organic farmers using drip irrigation for their tomatoes, whereas the majority of conventional farmers used furrow irrigation.

## Preliminary Results

Significant differences were observed between management categories based on soil, plant, and disease parameters. However, cultivar and planting and sampling date effects were also significant and were partially confounded with management types. The cultivar effect was mainly due to the fact that one field planted with a tomato cultivar differed substantially from the two more similar cultivars that were planted in all of the other fields. Based on these results, only one cultivar will be studied in the future. When necessary, transplants will be provided to the cooperating growers if they would not usually plant the cultivar selected by the project team. Differences in planting dates are unavoidable. However, confounding by transplant and sampling dates will be minimized in the future by equally dividing earlier and later planting dates over the farm types as much as possible.

Preliminary examination of the data suggests that differences exist between organic and conventional farms with regard to nitrogen mineralization potential, inorganic nitrogen pools, microbial activity, corky root severity, and insect damage. Yield differences were not evaluated in 1989 because of the confounding effects of cultivar and planting date. Average soil nitrate concentrations were higher in conventionally managed soils than in organically managed soils, whereas nitrogen mineralization potential was generally lowest in the conventional soils. Transitional soils tended to be intermediate in both respects. Microbial activity and organic carbon levels followed the same trend as nitrogen mineralization, but differences between organic and conventional farms in terms of organic carbon were less pronounced. In contrast, soil wet aggregate stability (an indication of soil porosity) appears to show no clear trends at this stage. The multivariate analyses will provide a clearer picture of the relationships among the soil variables and what differences exist among management types.

The leaf-baiting technique that was used to determine population levels of *Phytophthora parasitica* in soil detected the pathogen in both conventional and transitional farms but not in organic farms. However, during the tomato growing season, no phytophthora root rot symptoms were observed on any of the plants sampled, even on farms where *Phytophthora* propagules were detected in the soil. This discrepancy was probably due to the patchy distribution of the pathogen; disease symptoms were observed in some conventional and transitional fields outside of the sampling area. Corky root caused by *Pyrenochaeta lycopersici* was observed in all farms at low levels, but disease levels were generally higher in conventional than in organic farms. At this stage, given the low incidence of either disease studied, it is premature to draw conclusions regarding any relationship between soil parameters, such as microbial activity and nitrogen availability, and disease severity observed in the field. In the future, plants from 3 feet of the row will be sampled from each randomly selected location to increase the chance of disease detection. In addition, samples will be collected outside the designated sampling locations to target areas with visible foliar symptoms of phytophthora root rot and to aid in the determination of relationships between disease severity and soil properties.

Initial analyses of field-level data suggest that pairing of organic and transitional sites with conventional sites based on geographic proximity did not add robustness to comparisons of farm-level data on arthropod community abundance. Indeed, it is not clear which of many factors should be used as a basis for pairing sites for this kind of comparison or whether the categories of organic, transitional, and conventional farming themselves are useful in this context. In 1989, many conventional growers actually used insect pest control strategies similar to those of organic growers. Injury levels on tomato foliage tended to be approximately five times greater in organic than in either conventional or transitional fields. The levels of damage observed, however, were low and unlikely to cause significant yield loss, even on the organic farms. The cultural practice of early transplant dates usually results in low damage to fruit by insect pests. In future seasons, data will be obtained from more late plantings for comparisons among management practices and between years. An increase in sample size to 18 fields and more uniform representation of management categories will strengthen the ability to identify field-level trends in damage and abundance of pests and natural enemies.

## CONCLUSION

The results available to date show that the approaches taken in this study have proved to be effective. There are clear indications that interesting differences exist between the organic and conventional production systems.

Completion of the multivariate analyses will provide more extensive information than is presently available on interactions between many of the parameters that were measured. At the same time, the study was able to identify critical factors that need to be dealt with more effectively in the future, notably, planting date and cultivar. In light of these results, procedures for subsequent field seasons are being modified to provide greater uniformity. Transplants of the same tomato variety are being provided to all growers, and greater attention is being paid to ensuring that a similar range of planting dates exists among sites representing each management category. An increase in the number of sample units would clearly be beneficial for many aspects of the study. Furthermore, it became obvious that the advantages of having complete data sets for all sites far outweighed the disadvantage of sampling some sites that may not be ideal for addressing a particular question. In the future it is intended that complete information be collected for at least 18 sites, with essentially even representation of farms covering the spectrum, from large-scale conventional through smaller-scale conventional, transitional, and organic tomato producers.

This multiyear integrated interdisciplinary study of existing farms in one of the world's premier vegetable-producing regions promises to provide valuable insights into the effects of a variety of alternative and conventional management practices on the processes that function within this agroeco-system. While the results pertain most directly to central California agriculture, much of the data should also be relevant to similar irrigated production systems in semiarid irrigated regions of the world. Furthermore, the methods used to achieve an integrated interdisciplinary project with the meaningful involvement of farmers represent a powerful methodological approach for farming systems comparison research.

## ACKNOWLEDGMENTS

The authors acknowlege the assistance of R. O'Malley, S. van Nouhuys, V. Morrone, J. P. Mitchell, D. Haaf, and A. Wong for field sampling and technical assistance; P. Johnson for statistical consultations; and R. L. Bugg for stimulating discussions and for providing information on other farming systems studies. In particular, the authors acknowledge their debt to all the growers who have participated in this study and who provided feedback and ideas in addition to their valuable time. The authors are grateful to the LISA program of USDA for funding this research, and D. K. Letourneau also acknowledges receipt of an Academic Senate Faculty Grant from the University of California, Santa Cruz.

## REFERENCES

Afifi, A. A., and V. Clark. 1984. Computer-Aided Multivariate Analysis. New York: Van Nostrand Reinhold.

Altieri, M. A., and D. K. Letourneau. 1982. Vegetation management and biological control in agroecosystems. Crop Protection 1:405–430.

Altieri, M. A., and L. L. Schmidt. 1986. The dynamics of colonizing arthropod communities at the interface of abandoned organic and conventional apple orchards and adjacent woodland habitats. Agriculture, Ecosystems and Environment 16:29–43.

Baumgartner, J., F. Cerutti, W. Berchtold, and B. Graf. 1985. Multivariate statistical analyses of visual arthropod counts on apple leaf clusters. Schweizeriche Entomologishe Gesellschaft Mitteilungen 58:31–38.

Bethke, L. A., M. P. Parella, J. T. Trumble, and N. C. Toscano. 1987. Effect of tomato cultivar and fertilizer regime on the survival of *Liriomyza trifolii* (Diptera: Agromyzidae). Journal of Economic Entomology 80:200–203.

Bolton, H., Jr., L. F. Elliott, R. I. Papendick, and B. F. Bezdicek. 1985. Soil microbial biomass and selected soil enzyme activities: Effect of fertilization and cropping practices. Soil Biology Biochemistry 17:297–302.

Bottrell, D. R. 1980. Integrated Pest Management. Council on Environmental Quality. Washington, D.C.: U.S. Government Printing Office.

California Farmer. 1987. California at a Glance. A Comprehensive Analysis of Acreage and Dollar Value of California's Principal Crops by County and in State Totals. San Francisco: California Farmer.

Chancellor, W. J. 1977. Compaction of soil by agricultural equipment. Publication No. 1881. Oakland, Calif.: Division of Agriculture and Natural Resources, University of California.

Chaney, K., and R. S. Swift. 1984. The influence of organic matter on aggregate stability in some British soils. Journal of Soil Science 35:223–230.

Chen, W., H. A. Hoiting, A. F. Schmittner, and O. H. Tuovinen. 1988. The role of microbial activity in suppression of damping-off caused by *Pythium ultimum*. Phytopathology 78:314–322.

Cook, R. J. 1984. Root health: Importance and relationship to farming practices. In Organic Farming: Current Technology and its Role in Sustainable Agriculture, D. F. Bezdicek, J. F. Power, D. R. Kerney, and M. J. Wright, eds. Publication No. 46. Madison, Wis.: Agronomy Society of America.

Culik, M. N., J. C. McAllister, M. C. Palada, and S. Rieger. 1983. The Kutztown Farm Report: A Study of a Low-Input Crop/Livestock Farm. Regenerative Agriculture Library Technical Bulletin. Emmaus, Pa.: Rodale Press.

Daamen, R. A., F. G. Wijnands, and G. van der Vliet. 1989. Epidemics of diseases and pests of winter wheat at different levels of agrochemical input. A study on the possibilities for designing an integrated cropping system. Journal of Phytopathology 125:305–319.

Doran, J. W., D. G. Fraser, M. N. Culik, and W. Liebhardt. 1988. Influence of alternative and conventional agricultural management on soil microbial processes and nitrogen availability. Journal of Alternative Agriculture 2:99–106.

Fery, R. L., and F. P. Cuthbert. 1974. Effect of plant density on fruit worm damage in the tomato. HortScience 9:140–141.

Freidrich, A., and J. Zicarelli. 1987. Update on Water Quality in Monterey County, Monterey County's Progress in Public Health. Salinas, Calif.: Department of Health.

Gauch, H. G., Jr. 1982. Multivariate Analysis in Community Ecology. New York: Cambridge University Press.

Goldstein, W. A., and D. A. Young. 1988. An agronomic and economic comparison of a conventional and a low-input cropping system in the Palouse. Journal of Alternative Agriculture 2:51–56.

Groody, K. 1990. The variability in microtopography of soil crusts and the implications for cover crop residue incorporation and mineral fertilizer applications upon crust strength and seedling emergence. M.S. thesis, University of California, Davis.

Johnston, A. E. 1986. Soil organic matter, effects on soils and crops. Soil Use and Management 2:97–105.

Kareiva, P. M. 1983. The influence of vegetation texture on herbivore populations: Resource concentrations and herbivore movement. Pp. 259–289 in Variable Plants and Herbivores in Natural and Managed Systems, R. F. Denno and M. S. McClure, eds. New York: Academic Press.

Karlen, D. L., D. C. Erbach, T. C. Kaspar, T. S. Colvin, E. C. Berry, and T. R. Timmons. 1990. Soil tilth: A review of past performances and future needs. Soil Science Society of America Journal 54:153–161.

Kemper, W. D., and R. C. Rosenau. 1986. Aggregate stability and size distribution. Pp. 425–442 in Methods of Soil Analysis I: Physical and Mineralogical Methods, A. Klute, ed. Agronomy Monograph 9, Part I. Madison, Wis.: American Society of Agronomy, Inc., and Soil Science of America, Inc.

Leath, K. T., and R. H. Ratcliffe. 1974. The effect of fertilization on disease and insect resistance. Pp. 481–503 in Forage Fertilization, D. A. Mays, ed. Madison, Wis.: American Society of Agronomy, Inc., Crop Science Society of America, Inc., and Soil Science Society of America, Inc.

Letourneau, D. K., and L. R. Fox. 1989. Effects of experimental design and nitrogen on cabbage butterfly oviposition. Oecologia 80:211–214.

Lockeretz, W., G. Shearer, and D. H. Kohl. 1981. Organic Farming in the Corn Belt. Science 211:540–546.

Madden, L. V. 1989. Dynamic nature of within-field disease and pathogen distributions. Pp. 96–126 in The Spatial Components of Plant Disease Epidemics, M. J. Jege, ed. Englewood Cliffs, N.J.: Prentice-Hall.

Maidl, F. X., M. Demmel, and G. Fischbeck. 1988. Vergleichende Untersuchungen ausgewählter Parameter der Bod enfruchtbark eit auf konventionell und alternativ bewirtschafteten Standorten. Landwirtschaftliche Forschung 41:231–245.

Miller, A. 1983. Integrated pest management: Psychosocial constraints. Protection Ecology 5:253–267.

Moore, D. C., M. J. Singer, and W. H. Olson. 1989. Improving orchard soil structure and water penetration. California Agriculture 43:7–9.

Nelson, D. W., and L. E. Sommers. 1975. A rapid and accurate procedure for estimation of organic carbon in soils. Proceedings of the Indiana Academy of Sciences 84:456–462.

Niederbudde, E. A., and H. Flessa. 1989. Struktur, mikrobieller Stoffwechsel and potentiell mineralisierbare Stickstoffvorräte in ökologisch und konventionell bewirtschafteten Tonböden. Journal of Agronomy and Crop Science 162:333–341.

Niederbudde, E. A., H. Kaubrügger, and H. Flessa. 1989. Verändrungen von Tonböden bei alternativ-ökologischen und konventionellen Anbauverfahren. Journal of Agronomy and Crop Science 162:217–224.

Nolin, M. C., C. Wang, and M. J. Caillier. 1989. Fertility grouping of Montreal lowlands soil mapping units based on selected soil characteristics of the plow layer. Canadian Journal of Soil Science 69:525–541.

Oades, J. M. 1984. Soil organic matter and structural stability: Mechanisms and implications for management. Plant Soil 76:319–337.

Paul, E. A. 1984. Dynamics of organic matter in soils. Plant Soil 76:275–285.

Pedigo, L. P. 1989. Entomology and Pest Management. New York: Macmillan.

Pfender, W. F., and S. L. Wootke. 1988. Microbial communities of *Pyrenophora*-infested wheat straw as examined by multivariate analysis. Microbial Ecology 15:95–113.

Proebsting, E. L. 1952. Some effects of long continued covercropping in a California orchard. Proceedings of the American Society of Horticultural Science 60:87–90.

Proebsting, E. L. 1958. Fertilizers and cover crops for California orchards. Circular No. 466. Berkeley, Calif.: University of California Division of Agricultural Sciences.

Ratkowsky, D. A., and D. Martin. 1974. The use of multivariate analysis in identifying relationships among disorder and mineral element content in apples. Australian Journal of Agricultural Research 25:783–790.

Reganold, J. P. 1988. Comparison of soil properties as influenced by organic and conventional farming systems. American Journal of Alternative Agriculture 3:144–154.

Reganold, J. P., L. F. Elliott, and Y. L. Unger. 1987. Long-term effects of organic and conventional farming on soil erosion. Nature 330:370–372.

Ristaino, J. B., J. M. Duniway and J. J. Marois. 1988. Influence of frequency and duration of furrow irrigation on the development of phytophthora root rot and yield in processing tomatoes. Phytopathology 78:1701–1706.

Sahs, W. W., and G. Lesoing. 1985. Crop rotations and manure versus agricultural chemicals in dryland grain production. Journal of Soil and Water Conservation 40:511–515.

Scheuring, A. F., ed. 1983. A Guidebook to California Agriculture. Berkeley, Calif.: University of California Press.

Schmitthenner, A. F., and C. H. Canaday. 1983. Role of chemical factors in development of *Phytophthora* diseases. Pp. 189–196 in *Phytophthora*: Its Biology, Taxonomy, Ecology, and Pathology, D. C. Erwin, S. Bartnicki-Garcia, and P. H. Tsao, eds. St. Paul, Minn.: American Phytopathological Society.

Schnuerer, J., and T. Rosswall. 1982. Fluorescein diacetate hydrolysis as a measure of total microbial activity in soil and litter. Applied and Environmental Microbiology 43:1256–1261.

Sengonca, C., and K. U. Bruggen. 1989. Auftreten von Winter weizen Schädlingen und ihren natürlichen Feindeu in unterschiedlich bewirtschafteten Acherbaubetrieben. Zeitschrift fur Pflanzenkrankheiten Pflanzenschutz 96:100–106.

Steiner, H., A. El-Titi, and J. Bisch. 1986. Integrierter Pflanzenschutz im Ackerbau: Das Lautenbach Projekt. I. Experimental design. Zeitschrift fur Pflanzenkrankheiten Pflanzenschutz 93:1–18.

Stynes, B. A., and L. G. Veitch. 1981. A synoptic approach for crop loss assessment used to study wheat. IV. The description of cultural practices and the combination of cultural data with soil data. Australian Journal of Agricultural Research 32:1–8.

Stynes, B. A., and L. G. Veitch. 1983. A synoptic approach for crop loss assessment used to study wheat. VI. The pathogen data and their relationship to soil and cultural practice data. Australian Journal of Agricultural Research 34:167–181.

Stynes, B. A., H. R. Wallace, and L. G. Veitch. 1979. A synoptic approach for crop loss assessment used to study wheat. I. An appraisal of the physical and chemical soil properties in the study area. Australian Journal of Soil Research 17:217–225.

Stynes, B. A., L. G. Veitch, and H. R. Wallace. 1981. A synoptic approach for crop loss assessment used to study wheat. V. Crop growth and yield. Australian Journal of Agricultural Research 32:9–19.

Tate, R. L., ed. 1987. Soil Organic Matter: Biological and Ecological Effects. New York: John Wiley & Sons.

Thomas, C. S., and J. H. Hart. 1986. Site factors associated with Nectria canker on black walnut in Michigan. Plant Disease 70:1117–1121.

Tisdall, J. M., and J. M. Oades. 1982. Organic matter and water-stable aggregates in soils. Journal of Soil Science 33:141–163.

Tsao, P. H. 1983. Factors affecting isolation and quantitation of *Phytophthora* from soil. Pp. 219–236 in *Phytophthora*: Its Biology, Taxonomy, Ecology, and Pathology, D. C. Erwin, S. Bartnicki-Garcia, and P. H. Tsao, eds. St. Paul, Minn.: American Phytopathological Society.

University of California, Davis. 1984. Land, Air, and Water Resources: Technical Report. Water Penetration Problems in California Soils. Davis, Calif.: University of California.

University of California, Oakland. 1985. Integrated Pest Management for Tomatoes. University of California Statewide Integrated Pest Management Project. Publication No. 3274. Oakland, Calif.: Division of Agriculture and Natural Resources, University of California.

van Bruggen, A. H. C., and P. R. Brown. In press. Distinction between infectious and noninfectious corkey root of lettuce in relation to nitrogen fertilizer. Journal of the American Society of Horticultural Science.

van der Linden, A. M. A., J. A. Van Veen, and M. J. Frissel. 1987. Modelling soil organic matter levels after long-term applications of crop residues and farmyard and green manures. Plant Soil 101:21–28.

Veitch, L. G., and B. A. Stynes. 1979. A synoptic approach for crop loss assessment used to study wheat. II. Relationships between soil properties and traditional soil classifications. Australian Journal of Soil Research 17:227–236.

Veitch, L. G., and B. A. Stynes. 1981. A synoptic approach for crop loss assessment used to study wheat. III. Variables derived from soil properties and traditional soil classifications suitable for regression studies. Australian Journal of Soil Research 19:13–21.

Vereijken, P. 1989. Experimental systems of integrated and organic wheat production. Agricultural Systems 30:187–197.

Vereijken, P., and J. H. J. Spiertz. 1988. Geintegreerde en biologische landbouwproduktie Vergeleken. Landbouwkundig Tijdschrift 100:29–32.

Wallace, H. R. 1978. The diagnosis of plant diseases of complex etiology. Annual Review of Phytopathology 16:379–402.

Waring, S. A., and J. M. Bremner. 1964. Ammonium production in soil under waterlogged conditions as an index of nitrogen availability. Nature 201:951–952.

Weisskopf, P., W. G. Sturmy, E. R. Keller, and F. Schwendimann. 1989. Erhaltung der Ertragsfähigkeit des Bodens auf lange Sicht unter dem Einfluss van Fruchtfolgenestaltung, Dungung und Herbicideinsatz. VI. Entwicklung van Unterschieden bei bodenchemischen und bodenphysikalischen Merkmalen der Ertragsfahigkeit des Bodens im Verlaufe der Versuchsperiode 1973 bis 1983. Journal of Agronomy Crop Science 163:90–104.

Widden, P. 1987. Fungal communities in soils along an elevation gradient in northern England. Mycologia 79:298–309.

Williams, W. A. 1966. Management of nonleguminous green manures and crop residues to improve the infiltration rate of an irrigated soil. Soil Science Society of America Proceedings 30:631–634.

Williams, W. A., and L. D. Doneen. 1960. Field infiltration studies with green manures and crop residues on irrigated soils. Soil Science Society of America Proceedings 24:58–61.

Zeddies, J., G. Jung, and A. El Titi. 1986. Integrierter Pflanzenschutz im Ackerbau: Das Lautenbach Projekt. II. Oekonomische Auswirkungen. Zeitschrift Pflanzenkrankheiten Pflanzenschutz 93:449–461.

# 7

# STEEP: A Model for Conservation and Environmental Research and Education

*Robert I. Papendick*

The Pacific Northwest wheatlands are plagued with an erosion problem that has become recognized as among the most serious in the United States. Each year, erosion costs the region millions of tons of topsoil that are washed from its croplands. In some places where conventional farming practices are used, as much as 12 bushels of topsoil are eroded for each bushel of wheat produced. Environmental damage to land and water and loss of soil productivity have occurred despite conservation efforts by concerned farmers and government programs that have been implemented for erosion control. One major limitation of these efforts was that research and practice for erosion control were not always coordinated with research and practice for crop production. For the most part, each one was dealt with as a separate issue. As a result, control practices often increased the cost of crop production and thus were not acceptable to farmers.

More people are becoming sensitive to the serious consequences of soil erosion and its threat to the environment and the economic security of the region. In the future, the capability of maintaining high yields in the inherently fertile soils of the Northwest will depend largely on the ability to prevent the loss of topsoil and the depletion of soil organic matter.

Erosion is the result of a combination of factors, including (1) a winter precipitation climate with high amounts of frozen soil runoff, (2) the exceptionally steep, irregular topography, and (3) management and cropping systems that leave the soil bare as the winter rainy season approaches. Management is especially a problem when wheat and barley are seeded in the fall and when fall moldboard plowing is done in preparation for the following spring planting. Average annual erosion rates in the Palouse region of eastern Washington State range from 17 to 25 tons/acre (approximately 1/8 inch of topsoil) with conventional tillage practices. Each year approximately one-third of the eroded soil is washed into streams, rivers, lakes, and harbors. The sediment in water and the buildup of silt reduces irrigation efficiency and hydroelectric power generation. It also adversely affects many desirable ecosystems and the use of water for industrial and recreational purposes. Costs to remove silt from roadsides and ditches alone amount to several millions of dollars each year in the three-state region of Washington, Oregon, and Idaho.

## THE STEEP PROGRAM

STEEP (solutions to environmental and economic problems) is a multidisciplinary research and education program in the three Pacific Northwest states designed to focus scientific and extension efforts on the control of soil erosion on croplands (Miller and Oldenstadt, 1987; Oldenstadt et al., 1982). The central idea is that soil erosion and water pollution can be reduced significantly by integrating new and improved soil and crop management practices, plant types, pest control methods, and socioeconomic principles into farming systems to achieve sustainable crop production. The research approach is to reduce environmental damage while simultaneously maintaining or increasing agricultural productivity, which is important to the economic and social welfare of the region and the nation.

The motivation for this research approach came from wheat producer organizations in the region. They organized the initial discussions on the subject, obtained supplemental congressional funding for research support, and continue to support and monitor its progress. Funds for STEEP program research have been made available each year since 1976 by a special grant from the U.S. Department of Agriculture (USDA) to the Agricultural Experiment Stations in Washington, Oregon, and Idaho and by appropriations to the Agricultural Research Service (ARS) of USDA.

The STEEP program effort has worked to develop and use new and improved systems of conservation management in which tillage methods, crop rotations, plant types, and methods of plant protection are integrated into complete management systems that minimize erosion without adversely affecting costs or levels of production. There are two main research approaches for erosion control: (1) development of conservation cropping systems along with plant types that can produce economical yields in trashy, hard-soil seedbeds, and (2) development of planting methods for winter wheat and barley in the early fall to provide increased ground cover before winter. Also included is research on the erosion and runoff process and the prediction of erosion and runoff with emphasis on frozen and thawing soils. Emphasis is given to the control of diseases, weeds, insects, and rodents and to socioeconomic factors in the development of conservation tillage systems.

The development and tailoring of new conservation systems for the region has involved ARS of USDA and university scientists from as many as 10 disciplines. These scientists constantly interact to develop consistent sets of data for practices that are compatible with conservation management systems in the Pacific Northwest. The work has an impact on 10 million acres of cropland in the Columbia Plateau and Palouse-Nez Perce Prairies, Columbia Basin, Snake River Plains, and Willamette-Puget Sound Valleys in Washington, Oregon, and Idaho.

## STEEP Program Objectives

The overall research effort of the STEEP program is organized into six major objectives (Oldenstadt et al., 1982). Under each objective there are several subobjectives (not listed here), each of which covers a specific research topic that is to be integrated into the conservation management system being developed. The main emphasis of each major objective is given and discussed below.

1. *Tillage and plant management.* Develop combinations of tillage, cropping, residue management, and weed control systems to control erosion and increase crop production.

2. *Plant design.* Develop crop cultivars with morphological and rooting characteristics that reduce erosion and maintain food (feed) production when grown in conservation cropping systems.

3. *Erosion and runoff prediction.* Improve the understanding and prediction of erosion and runoff processes as they are affected by climate, topography, soils, tillage, and crop management for use as a decision-making tool for planning conservation applications.

4. *Pest management.* Integrate control of weeds, diseases, rodents, and insects into conservation tillage and plant management systems.

5. *Socioeconomics of erosion control.* Determine the impact of improved erosion control practices on farm organizations, cost, and net incomes and on maintaining agricultural productivity in the region.

6. *Soil erosion-productivity relationships.* Develop relationships that show how erosion affects crop production over both the short and long term.

All of the objectives have a common goal, that is, the development of farming systems for control of soil erosion. Each objective contributes in a special way to the achievement of this goal and is a necessary link in the integration approach. The objectives and/or subobjectives are revised as necessary as new problems arise in the process of changing tillage and cropping systems.

## ORGANIZATION AND MANAGEMENT OF THE RESEARCH PROGRAM

Members of the research team, numbering some 30 to 40 scientists, have repeatedly demonstrated their ability to work together over the past 14 years. Every effort is made to ensure that the solutions developed are practical and workable. The size of the effort needed to solve the soil erosion problem is greater than the resources of any single research institution or agency in the region. The combined effort, which is bound together by the common objectives, coordination of effort, and supplemental funding, is demonstrating a high degree of success.

A coordinating committee comprising six scientists, one extension specialist, and a representative from the USDA Soil Conservation Service has the responsibility for organizing annual reporting sessions. These sessions serve as a mechanism for monitoring progress for the wheat industry and interested federal and state agencies, and for facilitating interactions among the participating scientists. Scientists are encouraged to submit research proposals in their area of interest within the six objectives. These are passed through departmental channels, and the research administrators may call on the coordinating committee to review and prioritize the individual proposals and provide recommendations for funding. Research grants are usually made for a 3-year period, after which they are terminated or extensively revised. This turnover provides an opportunity for more scientists to participate in the program and also provides a means to maintain a balance in the needed disciplines.

### STEEP Program Extension

An extension component was added to the STEEP program in 1982 to help disseminate new research findings and to assist farmers in applying research results in the field. One specialist is located at the Columbia Plateau Conservation Research Center at Pendleton, Oregon, and the other is located on the University of Idaho campus at Moscow, Idaho. These specialists interact on a regular basis with scientists in their areas and host radio and television programs, write newsletters, organize grower information meetings, and conduct tours of fields. The STEEP program extension component is a vital link in narrowing the gap between the generation of research information and farm applications.

### Farmer Support

The STEEP program has often been labeled as a "growers' program," meaning that individual farmers and the wheat grower organizations have provided major inputs into the development and operation of the research and extension program. Not only do the growers seek support for STEEP program funding, they also assist in establishing research priorities and stimulating the needed research in other ways. Growers actively participate in the annual reviews, aid the coordinating committee in the evaluation of existing research projects, and suggest new problem areas and needs for future research based on their own experiences. Much of the guidance on the direction of research has come from the farmers themselves. The growers were largely responsible for the addition of the extension component to the STEEP program.

The wheat grower organizations of the three Pacific Northwest states have done much to obtain the special grants funding that supports the research. By pooling their resources, they have been able to convince the U.S. Congress of the seriousness of the erosion problem and the need for increased research on erosion control. Their efforts have largely been responsible for marshaling the resources of the various research agencies and of agri-industry into this high-priority research area.

## STEEP PROGRAM RESEARCH ACCOMPLISHMENTS

The STEEP program has contributed a number of major scientific and technical advancements in conservation farming since its inception. The following are a few examples of accomplishments for each of the six main objectives.

### Tillage and Plant Management

• The yield advantages and improved efficiency of band placement of nitrogen fertilizer in no-till planting of small grains were established (Koehler et al., 1987).

• No-till drills that have the capability to band fertilizer and sow small grains in moderate to heavy amounts of crop residues and in hard, dry soils have been developed (Hyde et al., 1987).

• A crop residue decomposition model that uses generated residue and soil temperature-moisture inputs was developed for predicting the rate at which surface residues disappear in the field (Stroo et al., 1989).

• A wheat growth model that predicts tillage and residue effects on developmental stages of wheat growth was developed (Klepper et al., 1987).

### Plant Design

• Wheat cultivars that perform best in conventional tillage management systems are the ones that perform best in conservation tillage systems (Allan and Peterson, 1987).

• Risks associated with seeding in the early fall to provide good ground cover over the winter have been reduced by developing wheat types that have increased resistance to rusts, foot and root rots, flag smut, and snow mold (Allan and Peterson, 1987).

• Spring wheats that produce only primary tillers were found to be more desirable than multiple tiller-producing varieties in areas with low amounts of rainfall. In these areas, more spring cropping instead of use of summer fallow is encouraged to control erosion (Konzak et al., 1987).

### Erosion and Runoff Prediction

• New factor relationships were developed for the Universal Soil Loss Equation to improve soil erosion prediction for the Pacific Northwest. These relationships are now being used by the Soil Conservation Service in farm planning applications to meet conservation compliance established in the Food Security Act of 1985 (McCool et al., 1987).

• A new method was developed for computing erosivity (R factor) values for rainfall characteristics using hourly rainfall data that are generally more available than the otherwise required 15-minute break-point rainfall data (Istok et al., 1987).

• A rill meter was developed for measurement of erosion in the field. This tool formed the basis of the new data collection needed to develop Universal Soil Loss Equation, length-of-slope (LS) factor relationships for the Pacific Northwest (McCool et al., 1981).

### Pest Management

• Conservation tillage practices and intensive cropping of small grains was found to increase root diseases of wheat and barley (Wiese et al., 1987).

• Studies determined that root diseases of wheat or barley in a conservation tillage system can be controlled by using a 3-year crop rotation, with the cereal being grown only 1 year in 3 (see R. J. Cook, "Challenges and Rewards of Sustainable Agriculture Research and Education," this volume).

• Improved methods of herbicide management were developed for control of annual grasses and broadleaf weeds in no-till wheat, barley, and chemical fallow (Rydrych, 1987; Thill et al., 1987).

### Socioeconomics of Erosion Control

• The diffusion of conservation practices in the Northwest will take considerable time and will depend to a large extent on changes in the context, the innovation, and the characteristics of potential adapters (Dillman et al., 1987).

• STEEP program research provided valuable insights for the highly erodible land protection provisions in the Food Security Act of 1985 (Hoag and Young, 1985).

• Studies that have surveyed farmers on their adoption of erosion control practices showed that (1) absentee landowners are not a major constraint to erosion control; (2) the major constraints to erosion control appear to be factors external to the farmer himself, especially the rules imbedded in government programs; and (3) over the past 10 years there has been a positive change in farmer attitudes toward erosion control as well as implementation of erosion control practices (Carlson et al., 1987; Dillman and Carlson, 1982).

### Soil Erosion-Productivity Relationships

• Wheat yield losses resulting from the loss of topsoil have been masked by technological progress, that is, improved varieties, fertilizer management, and weed control (Papendick et al., 1985; Walker and Young, 1986; Young et al., 1985).

• A computer modeling study supported by field data showed that the effect of erosion on the loss of wheat yields and the anticipated payoff from future technical progress is greater for deep topsoils than it is for shallow topsoils (Young, 1984).

### IMPACTS OF STEEP PROGRAM RESEARCH

To what degree have the objectives of the STEEP program been achieved? Has erosion on cropland been controlled? Has farm profitability been maintained or increased?

After 14 years of the STEEP program, erosion still occurs at unacceptable levels on Pacific Northwest croplands, and some farmers are making less money than they did in past years. Nevertheless, the benefits of STEEP program research are becoming evident. There has been a visible increase in the adoption of new soil and crop conservation management technologies developed or refined by STEEP program researchers. Much of the credit for this goes to the STEEP program research on fertilizer placement and the subsequent development of the fertilizer banding (placing fertilizer close to the seeds) capability of the no-till drills that became commercialized in the early 1980s. Conservation methods have enabled farmers to reduce the number of tillage operations on wheat-based rotations from five or more to less than three. These methods leave the seedbed with a rough surface and covered with residues that are effective in controlling erosion.

Much has been learned about the relationships between tillage and plant diseases and how these can be controlled with residue management and crop rotations. Methods have been devised to move pathogen-infested residues away from the seed row. Crop rotation systems have also been designed to reduce adverse biological effects in no-till systems. Economic studies have shown the long-term benefits of soil conservation and how cost-sharing aids in implementation of the best management practices by defraying the added short-term costs that otherwise could discourage their use by farmers. Overall, the STEEP program, more so than any other program, has created increased public awareness of soil erosion and its consequences and has made growers much more receptive to implementing conservation measures on their farms.

## FUTURE DIRECTIONS

Despite the technological advances made by the STEEP program, soil erosion is still a major environmental and economic problem for the Pacific Northwest region. Development of better conservation practices and achievement of more widespread application of such practices on the land remains an urgent, high-priority need. The mandatory conservation compliance requirements of the Food Security Act of 1985 will likely speed up the adoption of no-till and other surface tillage practices, and many growers will be trying these practices on their farms for the first time in the next several years. Many technical problems relating to the use of these conservation practices have not yet been solved, and consequently, their final economic and social acceptability are not known. For example, many farmers will have difficulty planting and harvesting crops and controlling weeds, diseases, and other pests in seedbeds that are rough-tilled or contain high amounts of surface residues. These obstacles can frequently be overcome in research plots; but they cannot always be overcome on farmers' fields for many reasons, including the lack of know-how, inadequate equipment, cost or time limitations, and variations in soil characteristics. Much remains to be done to accomplish the large-scale adoption of conservation tillage technology on the region's farms.

### Emerging Issues

In addition to soil erosion, other concerns have now placed land stewardship in a new context. Water quality will be a major national issue of the 1990s. In the future, increasingly severe restrictions on chemical and nutrient management are likely to be mandated legislatively. Fewer agricultural chemicals will be available to farmers, particularly for minor-use crops, because chemical manufacturers face increased registration restrictions and expenses. Other issues that farmers must face include energy conservation, environmental stability, fish and wildlife protection, farm worker health, food safety, and maintenance of net farm income. Conservation farming for erosion control affects all of these issues and concerns. For example, it appears that with current technology, conservation practices such as reduced tillage and no-till farming will usually require increased use of pesticides and nitrogen fertilizers. Because of the reduced runoff and evaporation under conservation tillage practices, this may increase infiltration and leaching of chemicals into groundwater. If this is true, the chemical-intensive practices being developed today likely will not be acceptable in the future. The same is true for the potential adverse impacts of chemicals on wildlife, farm worker health, and food safety. Thus, concerns for efficient crop production; conservation of soil, water, and energy; and maintaining net farm income must be integrated with the need to safeguard human health and protect the environment.

About 2 million acres of highly erodible land in the three Pacific Northwest states are now in the Conservation Reserve Program (CRP). This land is under 10-year contracts so that only grasses are grown or the land is used for other activities that do not require tillage. Much of this land had been seriously degraded because of excessive erosion rates as a result of previously used cropping systems. If these lands are converted back to crops after the expiration of CRP contracts, the challenge will be to conserve the accrued productivity benefits from 10 years of growing grasses by returning to cropping systems that are more sustainable. The outcome of this post-CRP transition will have a profound impact on soil erosion, water quality, farm income, and the future of agriculture in the region.

## STEEP II Program

The STEEP program has achieved an organizational structure and stance that makes the program ideally poised to meet the production, conservation, and environmental challenges of the 1990s. It has a proven ability to mobilize scientific and extension resources on relatively short notice to solve regional problems. A proposal has been drafted for a replacement project for the expiring STEEP program, known as the STEEP II program. This revised program will build on the progress and accomplishments of the STEEP program. STEEP II will seek to coordinate a regional research and information delivery system designed to provide growers in the Pacific Northwest with advanced technologies for simultaneously controlling soil erosion and protecting water quality while achieving more cost-efficient crop production and increased farm profitability.

It is proposed that the STEEP II program should be expanded to include water quality protection in addition to erosion control. The program will be organized around three main objectives:

1. Obtain and integrate new technical and scientific information on soils, crop plants, pests, energy, and farm profitability into sustainable, whole-farm management systems.

2. Develop tools for assessing the impacts of farming practices on erosion and water quality.

3. Develop and implement programs for dissemination of information and transfer of technology to the farm.

Within these objectives, considerable effort will be given to developing conservation cropping systems that will consistently produce acceptable crop yields, are adaptable to farmers in the different agronomic zones, and are environmentally safe. Attention will be directed toward overcoming factors that now limit crop growth with conservation planting. Some of the factors that have been identified include increased root diseases and weed infestations associated with increased surface residues, and increased pests in wheat planted early in the fall to enhance ground cover for erosion control in the winter and spring. Other limiting factors related to nutrient cycling and the physical properties of soil will also be examined.

Soil microbial properties of various tillage and cropping systems will be assessed to develop methods and principles that will prevent plant diseases, reduce the potential for nutrient leaching, and increase the use efficiency of applied fertilizers and organic amendments. Increased attention will also be given to new tillage methods that can complement the use of surface residues for erosion control. Research will be conducted to explore management strategies and options for farmers on how to maximize the benefits of soil productivity of CRP lands while in grass and how to return these highly erodible lands from grass to more sustainable cropping systems.

Development and testing of erosion and water quality control systems will include evaluations of economic impacts at the farm and regional levels and of social factors that limit or enhance the adoption of conservation practices by farmers. The STEEP II program will also give new emphasis to developing methodologies for on-farm research and providing scientific backup to research and education projects such as those supported by the USDA low-input sustainable agriculture (LISA) program that involve direct work with growers in testing treatments on large plots or whole fields.

Erosion and water quality models will be developed as tools to evaluate the impacts of conservation tillage and other management options on runoff and erosion, water conservation, and water quality. Physical and chemical process models will ultimately be incorporated with crop production models to provide an interactive link between crop production practices, erosion rates, nutrient use efficiency, and the potential for nutrient escape from the crop root zone.

A variety of innovative approaches will be used for information transfer in the STEEP II program. These include on-farm research and demonstration projects and associated field tours, decision-aid computer software, newsletters, farm magazine and newspaper articles, cooperative extension publications, meeting presentations, conferences and workshops, and audiovisual aids.

## CONCLUSION

The momentum for conservation is present and appears to be accelerating. The STEEP program continues to be the best vehicle for coordinating the research and extension efforts needed to ensure that national and regional goals for resource protection and economic viability are achieved.

## REFERENCES

Allan, R. E., and C. J. Peterson, Jr. 1987. Winter wheat plant design to facilitate control of soil erosion. Pp. 225–245 in STEEP—Conservation Concepts and Accomplishments, L. F. Elliott, ed. Pullman, Wash.: Washington State University Press.

Carlson, J. E., D. A. Dillman, and L. Boersma. 1987. Attitudes and behavior about soil conservation in the Pacific Northwest. Pp. 333–341 in STEEP—Conservation Concepts and Accomplishments, L. F. Elliott, ed. Pullman, Wash.: Washington State University Press.

Dillman, D. A., D. M. Beck, and J. E. Carlson. 1987. Factors affecting the diffusion of no-till agriculture in the Pacific Northwest. Pp. 343–364 in STEEP—Conservation Concepts and Accomplishments, L. F. Elliott, ed. Pullman, Wash.: Washington State University Press.

Dillman, D. A., and J. E. Carlson. 1982. Influence of absentee landlords on soil erosion control practices. Journal of Soil and Water Conservation 37:37–40.

Hoag, D., and D. Young. 1985. Toward effective land retirement legislation. Journal of Soil and Water Conservation 40:462–465.

Hyde, G., D. Wilkins, K. Saxton, J. Hammel, G. Swanson, R. Hermanson, E. Dowding, J. Simpson, and C. Peterson. 1987. Reduced tillage seeding equipment. Pp. 41–56 in STEEP—Conservation Concepts and Accomplishments, L. F. Elliott, ed. Pullman, Wash.: Washington State University Press.

Istok, J. D., J. F. Zuzel, L. Boersma, D. K. McCool, and M. Molnau. 1987. Advances in our ability to predict rates of runoff and erosion using historical climatic data. Pp. 205–222 in STEEP—Conservation Concepts and Accomplishments, L. F. Elliott, ed. Pullman, Wash.: Washington State University Press.

Klepper, B. L., R. W. Rickman, and P. M. Chevalier. 1987. Wheat plant growth in conservation tillage. Pp. 93–107 in STEEP—Conservation Concepts and Accomplishments, L. F. Elliott, ed. Pullman, Wash.: Washington State University Press.

Koehler, F. E., V. L. Cochran, and P. E. Rasmussen. 1987. Fertilizer placement, nutrient flow, and crop response in conservation tillage. Pp. 57–65 in STEEP—Conservation Concepts and Accomplishments, L. F. Elliott, ed. Pullman, Wash.: Washington State University Press.

Konzak, C. F., D. W. Sunderman, E. A. Polle, and W. L. McCuiston. 1987. Spring wheat plant design for conservation tillage management systems. Pp. 247–273 in STEEP—Conservation Concepts and Accomplishments, L. F. Elliott, ed. Pullman, Wash.: Washington State University Press.

McCool, D. K., M. G. Dossett, and S. J. Yecha. 1981. A portable rill meter for field measurement of soil loss. Erosion and sediment transport measurement. Proceedings of the Florence Symposium, June 1981. IAHS Publication No. 133. Wallingford, England: International Association of Hydrological Sciences.

McCool, D. K., J. F. Zuzel, J. D. Istok, G. E. Formanek, M. Molnau, K. E. Saxton, and L. F. Elliott. 1987. Erosion processes and prediction for the Pacific Northwest. Pp. 187–204 in STEEP—Conservation Concepts and Accomplishments, L. F. Elliott, ed. Pullman, Wash.: Washington State University Press.

Miller, R. J., and D. Oldenstadt. 1987. STEEP history and objectives. Pp. 1–7 in STEEP—Conservation Concepts and Accomplishments, L. F. Elliott, ed. Pullman, Wash.: Washington State University Press.

Oldenstadt, D. L., R. E. Allan, G. W. Bruehl, D. A. Dillman, E. L. Michalson, R. I. Papendick, and D. J. Rydrych. 1982. Solutions to Environmental and Economic Problems (STEEP). Science 217:904–909.

Papendick, R. I., D. L. Young, D. K. McCool, and H. A. Krauss. 1985. Regional effects of soil erosion on crop productivity—the Palouse area of the Pacific Northwest. Pp. 305–320 in Soil Erosion and Crop Productivity, R. F. Follett and B. A. Stewart, eds. Madison, Wis.: American Society of Agronomy, Crop Science Society of America, and Soil Science Society of America.

Rydrych, D. J. 1987. Weed management in wheat-fallow conservation tillage systems. Pp. 289–298 in STEEP—Conservation Concepts and Accomplishments, L. F. Elliott, ed. Pullman, Wash.: Washington State University Press.

Stroo, H. F., K. L. Bristow, L. F. Elliott, R. I. Papendick, and G. S. Campbell. 1989. Predicting rates of wheat residue decomposition. Soil Science Society of America Journal 53:91–99.

Thill, D. C., V. L. Cochran, F. L. Young, and A. G. Ogg, Jr. 1987. Weed management in annual cropping limited-tillage systems. Pp. 275–287 in STEEP—Conservation Concepts and Accomplishments, L. F. Elliott, ed. Pullman, Wash.: Washington State University Press.

Walker, D. J., and D. L. Young. 1986. Effect of technical progress on erosion damage and economic incentives for soil conservation. Land Economics 62:89–93.

Wiese, M. V., R. J. Cook, D. M. Weller, and T. D. Murray. 1987. Life cycles and incidence of soilborne plant pathogens in conservation tillage systems. Pp. 299–313 in STEEP—Conservation Concepts and Accomplishments, L. F. Elliott, ed. Pullman, Wash.: Washington State University Press.

Young, D. L. 1984. Modeling agricultural productivity impacts of soil erosion and future technology. Pp. 60–85 in Future Agricultural Technology and Resource Conservation, B. C. English, J. A. Maetzold, B. R. Holding, and E. O. Heady, eds. Ames, Iowa: Iowa State University Press.

Young, D. L., D. B. Taylor, and R. I. Papendick. 1985. Separating erosion and technology impacts on winter wheat yields in the Palouse: A statistical approach. Pp. 130–142 in Erosion and Soil Productivity, D. K. McCool, ed. ASAE Publication No. 8-85. St. Joseph, Mich.: American Society of Agricultural Engineers.

# 8
# Soil Moisture Monitoring: A Practical Route to Irrigation Efficiency and Farm Resource Conservation

*Gail Richardson*

Surface water and groundwater withdrawals for farm irrigation in the 17 western states constitute about 85 percent of the region's developed supplies. More efficient irrigation management could reduce aquifer depletion, protect water quality by curtailing runoff and drainage, help to control salinization and soil erosion, and make more water available for other uses. The potential for improvement is vast, but the required changes must occur field by field and farmer by farmer. There are no shortcuts.

In the past few years, because of research by INFORM (a nonprofit environmental research organization), training programs by the Soil Conservation Service (SCS) of the U.S. Department of Agriculture and support from private foundations, the California Energy Commission, and the Western Area Power Administration (WAPA), farmers in California and Colorado have adopted a low-cost, site-specific water management method on more than 70,000 irrigated acres. They use the information to reduce their water and energy use.

Corn and wheat farmers in Colorado are the largest and fastest-growing pool of users. Most of them irrigate with center-pivot sprinklers. Ranchers, field crop farmers, and specialty crop growers in California are also adopting the method and showing its applicability to diverse crops, soils, and farm operations—and especially to surface irrigation systems like those that predominate in irrigated agriculture systems in the United States.

Farmer testimony and field evidence now amply justify broader public support for soil moisture monitoring as a component of integrated conservation plans for irrigated farms. There is also growing evidence that soil moisture monitoring is effective in persuading farmers of dryland as well as irrigated crops in water-short regions (where wind erosion problems are concentrated) to convert to various forms of conservation tillage. Moisture monitoring is currently in use on 10,000 acres of dryland wheat in Colorado and is spreading.

## HOW SOIL MOISTURE MONITORING REDUCES GUESSWORK AND IMPROVES FARM RESOURCE MANAGEMENT

"Surface moisture, like beauty, is thin." These words of a Colorado farmer capture a common predicament. The field surface masks what lies beneath it. Irrigators must, therefore, guess about the effects of rainfall, snow, irrigation, sun, wind, and cultivation practices on moisture levels in the root zone. They follow instinct, the calendar, or their neighbors; and when in doubt, they run water.

Soil moisture monitoring reduces this uncertainty. It gives irrigators a record of the rise and fall in moisture levels at several soil depths in a field. It enables them to see, for example, whether an irrigation fully refills the root zone (often it does not, even if plenty of water is applied) and how long it sustains crop growth before another irrigation is needed.

Each field has unique soils and a unique history. Solutions to over- or underwatering on one field may be inappropriate on another. Site-specific trial-and-error experiments guided by soil moisture data help farmers solve such puzzles. They give farmers the direct proof they need to assess which of various possible remedies to irrigation problems actually increase their efficiency, reduce costs, and maintain or improve yields. Changes they are likely to adopt are described below.

### Scheduling Irrigations To Meet Crop Needs

Farmers who monitor moisture changes in the root zone often find that they can delay irrigation in the spring until crops deplete stored rainfall, space irrigations farther apart during the season, and on some crops, cut water off earlier toward the season's end. These changes are often straightforward and can bring sizable water and energy cutbacks and cost reductions.

### Managing Equipment More Efficiently

Colorado farmers typically run their center-pivot sprinklers as fast as they will go, often reapplying water on wet field sections before the last dousing has been absorbed. Such practices increase surface runoff and can damage equipment and fields. Soil moisture monitoring has shown that slower rotations result in better moisture penetration, fuller utilization of storage capacity in the root zone, and less runoff. Similarly, monitoring can help irrigators adjust pipeline valves, select siphons of appropriate size, and adapt all types of sprinkler systems to specific field and soil conditions.

### Controlling Distribution and Drainage

Surface irrigators often apply too much water at one field end and not enough at the other, as data collected by INFORM and SCS from California fields amply show. Heavy drainage and the leaching of chemicals can thus occur at the wet end, even though the total application may not exceed the volume needed to meet crop needs. Farmers who detect these problems by soil moisture monitoring are often motivated primarily to correct the underirrigation on the dry field end because of its potential to damage yields. Sometimes a change in the water application rate alone will improve distribution uniformity and reduce drainage.

### Reducing Soil Compaction and Infiltration Problems

Compacted soil layers lying beneath a field surface can restrict water infiltration below relatively shallow depths and cramp root growth into narrow soil bands. In addition to depressing yields, compaction is a major contributor to surface runoff. Soil moisture monitoring helps farmers locate these restrictive layers and evaluate the effectiveness of potential remedies. Other chemical or physical conditions that impede soil water movement and often contribute to runoff can be similarly analyzed and remedied by using soil moisture data to evaluate trial-and-error field experiments.

Heavy harvesting equipment creates more compaction on wet ground than it does on dry ground. Soil moisture monitoring helps irrigators improve the timing of irrigation cutoff dates to ensure a dry, less damageable field at the time of harvest. A drier field also reduces wear and tear on equipment.

### Converting to Conservation Tillage

Tillage systems that leave plant remains on field surfaces typically increase moisture retention at deep levels in the root zone and control soil erosion. With soil moisture monitoring, farmers on dryland as well as irrigated fields can see these moisture-saving benefits for themselves. This field proof can be a persuasive means of converting them to ridge-till, no-till, and other forms of conservation tillage. Good examples of this can be found in Colorado.

### Integrated Resource Management

After several seasons, farmers who use moisture data to manage water, equipment, crops, and soils may be drawn step-by-step into a virtual revolution in their practices. Since 1987, on Colorado's eastern plains overlying the Ogallala Aquifer, soil moisture monitoring is playing a central role in convincing mainstream grain producers to adopt an integrated conservation package, component by component.

The first component is typically pump tests (by nature of their work, farmers understand motors). Then come soil moisture monitoring, adjustments of irrigation schedules and equipment, the identification and analysis of soil compaction, year-round monitoring of dryland as well as irrigated fields, the substitution of no-till for clean fallow tillage methods, conversion to ridge-till methods on irrigated fields (with major reductions in herbicide use), and conversion to low-energy precision application sprinkler technology. The financial and environmental benefits of these interrelated changes far exceed those achieved through improved irrigation scheduling alone.

## TOOLS FOR MONITORING SOIL MOISTURE

The simplest form of soil moisture monitoring involves forcing a metal rod into the ground to see how deep it penetrates. (A dry soil layer will stop it.) Rods with special tips can be used to remove soil samples from various depths to examine their moisture content. These tools are useful for spot-checking wet or damp fields, but they are unusable after a field surface dries out.

For the systematic and continuous monitoring of soil moisture levels at deeper as well as shallower root zone depths, three other field tools are available. These tools are suited to different uses and vary in price.

The neutron probe emits neutrons from radioactive material into soil that has been previously analyzed for its water-holding capacity and prepared for testing. A counter records the neutrons that slow down after colliding with hydrogen atoms in water molecules. This count establishes the soil's water content with great precision. One unit costs about $4,000, and the user must be licensed by the Atomic Energy Commission.

The neutron probe is used primarily by researchers, private consultants, irrigation or water district staff, and farm advisers who are trained to do the tests and interpret the results for farmers. The probe can be used under all field conditions and on all irrigation systems.

Tensiometers are long plastic tubes (typically 2 to 4 feet in length) that are filled with water and equipped with a vacuum gauge. Their ceramic-tip ends are inserted into root zones and, as the root zone dries out, water that is "pulled" from the tubes creates a vacuum that indicates the soil's dryness: the drier the soil the stronger the vacuum. Tensiometers cost about $40 each.

Tensiometers are useful in fields with high moisture conditions, such as those maintained by certain trickle and drip systems. They are less useful for monitoring full wetting and drying cycles, because in heavier soils they become inoperative below a relatively wet range. Problems of breakage and maintenance make them impractical for use on field crops. They are more commonly used on permanent crops.

Gypsum blocks are small electrical sensors that are implanted in root zones. Wires attached to electrodes inside each one are drawn to the soil surface. By plugging the protruding wires into a hand-held meter, farmers get a reliable relative reading of moisture conditions surrounding the buried block: Conductivity is high under wet soil conditions and low under dry conditions.

Gypsum blocks are used in sets (stations) to monitor three or four soil depths at the same site. About six monitoring sites are required on surface-irrigated fields for analyzing patterns of water distribution, but as few as one or two sites can be sufficient on these or other fields where distribution is known to be relatively uniform.

Gypsum blocks cost from $3.50 to $14.00 each. Meters for reading them run from $150 to $600. Gypsum blocks are adaptable to nearly all crops and soils and all except high-moisture irrigation systems. They normally last 2 to 4 years in well-drained soils.

## Soil Moisture Monitoring and Other Approaches to Irrigation Management

For purposes of system evaluation and irrigation scheduling, soil moisture monitoring can often be combined with—and enhanced by—other tools and approaches. These include the use of evapotranspiration data to predict water consumption by crops and plan irrigation schedules, pump tests and other equipment evaluations, and advance-recession analyses of irrigation distribution on surface-irrigated fields that help farmers analyze and correct inefficient practices.

## INFORM'S RESEARCH: FORERUNNERS AND FINDINGS

Most of the scientific testing of gypsum blocks was done in the 1940s and 1950s and predates recent improvements in block design. An exception is the work of D. W. Henderson (now retired) of the University of California at Davis, whose decades of laboratory research provided, by the late 1970s, what may be the richest store of technical data ever accumulated. Unfortunately, Henderson published very few of his findings.

Now, after a 10-year hiatus, laboratory testing with gypsum blocks has resumed at the University of California at Davis. There, Larry Schwankl, an irrigation specialist with cooperative extension, is testing and comparing different models of gypsum blocks and meters. The results of his work will be published in 1991.

Outside the laboratory, gypsum blocks have been used at different times in several western states, but field results were sparsely documented until INFORM's field demonstrations in California in the 1980s. These provided the first comprehensive record of a systematic method for applying soil moisture data to a wide range of irrigation and other farm management decisions.

Paving the way to INFORM's study were several gypsum block field trials in the late 1970s that were organized by the Yolo County (California) Resource Conservation District to find a water management method that met farmers' practical needs. The director was Peter Mueller-Beilschmidt, an irrigation engineer and independent consultant from Davis, California. Mueller-Beilschmidt for many years had used blocks to evaluate sprinkler systems. He had also contributed to the technical improvement of gypsum block design and manufactured small quantities for his own engineering purposes.

During the course of these trials, Mueller-Beilschmidt developed a monitoring method specifically for surface irrigators, but it was equally applicable to sprinkler systems. Although the field trials went well, no resources were available to expand this initiative.

## Field Demonstrations, 1984–1986

In 1983, INFORM learned of Mueller-Beilschmidt's work. Perceiving the value to farmers and the public of documenting and publicizing a promising conservation method, INFORM organized 3 years of gypsum block trials on 31 surface-irrigated fields (chiefly alfalfa, cotton, and tomatoes) in northern and southern sections of the Central Valley of California. (One sprinkler-irrigated almond orchard was also studied in the project's pilot phase.)

Cooperators were recommended by technicians or their neighbors. The 16 farmers who participated ranged in age from their 20s to their 80s and had different backgrounds, philosophies, and managerial styles. They owned or managed farms that averaged 1,000 to 1,500 acres in size.

Fields were selected by farmers and accepted by INFORM for testing if they met a bare minimum of practical criteria. No prior soil analyses were conducted, nor did the testing methods or measurements exceed, at any point, a precision that was easily achievable by the farmers themselves. The aim was not to control variables or make statistical inferences of cause and effect, but to demonstrate how farmers could manage water more efficiently within existing limits of information, equipment, and established (and diverse) farm routines.

Each field trial began with a check for uneven root zone wetting, which is a major cause of irrigation inefficiency and drainage buildup. As re-

Soil erosion is agriculture's leading problem in the Pacific Northwest of the United States. (See Chapter 7.) Credit: Robert I. Papendick.

Cattle graze on crop residue, which represents one aspect of a low-input beef production system. (See Chapter 16.) Credit: Terry Klopfenstein.

The mowing of winter cover crops appears to have the potential to reduce herbicide and insecticide use and to increase yields and profits. (See Chapter 11.) Credit: John M. Luna.

A viable conservation tillage approach for corn production in the Southeast may be the planting of corn into ridges after a winter cover crop has been mowed. (See Chapter 11.) Credit: John M. Luna.

Surface mulch from a rye cover crop helps suppress weeds and conserve soil moisture in a low-input, ridge-till corn production system. (See Chapter 11.) Credit: John M. Luna.

Soil moisture changes in root zones can be monitored with buried electrical sensors known as gypsum blocks. The gypsum block on the right shows the two concentric electrodes to which two insulated wires will be attached before its construction is completed. The finished block (left), which is 1.25 inches in diameter, will be buried in a crop-root zone with its two insulated wires drawn to the surface for testing with an impedance meter. (See Chapter 8.) Credit: Gail Richardson.

The corn root on the left is from a plant grown using a low-input system that included manure. The root on the right is from a plant grown using a conventional cash grain system. (See Chapter 18.) The variation in root morphology may be explained by crop rotation, use of manure versus chemical fertilizer, differences in soil structure, and exposure to herbicide (in the conventional system). Spring moldboard plowing was used for tillage in both systems. Credit: Rhonda R. Janke and Steven E. Peters.

On the right, crabgrass has taken over a check plot, which required spraying with paraquat to prevent the weed from growing into the rows of blueberries. Crabgrass growth was completely suppressed on the left row, where pearl millet was grown as a living mulch cover crop. (See Chapter 12.) Credit: Kim Patten.

**TABLE 8-1** Water Distribution on 31 Surface-Irrigated Fields in California, Tested by INFORM, 1984–1986

| Patterns of Soil Wetness After Irrigation | Number of Fields |
|---|---|
| Uniform saturation to 4-foot depth | 7 |
| Too wet before irrigation to judge | 8 |
| Underirrigated top-middle section(s) | 9 |
| Underirrigated end-middle section(s) | 9 |
| Underirrigated depth | 2 |
| Underirrigation of entire field | 2 |
| Total | 37* |

*Six fields were counted twice because distribution patterns varied in different sections.

SOURCE: Richardson, G., and P. Mueller-Beilschmidt. 1980. Winning with Water: Soil-Moisture Monitoring for Efficient Irrigation. New York: INFORM.

quired for this analysis and later testing, a multiple-site monitoring system provided data from several points down the irrigation run on two strips in each field.

Of the 31 surface-irrigated fields studied, only 7 were found to be uniformly irrigated. Of the remainder, 16 had some sections that were underirrigated and some that were overirrigated. On these fields, locations near the water source were just as likely to be skipped by irrigation as locations near the drain ditch—which is contrary to common teaching and farmers' expectations. Eight fields were so heavily watered that distribution patterns were obscured altogether (the soils were continuously soggy) (Table 8-1).

Distribution patterns typically changed during the irrigation season and sometimes varied from one field section to another. Such complexities had diverse causes, including soil compaction and other poor soil conditions—as well as application rates that were poorly matched to soil types. These site-specific variations ruled out any single remedy for inefficient water use.

On several fields, where farmers' cooperation and logistics permitted, INFORM demonstrated how soil moisture data helped farmers spot and correct uneven distribution patterns. Then, on 21 fields, INFORM used soil moisture data to make scheduling changes and reduce water applications on test strips of under 1 to more than 17 acres. Water use, water cost, and yields on these strips were compared with the results of farmers' standard practices on each field's control strip.

## Findings

On 21 fields, INFORM found that the soil moisture method achieved water reductions ranging from 6 to 58 percent compared with farmers' standard practices. Most reductions fell in the range of 20 to 40 percent. The highest reductions occurred on strips where both the application rate and the irrigation frequency were reduced (Richardson and Mueller-Beilschmidt, 1988).

Yields improved on 10 fields, remained the same on 6 fields, declined on 4 fields, and could not be assessed on 1 field. Of the 4 fields with lower yields, 2 cases were due to serious infiltration problems and two cases were due to irrigation delays caused by confusion about instructions to farmers.

The financial benefits from lower water costs on 13 test strips ranged form $1 to $90 per acre, depending both on the cost of the water and the size of the reductions. The median cost of water per acre-foot (whether purchased or pumped) was $17. The median benefit of water reductions was $10 per acre. On eight fields irrigated through large valves in underground pipelines, there was no practical way to measure outflow. Hence, neither the volumes of water reductions nor their economic values could be estimated except in percentages.

The financial benefits from higher yields on eight fields ranged from $5 to $126 per acre. All but one of these fields were planted with alfalfa, on which irrigations were reduced from two to one per monthly cutting cycle. On two more fields (one alfalfa and one tomato), farmers reported improvements in yield quality on the drier test strips, but no quantitative evaluation was possible.

The combined benefits of water cost reductions and increased yield revenues ranged from $25 to $165 per acre. These occurred on the six fields where both benefits could be measured.

The documented financial benefits fell short of the potential for improvements indicated by INFORM's field data. On most fields, two or three seasons of gradual changes would be needed to fully realize this potential. Moreover, reduced labor time, better assessment of new field practices, better management of equipment, and other gains reported to INFORM by farmers could not be quantified, yet they appeared to be significant in persuading farmers of the benefits of monitoring soil moisture.

## GYPSUM BLOCK PROGRAMS IN CALIFORNIA AND COLORADO

Since the mid-1980s, when INFORM completed its research, the use of gypsum blocks has spread to more than 170 farmers in California and Colorado. Their field records and innovations provide the first broad-scale

**TABLE 8-2**  Gypsum Block Demonstrations
in California in 1989

| Field Trial Result | Number of Farmers |
|---|---|
| Total farmer participants | 45 |
| Water and energy savings achieved | |
|     High ($59/acre average) | 12 |
|     Moderate ($11/acre average) | 7 |
|     Low ($4/acre average) | 8 |
| Insufficient data to quantify | 3 |
| Full season used for | |
|     observation and analysis | 8 |
| Underwatering or infiltration problem | 3 |
| Farmer not interested | 4 |

SOURCE: California Association of Resource Conservation Districts. 1989. Gypsum block demonstration trial. Unpublished draft. Sacramento: California Association of Resource Conservation Districts.

documentation under commercial conditions of the economic and environmental benefits of systematic soil moisture monitoring on irrigated fields. Several private and public programs have contributed to this trend. Farmers' receptivity is largely due to rising energy and water costs.

### The California Program, 1989–1990

SCS technicians in California watched INFORM's research with interest. At its conclusion they sought INFORM's assistance in training SCS staff to use and teach the gypsum block method. After 2 years of technology transfer involving more than three dozen additional field trials, SCS concluded that the method met farmers' needs.

In 1988, with SCS's technical input, the California Association of Resource Conservation Districts won a 2-year grant of $90,000 from the California Energy Commission to teach moisture monitoring to 100 farmers in north central California (SCS Area IV). This program is developing thorough documentation of field results and farmers' reactions (California Association of Resource Conservation Districts, 1989).

In 1989, all but 4 of the 45 farmers who participated in the California Association of Resource Conservation Districts program in 1989 were persuaded of the benefits of soil moisture monitoring and reported their plans to continue using gypsum blocks. Of these, about half realized significant energy and water savings during the instructional period (see high and moderate water- and energy-savings categories in Table 8-2). Most of the remainder gathered useful information about fields, crops, and practices and said they would try scheduling changes in 1990.

The water reductions, where measurable, averaged nearly 1 acre-foot per acre. About 7,000 acres are currently managed with gypsum blocks in California.

### Colorado's Ogallala Program

Colorado's Ogallala region includes more than half a million irrigated farm acres in the state's eastern plains. Parts of this area contain soils with a high clay content. Other sections belong to the sand hill country where the leaching of farm chemicals into groundwater is of growing concern. Yuma County, one of the top corn-growing counties in the United States, is located in this region.

In contrast to the California program, which emphasizes surface irrigation and employs several monitoring sites per field, the program in Colorado has concentrated on center-pivot irrigation and uses only one or two monitoring sites per circle. This reduces the yearly cost of equipment and labor for monitoring to $0.50 per acre in most cases.

Beginning in 1986, the Ogallala Team of SCS initiated a soil moisture monitoring program for corn and wheat producers. The 1986 work on a few fields was followed by a broader educational effort in 1987. In that year, INFORM gave essential support in the form of equipment and technical advice. Simultaneously, WAPA began a pump-testing program that quickly expended to include soil moisture monitoring. Since then the SCS- and WAPA-supported programs have branched into integrated resource management (see above). WAPA considers these programs to be models for programs in neighboring states.

In three growing seasons, 1987 to 1989, moisture monitoring has spread to 65,000 irrigated acres in six Colorado counties (and 10,000 additional acres of dryland wheat). The affected acreage is probably much greater because many farmers use monitoring data from one or two center-pivot irrigation circles to manage several others. In general, farmers who monitor their fields cut back from a 90- to 100-day irrigation season to a 60- to 70-day season. Water and energy cutbacks exceeding 50 percent are not uncommon.

Bruce Unruh, a corn farmer who has been in the program since 1987, reports that he has had to "throw away all the things that you were taught about irrigation and start all over." Spurred by discoveries made with soil moisture monitoring, he has adopted a variety of new practices on irrigated corn. These include slower rotations of his center-pivot sprinkler system, the elimination of irrigation prior to planting, earlier cutoff dates in the fall, and ridge-till cultivation (which reduces herbicide use).

Unruh says that $70 per acre is a conservative estimate of his cost savings from reduced inputs alone. Compared with his pre-1987 practices, his per acre costs have dropped $30 because of energy use cutbacks, $30 because of lower fuel consumption, and $10 because of reduced applications of fertilizer and herbicides. When Unruh completes his present transition to low-energy precision application sprinkler technology, he can probably anticipate another $10 per acre savings in lower energy bills, according to a local technician. Moreover, Unruh identifies many other benefits of his "revolution" for which he does not have figures readily at hand. These include higher yields, lower labor costs, and less equipment damage.

Most irrigators on Colorado's eastern plains use electric pumps. In one county, Kit Carson, an added boost to the SCS- and WAPA-supported programs has come from the rural electric cooperative, K.C. Electric Association. A recent change in electricity rate structure gives farmers a powerful incentive to withhold irrigation as long as possible in the spring and to cut it off as early as possible in the late summer. The new system is most effective for both the utility and farmers when combined with soil moisture monitoring, so the K.C. Electric Association provides gypsum blocks to farmers to bring them into the SCS and WAPA programs.

The cooperative extension and state energy and soil conservation programs also have links with the SCS program. Governor Roy Romer has twice presented the state's farm energy conservation award to farmers who participated in this program.

A staff of four in six counties can no longer meet growing farmer demand for instruction in soil moisture monitoring. Both farmers and technical staff believe that at least half of the farmers in the Ogallala region of Colorado would adopt the method if they could observe its benefits on their own fields. These farmers would include the largest and best producers and would affect far more than half the region's irrigated acreage. The cumulative impact of their field-level changes could greatly reduce the region's energy use, slow the depletion of the Ogallala Aquifer, and curtail the runoff and leaching that degrade water quality.

## THE ROLE OF EDUCATIONAL ADVISERS AND AGENCIES

SCS technicians in California and Colorado agree that soil moisture monitoring with gypsum blocks sells itself. However, they emphasize the importance of initial instruction that includes the following components as being essential for getting farmers to the takeoff point over one or two seasons:

- demonstration on farmers' own fields;
- one-on-one teaching (several meetings a year);

• farmer participation as early as possible in installing gypsum blocks and collecting readings; and

• patience; farmers "look before they leap," and it sometimes takes a while for them to respond.

Soil moisture monitoring opens a new world to both farmers and technicians, and they learn together. This creates a strong foundation for the development of integrated farm management plans based on site-specific conditions and farmers' actual questions and needs (Richardson et al., 1989).

## PRODUCERS OF GYPSUM BLOCKS

There are several producers of gypsum blocks and meters in the United States, chiefly the following:

• Beckman Industrial Corporation, Cedar Grove, New Jersey
• Delmhorst Company, Towaco, New Jersey
• Electronics Unlimited, Sacramento, California
• Irrometer Company, Riverside, California
• Soilmoisture Equipment Corporation, Santa Barbara, California
• Soil Test Incorporated, Denver, Colorado

Different models of gypsum blocks and meters perform differently. No technical standards have yet been developed for evaluating and comparing their features.

Electronics Unlimited supplies the blocks and meters that were used in all of INFORM's work and continue to be used by SCS in California. The Delmhorst Company has been the leading supplier of the blocks and meters used in the program in Colorado.

## THE PUBLIC ROLE

Soil moisture monitoring is a practical route to lower production costs for irrigated farms and more effective protection of the environment in the western United States. It deserves increased public support of several kinds.

• Research is needed to establish technical standards to enable farmers and technicians to choose wisely among the available soil moisture sensors, which vary in sensitivity, uniformity, longevity, and price.

• Field demonstrations targeted in areas plagued by drainage and contamination problems could help to identify practical applications of soil moisture monitoring as components of regional water quality plans.

• The training of field staff in major federal and state agencies would give farmers much broader access to instruction in soil moisture monitoring.

• Modest financial assistance to farmers during an initial instructional period of 1 or 2 years would encourage more farmers to try soil moisture monitoring. Once farmers see its benefits on their own fields, they are generally willing to carry the method's full costs.

## REFERENCES

California Association of Resource Conservation Districts. 1989. Gypsum block demonstration trial. Unpublished draft. California Association of Resource Conservation Districts, Sacramento, Calif.

Richardson, G., and P. Mueller-Beilschmidt. 1988. Winning with Water: Soil-Moisture Monitoring for Efficient Irrigation. New York: INFORM.

Richardson, G., J. Tiedemann, K. Crabtee, and K. Summ. 1989. Gypsum blocks tell a water tale. Journal of Soil and Water Conservation 44:192–195.

# 9

# Reactors' Comments

## Research and Education in the Western Region

*Dale R. Darling*

Many of the processes proposed for changing agriculture, including low-input sustainable agriculture (LISA), sustainable agriculture, alternative agriculture, best management practices, integrated pest management, and integrated crop management, are used by many highly efficient, profitable farm operators today.

Agriculture was changing before LISA; however, the pace and intensity, as well as the focus, have quickened since LISA was introduced. Some of the products developed by DuPont involved in this change are described below:

• methomyl (Lannate) was used for insect control in soybeans based on economic thresholds in the early 1970s; it is now referred to as integrated pest management;
• linuron (Lorox) herbicide was used for no-till soybeans in the early 1970s;
• chlorsulfuron (Glean) herbicide, the first of a family of new low-level, low-environmental-impact herbicides, is a virtually nontoxic crop protectant for cereals; and
• low-level, low-environmental-impact pyrethroid insecticides, such as estenvalerate (Asana), are used along with low-level herbicides to reduce significantly the quantity of pesticides placed in the environment.

Through change comes opportunity. Industry looks forward to the challenges, changes, and opportunities for a more environmentally sound, socially acceptable, and more profitable agriculture system for U.S. farmers and their customers, the consumers.

In reviewing these comments, the reader must keep in mind my biases about agriculture. First, agriculture is fundamentally sound and very productive, with abundant opportunities for improvement through scientific discoveries. However, the starting point for initiating change should be with the highly efficient, profitable, environmentally conscientious farmers who have incorporated the results of 90 years of agriculture research and education into their profitable enterprises. Second, to create change efficiently and effectively, all of those involved in the input side of agriculture should be included in the process along with those who envision the need for dramatic changes.

## REACTIONS

Below are some of my general reactions to the chapters presented in the section, "Research and Education in the Western Region."

• In general, there is no argument with the concepts and principles presented in the sustainable agriculture focus.

• There appears to be more of a spirit of cooperation developing between most facets of agriculture and production; research and education; and government, industry, and producers.

• Some still need to understand that crop protectants are marketed for the purpose of managing excesses in pest populations; they are tools, like a plow, a cultivator, or a hoe.

• A team approach to research, demonstration, education, marketing, and production is the most economical and rapid method for creating change. The highly efficient, profitable farming enterprises most rapidly adopt new ideas.

The following are specific responses to the chapters in this section.

In their opening, John Gardner and colleagues ("Overview of Current Sustainable Agriculture Research") asked, "if sustainable agriculture is so good, why is it so controversial?" The concepts presented in that chapter are on target in relation to the title. However, if the leading question was answered, I missed the point. It is possible that the controversy Gardner and colleagues referred to is a result of the lack of communication between those involved in the established systems of research, education, and production and those desiring to change the systems.

Robert Papendick, in "STEEP: A Model for Conservation and Environmental Research and Education," showed that STEEP is truly a model of cooperation between producers, researchers, conservationists, extension personnel, commercial consultants, and retailers, as well as those involved in many production agriculture input industries, including the equipment, seed, fertility, and crop protection industries. The results are the proof.

The project described by Carol Shennan and colleagues in "Comparative Study of Established Organic and Conventional Tomato Production Systems in California: An Approach to On-Farm Systems Studies" could possibly benefit from involvement in the planning process by producers and input suppliers already involved in tomato production in the central coastal and interior Central Valley of California (Yolo, Sutter, and Sacramento counties). Those researchers who are currently involved in the project include a plant physiologist, a plant pathologist, a zoologist, an ecologist, as well as an economist. The model of cooperation between all entities presented by Papendick in the STEEP program should be considered.

I suggest that the project be discussed with commercial agronomists, consultants, and other specialists currently involved in commercial tomato production in these regions. They are valuable sources of experience and information.

Gail Richardson, in "Soil Moisture Monitoring: A Practical Route to Irrigation Efficiency and Farm Resource Conservation," provided an excellent demonstration of what appears to be a relatively low-cost, low-technology, practical approach to measuring soil moisture and plant response.

The intriguing thing she noted is the involvement of farmers, researchers, and educators with the suppliers of the technology in the field. The approach to developing the technology, as well as the simultaneous transfer of the technology, is similar to the approach used in the agricultural chemical, seed, and fertilizer industries, as well as that demonstrated in the chapter by Papendick on the STEEP program.

In the presentation of Kevin Gamble at the workshop (not included in this volume), "A National System for Sustainable Agriculture Information Dissemination," it was difficult to understand the technological elements of the concepts that he presented. It was easy to grasp the theories themselves. They should be carried out.

There is an important link in the technology transfer process that should be considered in the development of a sustainable agriculture system. Regardless of where farmers get their information, for example, media, extension, other farmers, commercial representatives, a local coffee shop, or field days, there is one vital last point where information is condensed and transferred: that is, the equipment, seed, fertilizer, and crop protection retailers; consultants; or contract applicators where farmers exchange their dollars for information, products, or services.

I hope all participants will find these suggestions constructive and helpful.

# A Farmer's Perspective

*Robert A. Klicker*

I am pleased to be able to react to some of the information presented in this volume (see the chapters by R. James Cook, Gail Richardson, Robert I. Papendick, Carol Shennan, and Kevin Gamble). The information in these chapters was presented clearly and accurately and can be incorporated into use on my farms, which are located in four areas with different soil types and rainfall amounts. I am concerned that results of the type of research discussed in those chapters take years before they are available to farmers. The results of this kind of research should be directed and incorporated into complete farm systems as soon as possible after they become available.

James Cook's documented microbial research and his opinions on soil microbial action are the major key to sustainable agriculture. The bulletin "Long-Term Management Effects on Soil Productivity and Crop Yield in Semi-Arid Regions of Eastern Oregon" (Columbia Basin Agricultural Research Center, 1989) expands on Cook's points. This bulletin explains in detail the long-term soil depletion and reduced crop yield problems for which there have been no corrective solutions since 1931. Cook's research is a major contribution to some of the corrective solutions that can be used in a complete farm system.

Gail Richardson's chapter documenting her 7 years of on-farm water conservation technology with the use of gypsum blocks is also a major key for sustainable agriculture. Richardson documented a 20 to 40 percent savings of water. She also discussed methods that can be used to inspire farmers to apply the information she has gathered to their irrigation techniques. She proved that these methods are an economic educational tool that can be widely accepted by farmers and that can help them to understand the movement of water, soil plow pans, or compaction and when there is too much or not enough water in the root zone.

STEEP coordinator Robert Papendick reviewed the excellent progress that has been accomplished in reducing soil erosion, which leads to improved water quality. I was also pleased to read a discussion of the plans for STEEP II research.

James Cook has stated it all correctly when he says, "Scientists are rewarded for discovering and working out mechanisms but not for carrying this technology into application." Experience has proven that piecemeal, conflicting data will not build a sustainable agriculture system. All sustainable agriculture could be lost if complete farm systems research does not zero in on water harvesting, organic matter, and soil tilth research. A complete sustainable agriculture system (including use of the correct amount of nutrients plus use of soil cation-exchange ratios, correct tillage, water, and air) promotes increased microbial biomass and increased organic matter. Organic matter has three times the water- and nutrient-holding capacity than clay does. Increases in organic matter promote higher yields and, often, higher-quality products. Because organic matter has a higher water-holding capacity, soils with good organic matter levels are less likely to erode. I know from my own and my neighbors' soil tests that organic matter levels can be increased slowly and systematically by using commercial fertilizers and incorporating stubble with the correct conventional tillage tools. For each farm system, of course, there are slight variations in the methods that are used to stop erosion and increase organic matter. However, many of the basic methods stay the same.

Soil tests were performed on 15 different fields at three separate farms. These fields have had a definite, substantial increase in organic matter in 4 to 5 years. The normal erosion on these fields, which have 10 to 40 percent slopes, has been completely controlled without the loss of yield.

It is my opinion that the entire U.S. agricultural community should move toward the complete farm research system concept, incorporating James Cook's research, Gail Richardson's water conservation research, farmers' own on-farm research, and other existing technologies. Successful sustainable agriculture can be achieved by working to increase organic matter and the microbial biomass, by correcting soil tilth, and by stopping erosion while using conventional fertilizers and tillage tools. To accomplish this, the valuable information from the scientific community that is presented in this volume must be coordinated and shared with U.S. farmers in a more timely fashion.

## REFERENCE

Columbia Basin Agricultural Research Center. 1989. Long-Term Management Effects on Soil Productivity and Crop Yield in Semi-Arid Regions of Eastern Oregon. Bulletin No. 675. Pendelton, Oreg.: Agricultural Research Service, U.S. Department of Agriculture, and Oregon State University Agricultural Experiment Station.

# PART THREE
# Research and Education in the Southern Region

# 10

# Southeastern Apple Integrated Pest Management

*Dan L. Horton, Douglas G. Pfeiffer, and Floyd F. Hendrix, Jr.*

Integrated pest management (IPM), in its commercially usable forms, is a synthesis of discrete management concepts. Sustainability is an inherent theme of pest management. Fruit crops, because of their high unit value and the demand for blemish-free products, are an especially challenging area for IPM research and implementation (National Research Council, 1989). Compromises between scientists from several disciplines focusing on growers' needs to manage pest populations in an optimum cropping system lead to good pest management. Successful pest management is also regionally adapted. It attempts to exploit any regional advantages while adopting and modifying the successful IPM practices of other regions as they are needed.

Low-input sustainable agriculture (LISA) funding in 1988 linked complementary apple pest management programs in Virginia and Georgia. Funding enabled both states to broaden and accelerate their long-standing commitments to pest management. Virginia is an important apple-producing state, while Georgia is not a major apple producer. The two are, however, on either end of the primary southeastern apple belt, which follows the Appalachian Mountains and includes production areas in North Carolina, South Carolina, Tennessee, and Alabama. Pest management programs in both Virginia and Georgia have sought to provide good regionally adapted control (Taylor and Dobson, 1974). The authors of this chapter have tried to provide growers with dependable, low-risk, preventative spray guidelines and to emphasize regional pest biology and selective, well-timed pesticide use. Through the years, significant effort has been made to encourage growers to adopt a pest management mentality. This chapter provides an overview of pest pressures and phenology with particular emphasis given to the differences between Georgia and Virginia.

## OVERVIEW OF SOUTHEASTERN ARTHROPOD
## PESTS AND DISEASES

Pests and beneficial arthropod complexes do not differ greatly from the mountains of north Georgia through North Carolina and on into much of Virginia. The abundance and severity of certain arthropod pests do vary within the region, however. Southeastern arthropod seasonality and pest severity were thoroughly evaluated and summarized in a 5-year study by Shaffer and Rock (1983). Management programs in Virginia and Georgia attempt to control primary pests (those that incur high economic cost if uncontrolled) early in the growing season.

Control of primary pests commonly begins with the use of dormant oil. Sampling to estimate the size of successful overwintering populations of these pests is not feasible. Control of mites and aphids becomes more challenging, more disruptive to predators and parasites, and more expensive as the crop progresses beyond the 0.5-inch green-tip stage of development. Use of superior oil treatment gives nondisruptive suppression of scales, primarily San Jose scale (*Quadraspidiotus perniciosus* Comstock), European red mite (*Panonychus ulmi* Koch), and a complex of aphids, with the rosy apple aphid (*Dysaphis plantaginea* Passerini) being of primary concern. The delayed dormant period around the 0.25-inch green-tip stage presents a second and somewhat more effective control window for the use of oil. The addition of an organophosphate insecticide such as chlorpyriphos (Lorsban) in this oil spray for the green-tip stage improves the control of rosy apple aphid, which is not controlled by oil alone (Hull and Starner, 1983b).

Tarnished plant bugs (*Lygus lineolaris* Palisot de Beauvois), spotted tentiform leafminers (*Phyllonorycter blancardella* F.), and green fruitworms (*Lithophane antennata* W. spp.) are injurious between the tight cluster and pink stages. Spotted tentiform leafminers are relatively new apple pests in the southeastern United States. Walgenbach et al. (1990) found that vigorous, prebloom control of overwintered leafminers generally provides acceptable season-long suppression. This minimizes the need for postbloom control, which often encourages mite outbreaks. Plant bugs are erratic, very mobile, and potentially damaging. The need for leafminer sprays at this stage makes integration of plant bug thresholds (Michaud et al., 1989) impractical for Georgia and Virginia growers since leafminer sprays provide plant bug control. Green fruitworms also are controlled at the pink and petal fall stages with sprays made for other pests. Prolonged cool springs with longer than normal periods between sprayings for the pink and petal fall stages make a scheduled spraying at the pink stage very worthwhile in current IPM practices.

The most important postbloom pests are codling moth (*Cydia pomonella* L.) and plum curculio (*Conotrachelus nenuphar* Herbst). Both of these primary pests attack the fruit. They are joined by tufted apple budmoth (*Platynota idaeusalis* Walker) and, in Virginia, variegated leafroller (*Platynota flavendana* Clemens) as pests that must be controlled for the production of a successful crop.

San Jose scale crawlers, white apple leafhoppers (*Typhlocyba pomaria* McAtee), Japanese beetles (*Popillia japonica* Newman), green June beetles (*Cotinis nitida* L.), and European red mites must also be monitored and controlled as necessary. San Jose scale crawlers remain in the susceptible crawler stage of development for a very brief period. Emergence normally occurs at about the time of the second cover spray; however, more precise timing can be had by using the earliest pheromone trap catches as a biological fix to better predict crawler emergence (Pfeiffer, 1985). Japanese beetles and green June beetles may be injurious, and caution should be exercised. Spraying for foliage feeding by these pests is sometimes warranted, but control of these pests carries the risk of inducing mite problems. White apple leafhopper infestations also raise concerns over mite infestations. They carry a greater risk of inducing mite outbreaks because the carbamate insecticides (carbaryl, Sevin; formetanate hydrochloride, Carzol) needed to control them are especially toxic to mite predators (Rajotte, 1988).

European red mites are challenging pernicious pests (Prokopy et al., 1980) whose leaf-feeding injury accumulates through a growing season. Mite management includes suppression with oil treatments early in the season. Conservation of natural enemies, primarily *Stethorus punctum* LeConte lady beetles and the predator mite *Amblyseius fallacis* Garman in the southeast (Farrier et al., 1980), is fostered by making every effort to avoid unnecessary pesticide use. Careful pest management minimizes all unwarranted pesticide use (Croft and Brown, 1975). This preserves mite predators and normally lowers pesticide inputs. Mite control decisions are based on regular monitoring and the use of thresholds.

Apple diseases in Georgia are unique compared with those in much of the rest of the United States. Scab is a minor problem, while the summer rots are of major importance. The opposite is true from the mountains of North Carolina through the northeastern United States. A thorough investigation and understanding of the epidemiology of Georgia apple diseases was a necessary prerequisite to beginning a LISA program for the crop in Georgia.

Black rot of apples in Georgia, caused by *Botryosphaeria obtusa* (Schw.) Shoemake, differs from the disease of northern areas in that conidia are the primary source of inoculum. Infestation of buds occurs in the winter, infection occurs in the early spring, and the disease can cause losses in excess of 50 percent. Conidia of *B. obtusa* are produced on dead wood in the trees and on the orchard floor throughout the year (Beisel et al., 1984). Conidia are found on and in buds from December through March. Even though conidia of the fungus are present in the buds for several months, infection does not occur until the silver-tip stage of bud phenology. Thus, a single spray at the  silver-tip stage instead of the five sprays suggested previously (Smith and Hendrix, 1984) can be used to control this fungus. The fungus does not rot fruit until about 6 weeks before harvest, even though infection occurs in March and April.

Apple scab, which is caused by *Venturia inequalis* (Cke.) Wint., is a minor problem in Georgia because high temperatures frequently preclude secondary disease cycles. In most years, there is a primary cycle early in the spring. In some years, there is one secondary cycle in the fall (Hendrix et al., 1978). If an orchard had scab in the previous year, it is suggested that three scab sprayings be applied. In the absence of scab the previous year, no sprays are needed.

Cedar apple rust (*Gymnosporangium junipera-virginianae* Schw.) and quince rust (*Gymnosporangium clavipes* Cke. & Pk.) occur to some extent in Georgia orchards (Hendrix et al., 1978). Two prebloom sprays are suggested for orchards where there is a history of a problem with these diseases, but no sprays are suggested in other orchards.

Fire blight of apples (*Erwinia amylovora* [Burr.] Winslow et al.) occurs sporadically in Georgia apple orchards (Hendrix, 1990). Temperatures between 70° and 80°F and rainy, humid weather are necessary for epidemic outbreaks (Van Der Zwet and Keil, 1979). Avoidance of excessive vegetative growth also aids in disease reduction. Fire blight control by spraying is suggested only in those years when the weather favors major outbreaks.

Brooks spot of apple (*Mycosphaerella pomi* [Pass.] Lindau) is a minor disease in Georgia.  Light infection may not be noticed at the time of harvest because of the small lesion size.  Infection occurs from late April to mid-June in North Carolina, but it is slightly earlier in Georgia because of warmer temperatures.  Sprayings from the time of petal fall until the second cover appears to provide adequate control (Sutton et al., 1987).  No sprays are suggested for use in orchards with no history of the disease.

Black pox of apple (*Helminthosporium papulosum* Berg) can cause severe losses.  The fungus reproduces in old bark lesions and spreads to new leaves, fruit, and bark.  Infections occur throughout the growing season (Taylor, 1970).  Because of the length of the period when infection can occur, orchards with a history of this disease require season-long preventative spraying and are not candidates for inclusion in a LISA-type program.

Bitter rot of apple (*Glomerella cingulata* [Stonem.] Spauld. & Schrenk.)

does not reach damaging levels every year in Georgia orchards. When it does occur, it can cause losses of up to 80 percent (Noe and Starkey, 1980, 1982; Taylor, 1971). While infection can occur anytime after bloom, it is considered a midsummer disease. Most infections occur after the fruit reaches full size (Hendrix et al., 1978). Temperatures of greater than 21°C and free moisture are necessary for disease development. The fungus overwinters primarily on dead wood in the tree and on the orchard floor. It also survives on a small percentage of mummified Georgia fruit. The Georgia spray guide suggests five to eight sprays for pest control from the time of bloom to harvest.

White rot on fruit and Bot canker on trees are caused by *Botryosphaeria dothidea* (Moug. ex. Fr.) Ces. & de Not. The fungus survives on dead wood in the tree and on the orchard floor. Conidia are produced throughout the summer, but fruit is susceptible to infection only after the soluble solids reach 10.5 percent (Kohn and Hendrix, 1982, 1983). This is usually about 6 weeks before harvest. Prior to the work of Kohn and Hendrix (1982, 1983), sprays were applied from the time of bloom to harvest.

*Botryosphaeria dothidea* infects apple stems primarily through improperly made pruning cuts (Brown and Hendrix, 1981). Stub cuts on which the bark dies are the most common point of infection. This disease can be controlled by making proper pruning cuts and maintaining proper sanitation.

Sooty blotch, which is caused by *Gloeodes pomigena* (Schw.) Copby, and flyspeck, which is caused by *Zygophiala jamaicensis* Mason, are fungi which grow on the surface of apples. They do not cause decay but do cause cosmetic blemishes. Efforts were made in 1986 and 1987 to monitor the development of these diseases and to control them with prescription-type sprays. None of the currently registered fungicides, however, is capable of arresting development of these diseases once they start. Control of sooty blotch and flyspeck in the orchard does not fit into any current IPM techniques for apple production. Data are not available for predicting these diseases.

## PATHOLOGY RESEARCH

University of Georgia LISA fruit research centered on the development of IPM-compatible controls for sooty blotch and fly speck. The heavy preventative spraying required to control these diseases was not conducive to further pest management implementation. This study is examining the efficacy of chlorine dips for postharvest removal of sooty blotch and flyspeck from apples. The removal of pesticide residues and effects on shelf life were also examined.

**TABLE 10-1**   Chlorine Removal of Sooty Blotch and Flyspeck from Apples with a 5-Minute Postharvest Chlorine Dip

| Chlorine (ppm) | Percentage Posttreatment | |
|---|---|---|
| | Sooty Blotch | Flyspeck |
| 0 | 100a | 100a |
| 100 | 97a | 96a |
| 300 | 16b | 70b |
| 500 | 6b | 56c |

NOTE: Fruits were not brushed. Values with the same symbol are not significantly different.

### Postharvest Removal of Sooty Blotch and Flyspeck

Initial testing of postharvest chlorine dips in 1988 showed that these treatments removed sooty blotch and reduced flyspeck. Several postharvest chlorine rinses are labeled for use on a variety of fruits and vegetables. Sodium hypochlorite, the active ingredient, volatilzes rapidly, eliminating chlorine residues. The U.S. Environmental Protection Agency (EPA) has exempted sodium hypochlorite from food tolerances, indicating its low risk. The rates found to be effective are above those currently listed by the EPA.

High levels of chlorine (940, 1,270, and 1,670 ppm) were found to remove completely sooty blotch from fruit at all concentrations tested. Flyspeck was reduced but not eliminated. This experiment was repeated with lower concentrations of chlorine (Table 10-1). At 300 and 500 ppm of chlorine, sooty blotch was reduced from 100 percent to 16 and 6 percent, respectively. Flyspeck was reduced from 100 percent to 70 and 56 percent, respectively. Fruit tested in both experiments showed no symptoms of phytotoxicity or damage to their finishes. This test was repeated five times, with similar results obtained each time.

In subsequent tests, fruit was treated in the dump tank of a commercial packing plant. Fruit was exposed to 0, 50, 100, 300, 400, or 500 ppm of chlorine for 5 to 7 minutes. It was then passed over a series of wet brushes and rinsed with nonchlorinated water. The addition of brushes improved the process. Sooty blotch removal by treatment with 200 ppm of chlorine (Table 10-2) was equivalent to that at 500 ppm without brushing (Table 10-1). Flyspeck removal by treatment with 300 ppm of chlorine with brushing was equivalent to that with 500 ppm without brushes. With 500 ppm of chlorine and brushes, flyspeck was reduced from 100 to 27 percent. At

**TABLE 10-2**  Chlorine Removal of Sooty Blotch and Flyspeck from Apples in a Commercial Packing Plant

| Chlorine (ppm) | Percentage Posttreatment | |
| --- | --- | --- |
| | Sooty Blotch | Flyspeck |
| 0 | 100a | 100a |
| 50 | 92b | 95a |
| 100 | 26c | 45c |
| 200 | 6d | 52b |
| 300 | 4d | 58b |
| 400 | 2d | 36c |
| 500 | 0d | 27d |

NOTE: Fruits were dipped postharvest in chlorine-treated, dump tank water for 5 minutes, brushed, and then rinsed with nonchlorinated water. Values with the same symbol are not statistically significant.

this level of chlorine, sooty blotch was reduced to 0 percent. This test was repeated in 1989 with similar results.

### Chlorine Treatment Effects on Pesticide Residues

Sample apples that were treated with 500 ppm of chlorine, brushed, and rinsed in nonchlorinated water were tested for pesticide residue in an EPA-approved laboratory at the University of Georgia. Pesticide residues were reduced by the postharvest chlorine treatment (Table 10-3). Captan resi-

**TABLE 10-3**  Effects of a 5-Minute Dip of 500 ppm of Chlorine Followed by Brushing and Rinsing with Non-chlorinated Water on Residues of Captan, Phosmet, and Maneb on Harvested Apples

| Treatment | Residues (ppm) | | | |
| --- | --- | --- | --- | --- |
| | Captan | Phosmet | | Maneb |
| | | Sample 1 | Sample 2 | |
| Water dip | 0.42 | 0.30 | 0.16 | 5.84 |
| Chlorine dip | 0.00 | 0.20 | 0.00 | 1.63 |

NOTE: Values are averages of 21 residue analyses.

**TABLE 10-4**   Effect of Postharvest Chlorine Treatment and Fruit Waxing on Weight Loss of Red and Golden Delicious Apples after 12 Weeks of Storage at 1° to 3°C

| Cultivar | Treatment Wax | Chlorine | Date 1 Weight (g) | Standard Deviation | Date 6 Weight (g) | Standard Deviation | Weight Loss (g) |
|---|---|---|---|---|---|---|---|
| Golden | − | + | 105.9 | 10.4 | 101.2 | 10.3 | 4.7 |
| Delicious | + | + | 102.1 | 11.1 | 97.8 | 10.8 | 4.2 |
| | − | − | 97.4 | 7.6 | 92.6 | 7.5 | 4.8 |
| | + | − | 102.4 | 10.8 | 98.3 | 10.4 | 4.1 |
| Red | − | + | 154.9 | 16.5 | 150.7 | 16.0 | 4.2 |
| Delicious | + | + | 142.1 | 27.2 | 138.5 | 24.4 | 3.6 |
| | − | − | 156.1 | 15.7 | 152.0 | 15.3 | 4.1 |
| | + | − | 164.9 | 14.1 | 156.5 | 14.2 | 8.4 |

dues were reduced to less than detectable levels, phosmet levels were reduced by 33 percent or more, and Maneb levels were reduced by 73 percent.

### Shelf Life of Chlorine-Treated Apples

Red Delicious and Golden Delicious apples were harvested and treated with chlorine in a commercial packing plant, and half of the fruit was waxed. The fruit was stored in boxes with dividers at 34° to 37°F for 12 weeks. Weight loss was determined by weighing individual fruit at 2-week intervals. Weight loss averaged about 4.5 percent for Golden Delicious and 3.6 percent for Red Delicious apples over the 12-week period (Table 10-4). Neither chlorine nor wax treatment affected weight loss.

### Phytotoxicity and Fruit Finish Trial

Fruit was treated with chlorine at levels of up to 4,100 ppm, with and without buffer, to determine phytotoxic levels. Fruit finish was not affected at 4,100 ppm of chlorine, with or without buffer, even when fruit was stored for 30 days in plastic bags in the presence of chlorine solution.

### Conclusions of Fruit IPM Research

Postharvest chlorine dips have been found to be an effective technique for the removal of sooty blotch and the reduction of flyspeck. Chlorine treatments allow growers to ignore sooty blotch and flyspeck. Postharvest use of chlorine complements existing IPM techniques, including sanitation;

scouting for insects, mites, and bitter rot; and prescription use of white rot control measures, as dictated by fruit-soluble solids and weather. A state label granting Georgia growers the right to elevate their chloride concentrations to 500 ppm has been obtained. Postharvest chlorine treatment has allowed Georgia growers to eliminate up to eight sprays that, in preventative programs, are dedicated, at least in part, to the control of sooty blotch and flyspeck. No chlorine treatment-induced phytotoxicity was observed, even when eight times the necessary concentrations or two times the necessary exposure durations were used. Apples can be stored for up to 3 months after treatment with minimal but acceptable weight loss. Postharvest chlorine treatment also reduces pesticide residues. This may be important, because many consumers feel that even minimal pesticide residues compromise food safety.

## ENTOMOLOGY RESEARCH

Virginia Polytechnic Institute and State University (VPI&SU) has provided the research lead in entomology. Objectives are (1) pheromone mating disruption to control codling moth and variegated leafroller, (2) inventory of orchard ground cover management practices and evaluation of the impacts of these practices on mites, (3) determination of the toxicity of herbicides to the predaceous mite *A. fallacis*, and (4) assessment of grower IPM expertise.

### Mating Disruption

Mating pheromones are chemical cues that many insects use to help them find mates. Pheromone mating disruption provides insect control without the use of conventional toxic insecticides. Saturation of an orchard with pheromone confuses the mate-finding process, which prevents mating and eliminates the damaging larval stages of these pests. The elimination or drastic reduction of reliance on conventional disruptive, toxic sprays to control these pests does a great deal to conserve natural enemies. This use of nondisruptive, behavior-altering chemicals may well usher in what has been called "second-stage IPM" (Prokopy, 1987).

Codling moth, a primary pest of apples, must be controlled each year. This entails considerable pesticide use. Rothschild (1982) reviewed codling moth biology and ecology and noted the characteristics of this moth that make it a good candidate for disruption of the mating process. The codling moth's narrow host range and relatively low fecundity, the females' limited dispersal capabilities, the low number of generations, and the apparent lack of nonolfactory mate-finding mechanisms were noted as factors that lend themselves to pheromonal control. Charmillot and Bloesch (1987) reported on an 11-year Swiss study that had considerable success. They also offered possible reasons for control failures by this approach. Variegated leafroller and tufted apple budmoth are also important fruit-feeding pests that are candidates for mating disruption.

Pheromone mating disruption of codling moth and leafrollers was performed very successfully in a 4.7-acre commercial block at Daleville, Virginia. The pheromone-treated block was isolated, but there was a small abandoned block nearby. The block in which pests were controlled by following VPI&SU's spray guide method was about 1 mile away, and was surrounded by additional commercial apple orchards. Leafroller pheromone dispensers were put into place on May 5; those for codling moth were put into place on May 8. Four hundred dispensers for each species were placed in per acre. The original target for leafroller disruption was variegated leafroller. The pheromone blend used was: $E,Z$-11-tetradecenyl acetate, 96 percent (ratio of $E:Z$ isomers, 70:30); $E,Z$-11-tetradecenol, 2 percent (ratio of $E:Z$ isomers, 70:30); and $Z$-9-dodecenyl acetate, 2 percent. It was hoped that this would provide disruption for several species with pheromones that are mainly of the $E$ isomer.

Pheromone traps for variegated leafroller, tufted apple budmoth, and redbanded leafroller were monitored weekly. Three traps for each species were placed in each block. Damage was also assessed weekly by examining at least 200 peices of fruit in each of the blocks (pheromone-treated, control by the Georgia spray guide method, and abandoned).

## Male Orientation to Traps

Codling moth traps were placed in the blocks before the arrival of dispensers; pretreatment trap counts demonstrated the presence of codling moth in the blocks. Placement of pheromone dispensers was followed by a 100 percent shutdown of codling moth attraction to traps. Trap captures were reduced for each of the leafroller species, but not by 100 percent. There was a working assumption that limited mating was taking place. Leafrollers usually feed by tying leaves together; occasionally, a leaf is tied to a fruit, causing damage. Because these leafrollers are not direct fruit feeders, control of fruit damage may be attained without complete mating disruption. Variegated leafroller capture has been suppressed to a greater degree than has tufted apple budmoth capture. An unexpected development involved another leafroller species. On August 9, obliquebanded leafrollers were caught in the redbanded leafroller traps in the control block (8.7 per trap). Only a single obliquebanded leafroller was detected in the pheromone block (0.3 per trap). Although this species has not been important in Virginia, it has been an important pest in New York State and New England.

*Damage Assessment*

Control of damage by mating disruption for both groups of pests appeared to be similar to pesticide control, with certain restrictions. Mating disruption of first generation codling moths did not result in detectable injury to the fruit. In the periphery of the block, codling moth damage at the time of harvest was about 0.6 percent in Red Delicious apples, and 5.0 percent in Golden Delicious apples. When injuries to fruit caused by first-generation codling moth were seen, the grower was asked to begin spraying the outer two rows in the pheromone-treated block. Injuries to fruit by second generation codling moth were higher in the interior of the disruption block (2 percent) than they were on the periphery (0.7 percent). This came about because of spraying of the outermost rows. Leafroller damage in the interior of the pheromone-treated block was 1 percent for Red Delicious apples and 3 percent for Golden Delicious apples. Leafroller damage to Red Delicious apples at the time of harvest was 2.7 percent on the periphery and 5 percent inside the orchard. Golden Delicious apples sustained more injuries, but injury counts were not taken at time of harvest. Damage from both codling moths and the leafroller on the periphery of the orchard resulted from immigration of mated females into this relatively small block. This has been found to be a common problem in mating disruption research. Spraying of the outer two rows of the pheromone-treated block effectively dealt with the immigration of mated females.

Damage in the abandoned block reached 39 percent for codling moth, 17.5 percent for leafroller, and 35.5 percent for plum curculio. Peak damage from two generations of codling moths could be discerned. Damage in the conventional control block was 0 percent for codling moth, 6.5 percent for leafroller, and 0 percent for plum curculio. Plum curculio is not controlled by mating disruption, but it is significant to note that a single, well-timed late petal-fall spray controlled this key pest in the pheromone disruption block.

## Ground Cover Management Inventory

A survey of the ground cover management practices of Virginia growers was expanded to all pesticides and plant growth regulators that are applied to fresh-market apple orchards. About 45 conventional growers responded. These growers accounted for about 76 percent of fresh-market apple producers in the state. Surveys were also mailed to organic growers, but this clientele was more diffuse and there was no single mailing list of growers. Even without complete survey information, it appeared that organic growers perceived a greater threat from direct pests, which attack the harvest product, like codling moth and plum curculio than did conventional growers; the reverse was true for indirect pests, which attack other plant parts.

European red mite was consistently rated as the most severe pest in conventional orchards. European red mite is a classic example of a secondary pest, which may be defined as one that is induced by the application of pesticides for primary pests.

Azinphosmethyl (Guthion) is the most widely used insecticide in Virginia. This organophosphate is not very disruptive to predatory populations. However, there are increasing problems with resistance in leafrollers, white apple leafhopper, and spotted tentiform leafminer. Methomyl, a carbamate, is the second most frequently applied insecticide and is often targeted against leafrollers, leafhoppers, and leafminers. Carbamates, particularly methomyl, are disruptive to predators. Propargite (Omite) is the most frequently applied acaricide; dicofol (Kelthane) follows.

Mancozeb (Manzate) is the most widely used fungicide in Virginia. Paraquat (Gramoxone) is the most widely used herbicide in Virginia, with an average of 1.5 applications per season. It is highly toxic to predatory mites while they are in the ground cover (Pfeiffer, 1986). In Virginia, chlorphacinone is the most frequently applied rodenticide for the control of voles (*Microtus* spp.). Rodenticides are frequently applied in bait stations and likely do not affect predatory populations.

## Pesticide Applications Toxic to Beneficial Predators

Some data are available from the literature on the toxicity of pesticides to *Amblyseius fallacis* (Hislop and Prokopy, 1981). Materials considered highly toxic are permethrin (Ambush, Pounce), formetanate hydrochloride, methomyl, oxythioquinox (Morestan), fenvalerate (Asana, Pydrin), carbaryl, oxamyl (Vydate), phosalone (Zolone), diazinon, dimethoate (Cygon), ammonium sulfamate (AMS), paraquat, glyphosate (Roundup), and benomyl (Benlate).

Insecticides considered highly toxic to *Stethorus punctum* are carbaryl, fenvalerate, and permethrin; moderately toxic insecticides are formetanate hydrochloride, dimethoate, diazinon, methomyl, parathion, phosphamidon (Dimecron), and endosulfan (Thiodan) (Colburn and Asquith, 1973; Hull and Starner, 1983a).

Pesticides that are highly toxic to *Amblyseius fallacis* are frequently applied to the majority of Virginia's apple acreage. This probably accounts for the low densities of this predator that have been observed.

## Assessment of Grower Expertise for IPM

Evaluations have revealed that most growers in Virginia are lacking in critical pest management skills. Similar weaknesses are thought to predominate in Georgia and may well exist in other southeastern states. Growers were unable to identify such important predators as predatory mites and the larvae of *Stethorus punctum*, lacewings, syrphid flies, and predatory midges. Most growers (80 percent) responded that they consider which predatory populations are present before selecting pesticides. There would surely be greater benefits if growers' predator identification skills were enhanced. Only 51 percent of growers responded that action thresholds are used when they decide whether sprays are needed.

## GEORGIA GROWER IPM TRIALS

Georgia grower IPM trials relied on the judicious use of currently labeled, readily available chemicals and cultural controls. Thorough control of overwintering populations was emphasized. This protected the crop from injury and suppressed the magnitude of pest pressure experienced from subsequent generations.

Insect control strategies in the Georgia trails were a regional adaptation of successful programs and research in Massachusetts, New York, Pennsylvania, North Carolina, and Virginia. Control of the primary pests in Georgia, codling moth and plum curculio were emphasized. Early-season suppression of other pests was sought. Historical data on pheromone catches from orchards in North Carolina (Rock and Yeargan, 1974), South Carolina (C. S. Gorsuch, personal communication), and Georgia (D. L. Horton, unpublished data) provided a framework for formulating the expected seasonal abundance of each species. Pheromone monitoring for codling moth, redbanded leafroller, tufted apple budmoth, and variegated leafroller was used to follow the population trends of each pest. Pheromone trap-based codling moth thresholds were adhered to (Rock et al., 1978). Rather than applying pheromones to the entire orchard, varieties were sprayed individually.

Dormant oil sprays were used. This standard practice provided for nondisruptive control of mites, scale, and aphids. Phosmet (Imidan) was the insecticide for all of the sprayings that followed. Sprayings that were solely for insect control were made to alternate row middles to hold down costs and conserve predators. Prepink and pink sprays were applied to the alternate row middles for spotted tentiform leafminer, tarnished plant bug, and green fruitworms in a scheduled preventative fashion. After bloom, sprayings were made for plum curculio, codling moth, and leafroller to the middles of alternate rows. Orchards were scouted once a week. Examination of fruit was emphasized, and at least 200 fruit were examined during each visit to check for insect injury or disease lesions.

Disease control in apples is based on sanitation and fungicides. In the spray guide approach, fungicides are applied on a weekly or biweekly schedule without regard for weather or disease activity. In a low-input pest

**TABLE 10-5**   Comparison of Sprayings in Four Orchards in 1988 Treated by Georgia Spray Guide and Low-Input Sustainable Agriculture (LISA) Methods

| | | Number of Sprays Applied Under | |
| | | Georgia | |
| Spray | Material | Spray Guide | LISA |
|---|---|---|---|
| Dormant | Superior oil | 1 | 1 |
| Silver tip | Captan | 1 | 1 |
| Prepink | Captan, Carzol | 1 | |
| | Imidan | | 1 |
| Pink | Captan, Thiodan | 1 | 0 |
| Bloom | Captan | 2 | 0 |
| | Agrimycin | 4 | 0 |
| Petal Fall | Captan, Guthion | 1 | |
| | Imidan, alternate row middle (ARM) | | 0.5 |
| Cover 1 | Captan, Guthion | 1 | |
| | Imidan, ARM | | 0.5 |
| Cover 2 | Captan, Guthion | 1 | |
| | Imidan, ARM | | 0.5 |
| Cover 3 | Captan, Guthion | 1 | |
| | Imidan, ARM | | 0.5 |
| Cover 4 | Captan, Guthion | 1 | |
| | Imidan, ARM | | 0.5 |
| Cover 5 | Captan, Guthion | 1 | |
| | Dikar (Early Blaze only) | | 1 |
| Cover 6 | Captan, Guthion | 1 | 0 |
| Cover 7 | Captan, Guthion | 1 | |
| | Benlate, Maneb (Early Blaze, Prima only) | | 1 |
| | Dikar (Red Delicious only) | | 1 |
| Cover 8 | Captan, Guthion | 1 | |
| | Maneb (Red Delicious only) | | 1 |
| Cover 9 | Captan, Benlate, Imidan | 1 | 0 |

NOTE: The common or chemical names of the brand names cited are as follows: Agrimycin, streptomycin; Benlate, benomyl; Captan, captan; Carzol, formetanate hydrochloride; Dikar,dinocap; Guthion, azinphosmethyl; Imidan, phosmet; Maneb, manganese ethylenebisdithiocarbamate; Thiodan, endosulfan.

management program, scouting, weather, and disease activity are considered. Fungicides are applied on a prescription basis.

Sanitation is practiced by pruning out and removing as much dead wood from the trees as possible and by either removing dead wood from the orchard and destroying it or mowing the wood on the orchard floor with a flail mower (Starkey and Hendrix, 1980). The flail mower chops the wood finely and removes the bark. The bark decays rapidly and no longer supports saprophytic growth of the fungi that cause black rot, bitter rot, white rot, and Bot canker. Each orchard participating in the Georgia IPM program is inspected for sanitation; those without superior sanitation are excluded from low-input management.

Using the low-input system under the conditions found in Georgia, a spray is applied at the silver-tip stage for black rot control. Only one spray is needed if it is properly timed. If scab, rust, and Brooks spot have not been a problem in the orchard in the past, it is not necessary to apply sprays for these diseases. If they have been a problem, the sprays suggested in the Georgia apple spray guide should be applied. Fire blight sprays should be applied only if the weather conditions are favorable for fire blight epidemics. Orchards with a history of black pox are not included in this system, since infection occurs throughout the summer.

Trees in the lowest, wettest part of the orchard are marked and scouted for bitter rot. This disease occurs in these areas first. No sprays are applied for bitter rot control until it occurs. One spray is then applied, diseased fruit is removed from the tree, and additional sprayings are used after 2 weeks if the disease becomes active again. Soluble solids are measured each week after the fruit is about half grown. When readings reach 10.5 percent, a spray is applied for white rot if the forecast calls for rain. In dry weather, no sprays should be applied.

Sooty blotch and flyspeck are not controlled in the orchard, which saves up to eight sprayings. These blemishes are removed by using chlorine as a postharvest treatment.

In 1989, three grower orchards and the Georgia Mountain Branch Experiment Station orchards were managed under the LISA system. Approximately 65 acres were involved. All orchards were examined for sanitation, and a history of insect and disease problems was established. Even though 1989 was an extremely wet year, total pesticide applications were reduced from 29 to 9.5 in the LISA orchards (Table 10-5). Many of the sprays in these orchards were only insecticides, applied to alternate row middles, or fungicides. This further reduced the amount of pesticides that were used. Because several varieties were present in the orchards and the varieties were sprayed individually, the amount of pesticide needed was also reduced. For example, the fifth cover spray was applied only to the cultivar Early Blaze, which made up about 5 percent of the trees in the orchards. Twenty-nine pesticide applications at a cost for materials of $246.85 per acre were applied by using the spray guide, while 9.5 sprays, at a cost for materials of $98.87 per acre, were applied to the LISA orchards. There were additional savings and reduced inputs in the LISA orchard because of reduced machinery costs (Table 10-6).

**TABLE 10-6**   Comparison of Georgia Spray Guide and Low-Input Sustainable Agriculture (LISA) Spraying Costs

| | Spray Guide | | LISA | |
|---|---|---|---|---|
| Type of Pesticide | Number of Sprayings | Cost ($) | Number of Sprayings | Cost ($) |
| Bactericides | 4 | 23.60 | 0.0 | 0.0 |
| Fungicides | 14 | 126.65 | 4.0 | 55.40 |
| Insecticides | 11 | 96.60 | 5.5 | 43.47 |
| Total | 29 | 246.85 | 9.5 | 98.87 |

The control of insect and disease injury in the LISA orchards was equal to or better than that in the Georgia spray guide blocks. It was not possible to make direct comparisons with spray guide treatments in adjacent orchards, because the cooperating growers began to adopt and mimic the IPM practices in their other orchards. Insect injury at the time of harvest in the IPM grower blocks is noted in Table 10-7. Grower 2 had heavy variegated leafroller pressure late in the season, which is an unusual occurrence in Georgia. Spray was applied to control this population. Despite the increased injury to the apples that was sustained, insect control was good. The IPM blocks were comparable to the control that could be expected with the Georgia apple spray guide. At the time of harvest, these growers had 1.75 percent bitter rot in Early Blaze apples and 2.4 percent bitter rot in Red Delicious apples. Injuries from tarnished plant bug, green fruitworms, and the leafroller complex were 5 to 6.5 percent. The percentage of apples that could be packed, after chlorine treatment for sooty blotch and flyspeck was equivalent to that in the Georgia spray guide blocks.

**TABLE 10-7**   Insect Injury to Harvested Apples in 1989 Grower Trials in Georgia

| | Loss (%) | |
|---|---|---|
| Pest | Grower 1 | Grower 2 |
| Tarnished plant bug | 4.0 | 3.0 |
| Green fruitworms | 0.5 | 0.5 |
| Leafrollers | 0.5 | 3.0 |
| Total | 5.0 | 6.5 |

# REFERENCES

Beisel, M., F. F. Hendrix, Jr., and T. E. Starkey. 1984. Natural inoculation of apple buds by *Botryosphaeria obtusa*. Phytopathology 74:335–338.

Brown, E. A., and F. F. Hendrix. 1981. Pathogenicity and histopathology of *Botryosphaeria dothidea* on apple stems. Phytopathology 71:375–379.

Charmillot, P. J., and B. Bloesch. 1987. La technique de confusion sexuelle: Un moyen specifique de lutte contre le carpocapse *Cydia pomonella* L. Rev. Suisse Vitio. Arboric. Hortic. 19:129–138.

Colburn, R., and D. Asquith. 1973. Tolerance of *Stethorus punctum* adults and larvae to various pesticides. Journal of Economic Entomology 66:961–962.

Croft, B. A., and A. W. A. Brown. 1975. Response of arthropod natural enemies to insecticides. Annual Review of Entomology 20:285–335.

Farrier, M. H., G. C. Rock, and R. Yeargan. 1980. Mite species in North Carolina apple orchards with notes on their abundance and distribution. Environmental Entomology 9:425–429.

Hendrix, F. F., Jr. 1990. Fire blight of apples in Georgia. Georgia Apple Grower 3(19):2–3.

Hendrix, F. F., W. M. Powell, and N. McGlohon. 1978. Apple diseases in Georgia and their control. Fruit of the South 2:112–116.

Hislop, R. G., and R. J. Prokopy. 1981. Integrated management of phytophagous mites in Massachusetts (U.S.A.) apple orchards. 2. Influence of pesticides on the predator *Amblyseius fallacis* (Acarina: Phytoseiidae) under laboratory and field conditions. Protection Ecology 3:157–172.

Hull, L. A., and V. Starner. 1983a. Impact of four synthetic pyrethroids on major natural enemies and pests of apple in Pennsylvania. Journal of Economic Entomology 76:122–130.

Hull, L. A., and V. R. Starner. 1983b. Effectiveness of insecticide applications timed to correspond with the development of rosy apple aphid (Homoptera: Aphididae) on apple. Journal of Economic Entomology 76:594–598.

Kohn, F. C., and F. F. Hendrix. 1982. Temperature, free moisture, and inoculum concentration effects on the influence and development of white rot of apple. Phytopathology 72:313–316.

Kohn, F. C., and F. F. Hendrix. 1983. Influence of sugar content and pH on development of white rot on apples. Plant Diseases 67:410–412.

Michaud, O. D., G. Boivin, and R. K. Stewart. 1989. Economic threshold for tarnished plant bug (Hemiptera: Miridae) in apple orchards. Journal of Economic Entomology 82:1722–1728.

National Research Council. 1989. Alternative Agriculture. Washington, D.C.: National Academy Press.

Noe, J. P., and T. E. Starkey. 1980. Effect of temperature on incidence and development of bitter rot lesions on apples. Plant Diseases 64:1084–1085.

Noe, J. P., and T. E. Starkey. 1982. Relationship of apple fruit maturity and inoculum concentration to infection by *Glomerella cingulata*. Plant Diseases 66:379–381.

Pfeiffer, D. G. 1985. Pheromone trapping of males and prediction of crawler emergence for San Jose scale (Homoptera: Diaspidiidae) in Virginia apple orchards. Journal of Entomological Sciences 20:351–353.

Pfeiffer, D. G. 1986. Effects of field applications of paraquat on densities of *Panonychus ulmi* (Koch) and *Neoseiulus fallacis* (Garman). Journal of Agricultural Entomology 3:322–325.

Prokopy, R. J. 1987. The second stage of IPM in Massachusetts. Fruit Notes (University of Massachusetts) 52(3):9–12.

Prokopy, R. J., W. M. Coli, R. G. Hislop, and K. L. Hauschild. 1980. Integrated management of insect and mite pests in commercial apple orchards in Massachusetts. Journal of Economic Entomology 73:529–543.

Rajotte, E. G., ed. 1988. Tree Fruit Production Guide 1988. University Park, Pa.: Cooperative Extension Service, The Pennsylvania State University.

Rock, G. C., and D. R. Yeargan. 1974. Flight activity and population estimates of four apple insect species as determined by pheromone traps. Environmental Entomology 3:508–510.

Rock, G. C., C. C. Childers, and H. J. Kirk. 1978. Insecticide applications based on Codlemone trap catchers vs. automatic schedule treatments for codling moth control in North Carolina apple orchard. Journal of Economic Entomology 71:650–653.

Rothschild, G. H. L. 1982. Suppression of mating in codling moths with synthetic sex pheromone and other compounds. Pp. 117–134 in Insect Suppression with Controlled Release Pheromone Systems, Vol. 2, A. F. Kydonieus and M. Beroza, eds. Boca Raton, Fla.: CRC Press.

Shaffer, P. L., and G. C. Rock. 1983. Arthropod abundance, distribution, and damage to fruit. In Integrated Pest and Orchard Management Systems for Apples in North Carolina, G. C. Rock and J. L. Apple, eds. Technical Bulletin 276. Raleigh: North Carolina Agricultural Research Service.

Smith, M. B., and F. F. Hendrix, Jr. 1984. Primary infection of apple buds by *Botryosphaeria obtusa*. Plant Diseases 68:707–709.

Starkey, T. E., and F. F. Hendrix, Jr. 1980. Reduction of substrate colonization by *Botryosphaeria obtusa*. Plant Diseases 64:292–294.

Sutton, T. B., E. M. Brown, and D. J. Hawthorne. 1987. Biology and epidemiology of *Mycosphaerella pomi*, cause of Brooks fruit spot of apple. Phytopathology 77:431–437.

Taylor, J. 1970. Incubation period of *Helminthosporium papulosum* on fruit and bark of apples. Photopathology 60:1704–1705.

Taylor, J. 1971. Epidemiology and symptomatology of apple bitter rot. Phytopathology 61:1028–1029.

Taylor, J., and J. W. Dobson. 1974. Minimum rates of pesticides on apples. Phytopathology 58:247–251.

Van Der Zwet, T., and H. L. Keil. 1979. Fire blight—a bacterial disease of rosaceous plants. USDA Agricultural Handbook No. 510. Washington, D.C.: U.S. Department of Agriculture.

Walgenbach, J. F., C. S. Gorsuch, and D. L. Horton. 1990. Adult phenology and management of the spotted tentiform leafminer (Lepidoptera: Gracillariidae) in the southeastern United States. Journal of Economic Entomology 83:985–994.

# 11

# Low-Input Crop and Livestock Systems in the Southeastern United States

*John M. Luna, Vivien Gore Allen, W. Lee Daniels,
Joseph P. Fontenot, Preston G. Sullivan, Curtis A. Laub,
Nicholas D. Stone, David H. Vaughan, E. Scott Hagood, and
Daniel B. Taylor*

During the past decade, increasing concerns for the economic viability and ecological sustainability of agriculture have produced an accelerated search for alternative farming systems. The concept of low-input sustainable agriculture (LISA) has emerged that addresses multiple objectives: increasing agricultural profitability, conserving energy and natural resources, and reducing soil erosion and loss of plant nutrients (Harwood, 1990; Schaller, 1989). The term *low-input* implies a reduction of external production inputs (i.e., off-farm resources such as fertilizers, pesticides, and fuels) "wherever and whenever feasible and practical to lower production costs, to avoid pollution of surface and groundwater, to reduce pesticide residues in foods, to reduce a farmer's overall risk, and to increase both short- and long-term farm profitability" (Parr et al., 1990, p. 52). Low-input farming systems seek to optimize the management and use of internal production inputs (i.e., on-farm, renewable resources) to utilize beneficial ecological processes such as nitrogen fixation, nutrient cycling, and biological pest control more efficiently (Luna and House, 1990).

This change of production paradigm, from managing industrial inputs to managing ecological processes, is fundamental for the transition to a more environmentally sound sustainable agriculture. Expansion and documentation of the scientific basis of agroecology and translation of this knowledge into site-specific, usable information form the challenge for agricultural researchers and educators. Also important is the acquisition, evaluation, and horizontal distribution of the indigenous, local knowledge that exists among farmers (Berry, 1984; Ehrenfeld, 1987).

## LOW-INPUT CROP AND LIVESTOCK SYSTEMS

Livestock production is a major agricultural enterprise in the southeastern United States. In Virginia, for example, beef cattle is the leading agricultural commodity, averaging $360 million annually (Virginia Agricultural Statistics Services, 1988). Current livestock production systems in this region rely heavily on forage and grain crops produced on the farm. A major portion of the variable production costs for most farms in the region is for the purchase of synthetic fertilizers, insecticides, and herbicides. Various alternative production practices are available, however, that have the potential for reducing the quantity of these inputs while maintaining or increasing yield (National Research Council, 1989).

Forages play a unique role in creating sustainable and profitable agricultural systems (Parker, 1990). Their dense canopy and extensive root systems stabilize soils, largely preventing soil erosion and fertilizer chemical movement into groundwater and surface waters. Nitrogen can be supplied to the soil-plant-animal system through legumes, providing a high-quality diet to grazing animals and avoiding purchases of nitrogen fertilizers. Properly managed, grazed perennial pastures require fewer inputs of seed, fertilizers, and lime and have lower planting and harvesting costs compared with those of annual row crop systems. Forage provides the major feed for beef cattle throughout the South, particularly for cow-calf and stocker cattle operations. Because of its favorable soil and climate, the South has a comparative advantage for forage production over many other areas of the country, and high-quality forages can, through proper management, be utilized in grazing systems nearly year-round.

The recent availability of economical and practical fencing materials has renewed interest in intensified grazing systems that allow for increased animal numbers per unit of land area. Grazing management systems can promote efficient recycling of nutrients for optimum plant and animal production, and can avoid contamination of surface water and groundwater. Controlled grazing can also promote vigorous growth of desirable forage species and reduce growth and encroachment of weedy species, thus reducing or eliminating the need for herbicides. These systems, however, must be compatible with the management capabilities of farmers who are often employed in off-farm enterprises.

Economically viable all-forage systems have been developed for cow-calf and stocker production (Allen et al., 1987, 1989c; Blaser, 1986). Recent research has demonstrated that stocker cattle grazed on tall fescue-alfalfa (*Festuca arrundinacea-Medicago sativa*) during fall and winter made higher daily gains both during the growing phase and again 1 year later when these cattle were finished on corn silage (Allen et al., 1989b). These pastures were maintained during 6 years with no nitrogen fertilizer. Yields were similar to all fescue pastures where nitrogen was applied twice yearly at 143 pounds/acre (lbs/acre) each year. Soil nitrate and ammonium levels were similar between pastures where nitrogen was applied and those where alfalfa was grown, but nitrogen utilization by sheep was improved by inclusion of the legume compared with that of the nitrogen-fertilized grass (Absher et al., 1989). Systems for rotational or continuous grazing for cows, calves, and 1stockers have also been developed (Allen et al., 1989a). Recently, research has been initiated to develop systems for taking stocker steers to a finished weight and grade to meet present market demands (Fontenot et al., 1985).

In addition to forages, corn (*Zea mays*) production plays a central role in southeastern livestock systems. This crop was chosen as a major focus for the study described in this chapter to evaluate low-input practices because of (1) the large acreage in the southeast; (2) the high-input requirements of fertilizers, herbicides, and insecticides in corn production; and (3) the high rates of soil erosion commonly associated with corn production, particularly where conventional tillage is used.

Growing winter-annual legumes is an important alternative practice for reducing nitrogen (N) fertilizer use in corn production in the southeast and for improving soil conservation and productivity (Hargrove and Frye, 1987). Legume cover crops can contribute more than 80 lbs of N per acre to the succeeding crop (Corak et al., 1987; Ebelhar et al., 1984; Hargrove, 1986; Mitchell and Teel, 1977; Neely et al., 1987). Extensive work on the use of legumes in conservation tillage systems has been summarized by Power (1987). However, very few farmers in the mid-Atlantic region and other areas of the South utilize these legumes in their production systems.

No-till planting of corn into cover crops or previous crop residue without primary tillage has been widely used by farmers in an effort to reduce soil erosion as well as production costs. A common practice in the mid-Atlantic region is to plant rye (*Secale cereale*) in the fall as a winter cover crop and then desiccate the rye in the spring with a herbicide prior to corn planting. The rye protects the soil during winter, recycles some soil nitrogen, and contributes a moisture-conserving mulch for the corn crop.

Insect pest problems, however, can be increased by using a rye cover crop. No-till corn planted into a rye cover crop has a higher incidence of damage from the common armyworm (*Pseudaletia unipuncta* Haworth), however, than do conventionally tilled fields or fields without a rye cover crop. Adult armyworm moths lay eggs in the rye, which serves as host for the developing larvae. When the rye is desiccated with the herbicide, the armyworm larvae move onto the corn seedlings, frequently causing severe defoliation and economic damage. To control this pest, growers commonly use a prophylactic application of an insecticide mixed with the herbicide when the cover crop is killed.

One conservation tillage alternative to no-till planting is that of ridge-till planting. Ridge-till systems have been used quite successfully for corn and soybean production in the Midwest (Behn, 1982; Little, 1987; National Research Council, 1989), but little work with ridge-till systems has been reported under southeastern conditions. While functioning as effective conservation tillage systems, ridge-till systems permit mechanical cultivation, reducing or eliminating the need for weed control with herbicides. In the low-input corn research project reported here, the central focus is the integration of winter-annual cover crops into ridge-till systems for corn production. Particular emphasis is placed on weed management, evaluation of mechanical cultivation, the role of the cover crop mulch in weed suppression, and banded herbicide applications.

## RESEARCH PROJECT ORGANIZATION

The research and education project described here comprises four distinct yet closely interrelated components: (1) establishment of a long-term crop and livestock farming systems comparison study, (2) development of a low-input corn production system, (3) development of a prototype expert system for whole-farm crop rotation planning, and (4) implementation of an extension education program.

An interdisciplinary research group involving faculty and graduate students from six academic departments at Virginia Polytechnic Institute and State University (VPI&SU) was involved in the design and operation of these projects. Two farmers and several extension faculty members participated in discussions of project design and field implementation. This degree of interdisciplinary cooperation is essential for this type of project to be successful and is a unique attribute of this effort. Some components of this project were initiated in 1987; however, major work began in 1988 with funding from the U.S. Department of Agriculture (USDA) low-input sustainable agriculture program.

## CROP-LIVESTOCK SYSTEMS COMPARISON STUDY

In order to examine the long-term productivity and ecological interactions associated with whole farming systems, a replicated crop-livestock farming systems comparison study was established. This interdisciplinary project compares a conventional crop-livestock system typical of the mid-Appalachian region with an experimental, low-input system. The conventional system utilizes the best management production practices most commonly used by growers within the region. The low-input system involves a different integration of farming practices, with specific emphasis on the minimization of soil erosion and the use of agricultural chemical

inputs while maintaining or improving economic viability. Both conventional and low-input systems consist of four replicates of 10 acres of land and 6
steers each, for a total of 80 acres and 48 steers. Although the land area and
total animal stocking rate remain the same in both systems, the crop mix,
rotation plan, and production methods differ between the systems. Grazing is
also used more extensively in the low-input system in an effort to reduce
harvest costs, optimize forage utilization, and recycle animal manures.

Each 10-acre replicate of the conventional system consists of 4 acres
of N-fertilized fescue (for winter stockpiling), 3 acres of a fescue-clover
mixture for grazing and hay, 1.5 acres of alfalfa, and 1.5 acres of corn.
All crops are grown in the same fields for 5 years, and then the corn and
alfalfa fields are rotated. In the low-input system, each 10-acre replicate is
divided into a 4-acre field of fescue-alfalfa mixture for stockpiling, grazing,
or hay, and four 1.5-acre fields are rotated among the following crops in a
4-year rotation: corn, wheat (*Triticum aestivum*) and foxtail millet (*Setaria
italica*) (double-cropped), and alfalfa (2 years). Insect pest management
practices in the low-input system rely on rotational effects, and insecticides are applied based on pest population sampling and economic thresholds. Use of preemergent herbicides is minimal, with postemergent herbicides and rates based on the weed species present and their densities and
stages of growth.

It is hypothesized that the low-input crop rotation and grazing system
will reduce need for insecticides, herbicides, and fertilizers. Specifically,
(1) spring and fall grazing of alfalfa will reduce the need for insecticides
for control of alfalfa weevil, (2) planting of alfalfa following millet will
reduce the need for herbicides and insecticides during establishment, (3)
first-year corn following alfalfa will not require insecticides for corn rootworm or armyworm control and will have reduced N fertilizer needs,
(4) inclusion of alfalfa in the fescue pasture will eliminate N fertilizer
needs for the grass, (5) grazing of all land areas will aid in recycling nutrients and reducing fertilizer needs, and (6) scouting for weeds and insect
pests and using pesticides based on economic thresholds should also
reduce pesticide inputs.

## Plot Location and Sampling

This experiment was established at the VPI&SU Whitethorne Research
Farm, near Blacksburg, Virginia, during 1988 and 1989. Each crop or
pasture block was located on a uniform soil landscape with a uniform cropping and cover history. A survey of the dominant soil types was made to
ensure uniformity within experimental blocks. Large bulk soil samples
were taken in the early spring of 1989 before the addition of fertilizers and
pesticides. An additional sampling was made in the late summer of 1989,

and the soil within each plot will be sampled annually. Soil samples have been analyzed for pH and nutrient levels and are undergoing analysis for organic matter, aggregate stability, and several other parameters. Changes in important soil parameters will be carefully documented in each plot over time, as will all nutrient additions and removals.

## Pasture Establishment

Four replications of pastures for the LISA and conventional systems were established during the summer and fall of 1989. Pastures for the LISA system are tall fescue-alfalfa (*Festuca arundinacea-Medicago sativa*). Pastures for the conventional system are (1) tall fescue and (2) tall fescue-red clover (*Trifolium pratense*). To prepare the land for pasture establishment, the entire area was seeded in millet (*Seteria italica*) in June 1989. Millet was harvested for hay in August, and the entire area was sprayed with paraquat. Pastures were established by drilling seed into the residual sod. Lime was applied as indicated by soil analyses in the early spring of 1989. Pastures were fertilized at the time of crop establishment with N, phosphorus, and potassium according to soil test recommendations from the VPI&SU Soil Testing Laboratory.

## Establishing the Crop Rotation Sequence

*LISA System*

In order to initiate the crop rotation sequence required in this experiment, Cimmaron alfalfa was seeded into three blocks and corn was seeded into the fourth block in the spring of 1989. Pioneer 3192 corn was no-till drilled into herbicide-killed sod in 1.5-acre blocks with four replications. Alfalfa was drilled into herbicide-suppressed sod with four replications of each of the three blocks. Fertilizer and lime were applied at the time of establishment of the respective crops based on soil test recommendations. Acceptable stands were achieved. Abruzzi rye was drilled into one alfalfa block in the fall of 1989. This rye was killed by mowing in the spring of 1990 for no-till seeding of corn. Corn was harvested as silage in September 1989, and the plots were overseeded with Massey wheat. Wheat was harvested in 1990 and was followed by millet preparatory to the planting of alfalfa. This completed the establishment of the rotation sequence for the LISA system of corn, wheat, millet-alfalfa, and alfalfa.

## Conventional System

Pioneer 3192 corn and Cimmaron alfalfa were established by the no-till method into herbicide-killed sods in the spring of 1989 in blocks of 1.5

acres each with four replications. Alfalfa establishment was successful, beginning the 5-year stand life for this crop. Corn was harvested for silage in September 1989, and Wheeler rye was drilled into the stubble. Rye was killed with paraquat in the spring of 1990, preparatory to no-till establishment of the second-year corn crop. Lime and fertilizer were applied following soil test recommendations, as described above for LISA systems.

Establishment of all crops was successful; thus, no delays are anticipated in the progress of this project. Cattle began grazing the systems in November 1990, and fencing and water systems were completed by the time the cattle entered the project.

### Sampled Variables

System productivity will be compared by measuring animal gains (both per animal and per area) and carcass quality, and excess forage will be harvested for hay. All systems will provide maximum grazing and minimum hay harvesting and feeding. The required inputs of labor, fertilizer, seeds, fencing, water, and shade will be compared among the systems. The influence of the grazing system on uniformity of nutrient recycling will be determined. Changes in soil physical and chemical properties will be monitored over time. Shifts in botanical composition within pastures will be measured to determine weed percentages and types and the dominance of desirable forage species. Percentage ground cover will be measured, and erosion potential will be monitored. The effects of animal impacts on soils will be determined by measurements of bulk density and characterization of traffic patterns and camping sites. The distribution of manure within pastures as influenced by the grazing system and the presence of shade will be evaluated.

An economic analysis of inputs, fixed costs, labor, and profitability will be conducted for each system. The potential of systems to provide economically viable forage and beef cattle systems for farmers in the region will be determined. The potential impact of the new systems (if adopted) on local employment and income profiles will be estimated.

### LOW-INPUT CORN PRODUCTION SYSTEMS

Three separate research subprojects have been conducted under the objectives of this project: (1) evaluation of the contribution of various winter-annual legume and small grain combinations to silage corn production, (2) evaluation of alternative cover crop management practices for winter-annual cover crops in no-till corn (comparing the effects of rotary mowing with those of conventional herbicide desiccation), and (3) evaluation of a ridge-till production system which uses winter-annual cover crops.

## Contribution of Winter-Annual Legume and Small Grain Cover Crop Combinations to Silage Corn Production

This experiment was conducted near Blacksburg, Virginia, in 1987 to 1989 on a Hayter cobbley loam soil. A 2-by-6 factorial experimental design with four replications was used to examine contributions of several winter-annual cover crops to corn production. Fall-seeded cover crops included rye, hairy vetch (*Vicia villosa*), bigflower vetch (*Vicia grandiflora*), rye-hairy vetch, and a hairy vetch-bigflower vetch mixture.

Two tillage practices were used: (1) no-till, with corn slot-planted directly into the herbicide-killed cover crops, and (2) minimum-till, in which the cover crops were disked prior to corn planting. For each tillage practice, crops were grown in control plots by standard practices. This consisted of a fallow plot with 125 lbs of N per acre for the minimum-till plots and a rye cover crop with 125 lbs of N per acre for the no-till plots. The individual plot size was 12 feet in width (four corn rows, 36 inches apart) and 50 feet in length.

Cover crops in the minimum-till plots were incorporated about 2 weeks before the corn was planted. Cover crops in the no-till treatments were desiccated with paraquat herbicide immediately before the corn was planted. All plots were treated with a residual herbicide mixture of metolachlor and cyanazine. Pioneer 3233 corn was planted on May 19, 1988, and on May 22, 1989, following cover crop kill and harvested as for silage (35 to 42 percent dry matter).

Cover crop yields, total nitrogen, and carbon:nitrogen (C:N) ratios were determined prior to disk incorporation or desiccation. The effects of cover crops and tillage practices on weed densities, water infiltration rates and soil moisture, seasonal N uptake, and corn silage yields were determined. Only the effects on corn silage yields and soil moisture are reported here.

### Results

The results reported here are summarized from the work of Sullivan (1990). Soil moisture during the corn growing season was correlated ($p < 0.05$) to the biomass of the cover crop on the soil surface, confirming the importance of soil mulches in reducing moisture stress to crops. In late summer, soil moisture was highest under the rye cover crop mulch. This was possibly due to the high C:N ratio of the rye (58:1) compared with the much lower C:N ratio of the vetch cover crops (11:1), which reduced the rate of microbial decomposition of the rye cover crop. In addition, reduced water uptake by nitrogen-deficient corn plants growing in the rye-only plots could have influenced the soil water content.

In both years in both tillage systems, corn that was grown following a mixture of hairy and bigflower vetch cover crops produced yields similar ($p > 0.05$) to those of corn that was grown without cover crops plus 125 lbs of N per acre or with a rye cover crop plus 125 lbs of N per acre (Figure 11-1). Pure stands of either bigflower or hairy vetch produced corn silage yields comparable to those in 1988, but produced less corn ($p < 0.05$) in 1989 than did the vetch mixtures or the controls (Table 11-1). Pure stands of rye cover crop without any nitrogen fertilizer produced the lowest corn yields in the no-till treatments in both years. Mixtures of rye and vetch produced corn yields similar to those of the control in 1988, but they yielded significantly less in 1989.

A rye-vetch mixture may be a more effective cover crop, however, since the rye produces a ground cover more quickly in the fall. However, additional N fertilizer will be needed for the rye-vetch cover crop mixture to provide economically optimum yields (Frye et al., 1985; Ott and Hargrove, 1989). Although the purpose of this experiment was not to evaluate cover crops as possible forage crops, the rye-hairy vetch mixture produced significantly more biomass than any of the other cover crops did in both years under both tillage systems (Table 11-2). The total N in the rye-vetch mixture was significantly higher than that in the rye alone. Since many farmers in the mid-Atlantic region grow rye as a silage double crop followed by no-till corn, the addition of hairy vetch to the rye could increase the yields and protein content of the forage. The delayed growth of the vetch in the spring, however, may reduce this benefit.

Corn establishment problems were encountered when no-till planting into the cover crops was used. The rolling coulter in front of the no-till corn planter tended to push the cover crop into the planted slot, reducing seed-to-soil contact. The cover crop residue also impeded seedling emergence in places.

All cover crop treatments were analyzed under the two tillage regimes for economic feasibility; however, only the hairy vetch and control crops grown by standard practices are discussed here. Variable costs for labor, machinery, seeds, pesticides, lime, and fertilizer were based on 1989 prices (Maxey et al., 1989). Differences in net returns occurred among treatments and among years (Table 11-3). Averaged across both years and cover crops, the no-till system produced a $44 greater net return per acre than that of the disk tillage system. Within the no-till system, the hairy vetch cover crop produced net returns of about $22/acre greater than that of the rye cover crop with 125 lbs of N per acre. Under disk tillage, net returns were similar, with fallow plus fertilizer producing an average of $4/acre more net return than that of the vetch cover crops.

These data augment a growing body of published literature confirming the economic advantages of using winter legume cover crops in corn and other cropping systems (Frye et al., 1985; Ott and Hargrove, 1989). How-

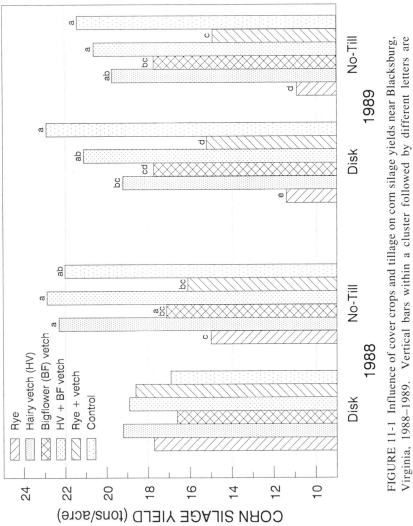

FIGURE 11-1  Influence of cover crops and tillage on corn silage yields near Blacksburg, Virginia, 1988–1989.  Vertical bars within a cluster followed by different letters are statistically different (F-protected LSD; alpha = 0.05).

**TABLE 11-1**  Corn Silage Yields Following Various Winter Annual Crops and Fallow Plus N Fertilizer, Blacksburg, Virginia, 1988 and 1989

| Cover Crop | Corn Silage Yields[*] (tons/acre) | |
| --- | --- | --- |
| | 1988 | 1989 |
| Fallow + 125 lbs of N/acre | 19.4ab | 22.2a |
| Hairy + bigflower vetch[†] | 20.8a | 20.9ab |
| Hairy vetch | 20.7a | 19.5bc |
| Bigflower vetch | 16.9b | 17.7c |
| Rye + hairy vetch | 17.3b | 14.9d |
| Rye | 16.3b | 11.3e |

NOTE: N, nitrogen. Means within a column followed by different letters are different (F-protected LSD; alpha = 0.05).

[*]Tons of silage per acre at 35 percent dry matter.
[†]No additional N fertilizer was used to grow the corn following the cover crops.

ever, a critical underestimation of the value of cover crops has occurred in virtually all economic analyses of cover crops, including the net benefit analysis of this project described above. Net benefit analyses, even though multiyear in scope, are generally based on single-year crop yield responses. The potential long-term, cumulative beneficial effects resulting from soil conservation, improvement in soil organic matter content and soil tilth, and delayed nutrient release from the mineralized cover crops are not factored into the analyses because the published studies have been short term in nature (Allison and Ott, 1987). Additional long-term research is needed to incorporate these variables into economic analyses of cover crops.

### Evaluation of Alternative Cover Crop Management Practices for Winter-Annual Cover Crops in No-Till Corn

The second subproject for developing a low-input corn system involved the evaluation of alternative cover crop management practices for rye cover crops in no-till corn, comparing the effects of mowing with those of conventional herbicide desiccation. Demonstration experiments were conducted on five farms in two Virginia counties in 1988 and 1989. Rye cover crops were planted in September and October 1988. In early May, two experimental treatments were established in a randomized block design: (1) conventional herbicide desiccation with paraquat and (2) mowing with a tractor-mounted rotary mower. Plots were 60 by 50 feet, and

**TABLE 11-2**   Cover Crop Biomass and Total N Production Averaged Across Two Tillage Methods

| Year | Cover Crop | | | | |
|------|-----|----------------|---------------------|----------------------------------|----------------|
| | Rye | Hairy Vetch | Bigflower Vetch | Hairy and Bigflower Vetch | Rye-Vetch |
| | | | *Biomass tons/acre* | | |
| 1988 | 3.1a | 1.6a | 1.3a | 2.0a | 4.3a |
| 1989 | 2.1b | 1.9a | 1.6a | 1.8a | 2.6b |
| | | | *Total lbs of N/acre* | | |
| 1988 | 60a | 85b | 45b | 124a | 133a |
| 1989 | 36b | 147a | 112a | 133a | 109b |

NOTE: N, nitrogen.  Means within a half column followed by the same letter are not significantly different (F-protected LSD; alpha = 0.05).

each treatment was replicated four times.  Corn was planted with the various no-till corn planters used by the cooperating farmers.  Residual herbicides and fertilizers were applied based on the cooperating farmers' practices.

Densities of armyworm larvae were estimated from the time of corn seedling emergence until armyworm larvae were no longer found.  Sampling was conducted every 4 to 5 days.  The sampling unit was a 2-by-5-foot quadrat placed lengthwise over a corn row.  Other variables examined included armyworm parasitoids, ground-dwelling predator populations, cover crop regrowth, and corn silage yields.

**TABLE 11-3**   Estimated Net Return to Management from Corn Silage Following Two Cover Crop Treatments and Two Tillage Methods

| Year | Estimated Net Return ($/acre)[*] | | | |
|------|---------|------|---------|------|
| | Vetch | | Rye + 125 lbs N/acre | |
| | No-Till | Disk | No-Till | Disk |
| 1988 | 436 | 345 | 389 | 284 |
| 1989 | 373 | 350 | 376 | 421 |
| Average | 405 | 348 | 383 | 352 |

NOTE: N, nitrogen.

[*]Based on corn silage at $25.00/ton and N fertilizer at $0.24/lb.

*Results*

The results reported here are summarized from the work of Laub (1990). In four of the five fields in both years, mowing significantly reduced armyworm population densities in the early stages of corn growth (Figure 11-2). In the fifth (Bishop) field in 1989, a very poor corn stand was produced in both treatments because of excessively wet conditions and feeding by the common garden slug. In the Bishop field, armyworm numbers were higher in the plots that were treated by mowing, although the densities in plots that were treated by both methods were very low. Increased numbers of certain species of predacious ground beetles and spiders were also higher early in the season in the mowed treatments in some of these fields, although it is not known whether these predators influenced armyworm abundance. Reduction in armyworm densities in the mowed plots could also be due to mechanical destruction during the mowing process.

Mowing also adequately suppressed rye cover crop regrowth in all fields, but it should be noted that the rye was mowed after the initiation of flowering. Mowing before this stage of growth does not kill the rye plant. Corn silage yields tended to be slightly higher in the mowed treatments than the herbicide-treated rye in all fields, although these differences were not statistically significant ($p < 0.05$) (Figure 11-3).

Costs of the cover crop management methods are calculated to be ca. \$6.00/acre for mowing and \$10.00/acre for paraquat spraying. Calculated costs include fuel, maintenance, labor at \$5/hour, and depreciation on equip-ment. Herbicide cost was calculated at 2 pints of paraquat (Gramoxone Super) per acre, which cost \$40.35/gallon in 1990. The recommended application rate for paraquat for contact killing of rye cover crops is 1.5 to 2.5 pints/acre (Webb et al., 1988). The net economic benefits of the two cover crop management practices were compared by using mean silage yields and calculated costs of the two treatments (Table 11-4). Averaged across both years and all fields, mowing of the cover crop produces an estimated net return of \$40/acre more than the use of paraquat does.

Although mowing winter cover crops appears to have the potential both to reduce herbicide and insecticide use and to increase yields and profits, additional research is needed to determine proper timing of mowing in relation to cover crop phenology to ensure cover crop kill. Also, the requirements for labor during a critical time of year may restrict the adoption of this practice. Spraying of a herbicide could be accomplished at a rate of 15.3 acres/hour (using a 35-foot boom traveling at 6 miles/hour and with a 60 percent efficiency), whereas only 4.9 acres could be mowed per hour (using a 12-foot-wide mower traveling at 4 miles/hour and with an 85 percent efficiency).

Future on-farm research will continue to evaluate the potential of rotary

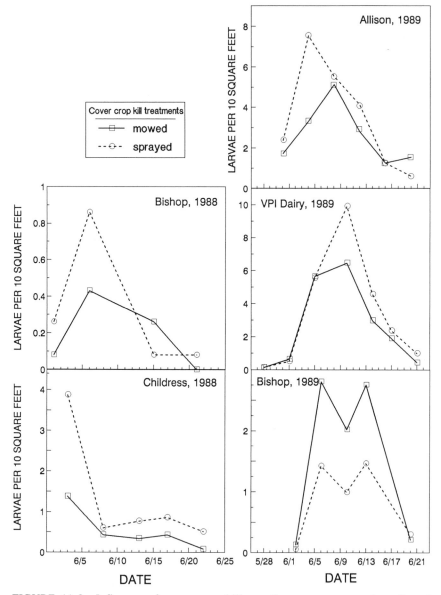

FIGURE 11-2  Influence of cover crop kill practices on mean number of total armyworm larvae in no-till corn in southwest Virginia, 1988–1989.

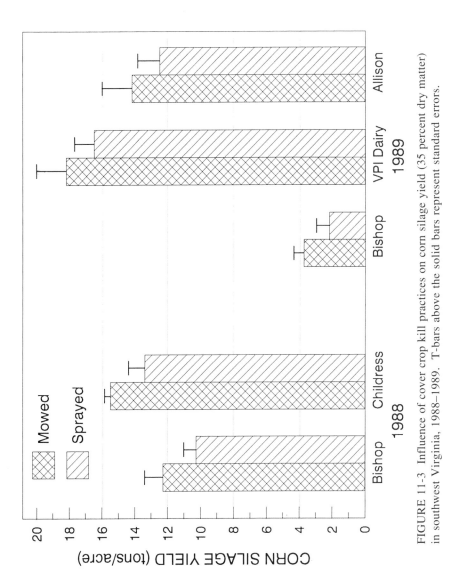

FIGURE 11-3 Influence of cover crop kill practices on corn silage yield (35 percent dry matter) in southwest Virginia, 1988–1989. T-bars above the solid bars represent standard errors.

**TABLE 11-4**   Estimated Net Benefit of Mowing a Winter Cover Crop Compared with Spraying with Paraquat for No-Till Silage Production, 1988 and 1989

| | Fields | | | | |
|---|---|---|---|---|---|
| | 1988 | | 1989 | | |
| Cover Crop Kill Treatment[*] | Bishop | Childress | Bishop | VPI&SU | Allison |
| *Average Value ($) of Silage/Acre*[†] | | | | | |
| Mow | 255 | 267 | 77 | 376 | 293 |
| Spray | 213 | 231 | 44 | 342 | 258 |
| Silage benefit/acre of mowing | 42 | 36 | 33 | 34 | 35 |
| Savings/acre from mowing the cover crop[‡] | 4 | 4 | 4 | 4 | 4 |
| Total benefit from mowing | 46 | 40 | 37 | 38 | 39 |

NOTE:  VPI/SU, Virginia Polytechnic Institute and State University.

[*]Mow indicates mowing with a rotary mower (bushog); spray indicates spraying with paraquat at 2 pints/acre.

[†]Calculated for corn silage at 35 percent dry matter, valued at $25.00/ton.

[‡]Costs of cover crop kill treatments:  spraying, $10.00/acre; mowing, $6.00/acre.

mowing as an alternative to herbicide desiccation of the cover crop. Since no insecticides were applied to any of the treatments in 1988 or 1989, an additional treatment variable, with and without insecticide, will be included to determine whether mowing alone will replace the need for insecticide to control armyworm populations.  Difficulty in planting into the mowed cover crop residue was also encountered in a few fields in this study. Corn planters will need to be equipped with residue-clearing attachments, although these are readily available for most modern conservation-tillage planters. Flail mowers will also be evaluated as an alternative to rotary mowers to improve the uniformity of distribution of the cover crop residue on the soil surface after mowing.

## Evaluation of Ridge-Till Corn Production Systems Using Winter-Annual Cover Crops

Weed management practices were evaluated in a ridge-till corn production system using winter-annual cover crops.  The experiment was conducted at the VPI&SU Whitethorne Research Farm, near Blacksburg, Virginia.  Ridges were established in September 1988 by using a Buffalo cultivator (Fleischer Manufacturing, Inc., Columbus, Nebraska).  Ridges were on 36-inch centers and were approximately 7 inches tall. A cover crop of rye (90 lbs/acre) and hairy vetch (20 lbs/acre) was planted with a spin spreader on September 29, 1988. Very little hairy vetch cover crop was produced; thus, the cover crop consisted primarily of rye. The cover crop was killed with a tractor-drawn rotary mower (bushog) on May 24, 1989. Pioneer 3295 corn was planted on June 2, 1989, by using a Buffalo ridge-till planter (Fleischer Manufacturing, Inc.). Seventy pounds of actual N fertilizer was broadcast per acre on June 26, 1989, with an additional 30 lbs of N per acre broadcast on July 28.

A 2-by-4 factorial experiment was used to evaluate various combinations of mechanical and banded herbicide weed management practices. Ridges and furrows were treated separately, with four weed control treatments used on the ridges: preemergence herbicide, postemergence herbicide, cultivation, and control (no weed control). Two furrow treatments were used: preemergence herbicide and control (no weed control). A third weed control variable for the furrow, cultivation, was planned for the experiment, but excessively rainy weather delayed cultivation until the corn was too tall to avoid plant damage by the cultivator tool bar. This treatment will be included in the 1990 second-year replication of this experiment.

Weed biomass was sampled on August 8 and 9, 1989, by using 18-by-60-inch quadrats. Twelve randomly selected samples were taken in each treatment plot, six samples each in the furrow and on the ridges. Weeds were sorted into four categories: broadleaves, grasses, yellow nutsedge (*Cyperus esculentus* L.), and Pennsylvania smartweed (*Polygonum pennsylvanicum* L.). Samples were oven dried at 130° F for 48 hours, and estimates of total weed biomass per acre were calculated. Grain yields were estimated by removing ears within two 30-foot rows of corn within each plot. Corn was shelled in the field, and subsamples were oven dried to determine dry weight.

## Results

Preemergence herbicides provided the best overall weed control in both ridges and furrows, although because of fairly high coefficients of variation in the data, no statistically significant differences ($p < 0.05$) were detected between the control (no weed control) and the preemergence herbicide treatment for total weed biomass (Table 11-5). Cultivation apparently stimulated germination of smartweed, with significantly more smartweed biomass than when no weed control treatment was used. Weed biomass in the furrows, where there was a thick mulch of cover crop residue, was considerably less than that on the ridges, where the soil was disturbed and no mulch existed.

Although weed control varied among treatments, all treatments produced

**TABLE 11-5**   Effects of Weed Control Practices on Weed Biomass in a Ridge-Till Corn Production System, 1989[*]

| Weed Control Practice | Pounds of Weeds/Acre[†] | | | | |
|---|---|---|---|---|---|
| | BL | PSW | GR | YNS | Total |
| Ridge | | | | | |
| None | 149b | 19b | 224ab | 1,648 | 2,040ab |
| Cultivation | 163b | 1,784a | 117b | 994 | 3,056a |
| Preemergence herbicide | 110b | 38b | 22b | 474 | 645b |
| Postemergence herbicide | 502a | 17b | 469a | 484 | 1,472ab |
| Furrow | | | | | |
| None | 52 | 277 | 43a | 203 | 574 |
| Preemergence herbicide | 53 | 3 | 2b | 128 | 185 |

NOTE:   Means within a half column followed by the same letter arenot significantly different (Duncan multiple range test; alpha = 0.05).

[*]Samples were taken on August 8 and 9, 1989. Calculated on an oven-dried (135° F) basis.
[†]BL, broadleaves; PSW, Pennsylvania smartweed; GR, grasses; YNS, yellow nutsedge.

statistically similar ($p > 0.05$) corn yields (Table 11-6).  Areas surrounding the experimental plots were covered with a rank growth of Pennsylvania smartweed; however, within the ridge-till experiment, smartweed levels were relatively low. Although these first-year results are preliminary, they indicate that the ridge-till system, in which a mowed cover crop was used for mulch and a very shallow skimming of the ridge was used during planting, may provide significant levels of weed control, reducing the need for both mechanical and chemical weed controls.  A second-year replication of this experiment is in progress, with rye-hairy vetch cover crops established in the fall of 1989.  Economic analysis of both years' results will be conducted at the end of the 1990 season.

## EXPERT SYSTEM DEVELOPMENT

Over the last decade, expert systems have been used extensively in industries from medicine to defense to solve complex problems that normally require human expertise.  They are particularly well suited to problems in which the solution requires judgment, dealing with uncertainty, qualitative assessments, and rules of thumb rather than solutions to mathematical equations.  When expert systems are integrated with conventional computer decision-making aids like simulation models and data bases, they become even more powerful, acting like a cadre of experts with access to sophisticated prediction tools and data.

Development of a low-input farming plan for a specific farm is an ideal

**TABLE 11-6**  Effects of Weed Management Practices on Corn Grain Yield in a Ridge-Till System Using a Winter Rye Cover Crop, Blacksburg, Virginia, 1989

| Weed Control Treatment | | Mean Corn Yield (bushels/acre) |
| --- | --- | --- |
| Ridge | Furrow | |
| None | None | 99.5 |
| None | Preemergence herbicide* | 93.2 |
| Preemergence herbicide | None | 99.8 |
| Preemergence herbicide | Preemergence herbicide | 84.8 |
| Cultivate† | None | 94.2 |
| Cultivate | Preemergence herbicide | 89.1 |
| Postemergence herbicide‡ | None | 94.7 |
| Postemergence herbicide | Preemergence herbicide | 100.1 |

*Atrazine (AAtrex) was used at 3 pints/acre with metolachlor (Dual) at 2 pints/acre.

†One cultivation with a modified V-sweep cultivator (set 5 inches from the centerline of the corn row) on June 30, 1989. Corn was approximately 12 inches tall and was at the six- to seven-leaf stage.

‡Bentazone (Basagran) was used at 2 pints/acre with crop oil (Dash) at 2 pints/acre.

problem for expert system techniques. There is an extensive qualitative knowledge base concerning the effects of crop rotations, tillage practices, legume N, and other practices on soil properties and crop productivity. Much of this knowledge cannot be put into mathematical equations because it is imprecise or qualitative. It is not possible or useful in a mathematical simulation, for example, to say that the incorporation of manure into soil increases the soil's tilth. Unless it can be transcribed to a rate function, the model cannot use this information. An expert system, on the other hand, like the one described here, is constructed from sets of statements. These statements make up an expert system's knowledge base, the knowledge that lets the computer solve a problem.

A prototype computer-aided decision-making system called CROPS (crop rotation planning system) has been developed for farm-level planning. This program uses artificial intelligence techniques to generate crop rotation plans for individual farms, implementing low-input sustainable practices and comparing these plans with conventional alternatives. It answers a fundamental need in the pursuit of a sustainable agriculture because it is impossible to implement low-input sustainable practices without addressing the whole-farm planning problem. Planning crop rotations involves or influences (1) the entire acreage of the farm, (2) tillage and soil conservation plans, (3) pest management, (4) use and purchase of fertilizers and lime, (5) farm economics, (6) farm diversification, and (7) livestock requirements and operation.

CROPS is now under further development, having received additional financial support from the 1990 USDA LISA program. The final version will not only generate crop rotation plans that implement low-input practices, it will also analyze the plans generated and allow the farmer-user to compare the generated plans with alternatives. The system will include simulation models for estimating soil erosion and for analyzing the financial status of the farm under various alternative combinations of crop mixes, farm program participation, and machinery complement.

## EXTENSION EDUCATIONAL PROGRAMS

Several educational programs for farmers and extension personnel were conducted through the Virginia Cooperative Extension Service to provide practical information of low-input sustainable farming practices and systems. These included the following:

1. A 1-day training session on low-input farming systems was held for extension agents as part of their annual in-service training. Extension agents learned new low-input practices that can be used across a wide array of cropping systems, as well as additional sources of information to serve interested clientele.

2. The statewide Virginia Conference on Sustainable Agricultural Systems, March 13–14, 1989, in Charlottesville, was cosponsored by the Virginia Cooperative Extension Service, the Virginia State Horticultural Society, and the Virginia Division of Soil and Water Conservation.

3. A multicounty farmer educational meeting on sustainable agriculture in Amelia County was cosponsored by the Virginia Cooperative Extension Service and the Virginia Farm Bureau Federation. Multicounty grower meetings on sustainable agriculture were also conducted at two other locations in 1989.

4. A research update in-service training session on LISA projects for extension agents of the Virginia Cooperative Extension Service West-Central District was conducted by LISA project personnel.

5. A low-input sustainable agriculture field day was held in August 1989 at the Whitethorne Research Farm, Blacksburg, Virginia.

6. A conference entitled "Farming for Profit and Stewardship" was held on March 15–16, 1990; it was cosponsored by the Virginia Cooperative Extension Service, the Virginia Department of Agriculture and Consumer Services, the Soil Conservation Service, and other groups.

## CURRENT PROJECT STATUS AND FUTURE PLANS

The long-term crop and livestock systems comparison study is now established, with all rotational sequences in place. Cattle will be introduced into the experiment in the fall of 1990. Additional funding is currently being sought to continue operation of this study. As a long-term study, results of this work are not anticipated before completion of the first rotational cycle in 1995. In the interim, the project will serve as a demonstration model to teach farmers, extension personnel, and other agricultural professionals the concepts of farming systems design to maximize beneficial agroecological processes and to reduce off-farm inputs.

On-farm research on the use of mowing as an alternative to herbicides for cover crop desiccation and insecticides for armyworm control is continuing. The Tennessee Valley Authority is providing additional funding for this work. Research on the integration of cover crops into low-input ridge-tillage systems will also continue. Development of the computer-aided crop rotation farm planning system will be continued, with on-farm testing of the system anticipated in 1991.

Funding from the USDA LISA program has provided for a significant level of research and extension education programs that would have been impossible otherwise. Establishment of the long-term crop and livestock systems study will provide a catalyst to obtain additional funding to continue this work. In addition, the projects have stimulated a broad interest in the general area of sustainable agriculture within the university, extension field staff, and agricultural communities.

## REFERENCES

Absher, K. L., V. G. Allen, and J. P. Fontenot. 1989. Effect of nitrogen fertilization of legumes on digestibility and nitrogen utilization of stockpiled fescue. Journal of Animal Science 67(Suppl. 1):270.

Allen, V. G., J. P. Fontenot, W. P. Green, R. C. Hammes, Jr., and H. T. Bryant. 1987. Year-round forage systems for beef cow-calf production. Journal of Animal Science 65(Suppl. 1):347.

Allen, V. G., J. P. Fontenot, and R. F. Kelly. 1989a. Performance and carcass characteristics of beef cattle on forage-based systems. Journal of Animal Science 67(Suppl. 1):270.

Allen, V. G., J. P. Fontenot, and W. H. McClure. 1989b. Intensive grazing systems for beef cattle. Pp. 160–164 in Proceedings of the 1989 American Forage and Grassland Conference. Belleville, Pa.: American Forage and Grassland Council.

Allen, V. G., J. P. Fontenot, W. P. Green, and R. C. Hammes, Jr. 1989c. Year-round grazing systems for beef production from conception to slaughter. Pp. 1197–1198 in Proceedings of the XVI International Grasslands Congress, Nice, France.

Allison, J. R., and S. L. Ott. 1987. Economics of using legumes as a nitrogen source in conservation tillage systems. Pp. 145–150 in The Role of Legumes in Conservation Tillage Systems, Proceedings of a National Conference, J. F. Power, ed. Ankeny, Iowa: Soil and Water Conservation Society.

Behn, E. E. 1982. More Profit with Less Tillage. Des Moines, Iowa: Wallace-Homestead Press.

Berry, W. 1984. Whose head is the farmer using? Whose head is using the farmer? Pp. 19–30 in Meeting the Expectations of the Land, W. Jackson, W. Berry, and B. Colman, eds. San Francisco: North Point Press.

Blaser, R. E. 1986. Forage-animal management systems. Virginia Agricultural Experiment Station Bulletin No. 86-7. Blacksburg, Va.: Virginia Polytechnic Institute & State University.

Corak, S. J., W. W. Frye, M. S. Smith, J. H. Grove, and C. T. MacKown. 1987. Fertilizer nitrogen recovery by no-till corn as influenced by a legume cover crop. Pp. 43–44 in The Role of Legumes in Conservation Tillage Systems, Proceedings of a National Conference, J. F. Power, ed. Ankeny, Iowa: Soil and Water Conservation Society.

Ebelhar, S. A., W. W. Frye, and R. L. Blevins. 1984. Nitrogen from legume cover crops for no-tillage corn. Agronomy Journal 76:51–55.

Ehrenfeld, D. 1987. Sustainable agriculture and the challenge of place. American Journal of Alternative Agriculture 2:184–187.

Fontenot, J. P., F. P. Horn, and V. G. Allen. 1985. Forages and slaughter cattle. Pp. 570–578 in Forages, the Science of Grassland Agriculture, 4th ed., M. E. Heath, R. F. Barnes, and D. S. Metcalfe, eds. Ames, Iowa: Iowa State University Press.

Frye, W. W., W. G. Smith, and R. J. Williams. 1985. Economics of winter cover crops as a source of nitrogen for no-till corn. Journal of Soil and Water Conservation 40:246–249.

Hargrove, W. L. 1986. Winter legumes as a nitrogen source for no-till grain sorghum. Agronomy Journal 78:70–74.

Hargrove, W. L., and W. W. Frye. 1987. The need for legume cover crops in conservation tillage production. Pp. 1–4 in The Role of Legumes in Conservation Tillage Systems, Proceedings of a National Conference, J. F. Power, ed. Ankeny, Iowa: Soil and Water Conservation Society.

Harwood, R. R. 1990. A history of sustainable agriculture. Pp. 3–19 in Sustainable Agricultural Systems, C. Edwards, R. Lal, P. Madden, R. Miller and G. House, eds. Ankeny, Iowa: Soil and Water Conservation Society.

Laub, C. A. 1990. Influence of Cover Crop Management on Armyworm, *Pseudaletia unipuncta* (Haworth) Seasonal Abundance, Natural Enemies, and Yield in No-Till Corn, and Diurnal Abundance and Spatial Distribution of Armyworm. M.S. thesis. Virginia Polytechnic Institute & State University, Blacksburg.

Little, C. E. 1987. Green Fields Forever; The Conservation Tillage Revolution in America. Washington, D.C.: Island Press.

Luna, J. M., and G. J. House. 1990. Pest management in sustainable agricultural systems. Pp. 157–173 in Sustainable Agricultural Systems, C. Edwards, R. Lal, P. Madden, R. Miller, and G. House, eds. Ankeny, Iowa: Soil and Water Conservation Society.

Maxey, H., T. Covey, B. McKinnon, and A. Allen. 1989. West Central District Crop Budgets. Special Publication. Blacksburg, Va.: Department of Agricultural Economics, Virginia Polytechnic Institute & State University.

Mitchell, W. H., and M. R. Teel. 1977. Winter annual cover crops for no-tillage corn production. Agronomy Journal 69:569–573.

National Research Council. 1989. Alternative Agriculture. Washington, D.C.: National Academy Press.

Neely, C. L., K. A. McVay, and W. L. Hargrove. 1987. Nitrogen contribution of winter legumes to no-till corn and grain sorghum. Pp. 48–49 in The Role of Legumes in Conservation Tillage Systems, Proceedings of a National Conference, J. F. Power, ed. Ankeny, Iowa: Soil and Water Conservation Society.

Ott, S. L., and W. L. Hargrove. 1989. Profits and risks of using crimson clover and hairy vetch cover crops in no-till corn production. American Journal of Alternative Agriculture 4:65–70.

Parker, C. F. 1990. Role of animals in sustainable agriculture. Pp. 238–248 in Sustainable Agricultural Systems, C. Edwards, R. Lal, P. Madden, R. Miller, and G. House, eds. Ankeny, Iowa: Soil and Water Conservation Society.

Parr, J. F., R. I. Papendick, I. G. Youngberg, and R. E. Meyer. 1990. Sustainable agriculture in the United States. Pp. 50–67 in Sustainable Agricultural Systems, C. Edwards, R. Lal, P. Madden, R. Miller, and G. House, eds. Ankeny, Iowa: Soil and Water Conservation Society.

Power, J. F., ed. 1987. The Role of Legumes in Conservation Tillage Systems. Proceedings of a National Conference. Ankeny, Iowa: Soil and Water Conservation Society.

Schaller, N. 1989. Low input sustainable agriculture. Pp. 216–219 in 1989 Yearbook of Agriculture. Washington, D.C.: U.S. Department of Agriculture.

Sullivan, P. 1990. Rye and Vetch Intercrops for Reducing Corn Nitrogen Fertilizer Requirements and Providing Ground Cover in the Mid-Atlantic Region. Ph.D. dissertation. Virginia Polytechnic Institute & State University, Blacksburg.

Virginia Agricultural Statistics Service. 1988. Virginia Agricultural Statistics. Richmond, Va.: Virginia Agricultural Statistics Service.

Webb, F. J., R. L. Ritter, E. S. Hagood, J. W. Wilcut, and H. P. Wilson. 1988. Weed control in field crops. Pp. 39–90 in 1988–89 Pest Management Recommendations for Field Crops. Publication No. 456-015. Blacksburg, Va.: Virginia Cooperative Extension Service.

# 12

# Solarization and Living Mulch to Optimize Low-Input Production System for Small Fruits

*Kim Patten, Jeff B. Hillard, Gary Nimr, Elizabeth Neuendorff, David A. Bender, James L. Starr, Gerard W. Krewer, Randall A. Culpepper, Mike Bruorton, and Barbara J. Smith*

In the South, production of most horticultural crops is chemical and labor intensive. Disease, insect, and weed pressures can be major factors that limit successful fruit production. However, some fruit crops, such as blueberries and strawberries, lend themselves to low-input farming systems.

Blueberries can be grown with reduced or no chemical inputs if suitable management alternatives for soil fertility and weed control are available, because disease and insect pressures are minimal. One alternative is the use of living mulches. Under this scenario, a series of cover crops are grown in an all-year rotation between blueberry rows. The cover crops are mowed and windrowed under the blueberry plants for use as a mulch. Through proper selection of living mulch cover crops, weed competition may be eliminated because of allelopathic or smothering effects from the mulch barrier (Putnam, 1988), and nutrient inputs could be supplied by the decomposing mulch (Wagger, 1989). An improved edaphic environment (Patten et al., 1989) and reduced erosion would be additional benefits to the blueberry grower from this agroecosystem.

Chemical fumigation for the control of weeds and soil pathogens is routinely practiced for strawberry production. Fumigants, like methyl bromide, are effective in controlling many soil pests, but they are also expensive, require special licenses and equipment for application, are hazardous to agricultural workers, and may damage the environment. For these reasons a need exists to develop safe and economical alternatives to chemical fumigation.

Soil solarization is a nonchemical pest management practice that can be used to control a plethora of soil pests (Katan, 1981; Pullman et al., 1984;

Stapleton and Devay, 1986). This technique relies on increased soil temperatures, using a cover of clear polyethylene film to trap solar energy, to thermally deactivate soil pests. Solarization, however, is not a panacea and is not a widely used farming practice in the South. This lack of adoption partly results from the expense and difficulty of integrating solarization into routine cultural practices.

The cost of integrating solarization into a routine cultural practice for bedded row crops may be reduced by solarizing individual beds. After solarization, the clear plastic is pigmented with a latex paint to cool the soil and to allow planting directly through the plastic (Hartz et al., 1985). This method has been reported to be more cost-effective and require less specialized machinery than the conventional wide-tarp solarization method.

Annual strawberry production is ideally suited for bedded row solarization for several reasons. The strawberry off-season occurs during the hottest time of year (July and August), when solarization is most effective; and soil bedding, fumigation, and plastic mulch are standard practices that are used in the annual production system. This low-input system may be further optimized when it is used in combination with legume cover crops or manure application that enhance soil fertility.

Management-intensive, but chemically input-free, production systems for small fruit in the South not only provide for an opportunity to capitalize on a new market niche but also foster long-term soil productivity and reduce environmental hazards.

The low-input sustainable agriculture (LISA) project that is the subject of this chapter has two objectives:

• to investigate the feasibility of eliminating fertilizer and herbicide inputs on blueberries grown in the South by using legumes and annual forage rotations for living mulches, and

• to evaluate solarization and cover crops or manures as replacements for fumigation and fertilizers in annual strawberry production in the South.

## OBJECTIVE ONE: BLUEBERRY LIVING MULCHES

### General Farm Background

Experimental plots were established at seven growers and one experiment station to demonstrate the efficacy of living mulch systems for blueberries. The planting locations ranged from northeast Texas (Winnsboro, Tyler, and Overton), central east Texas (Nacogdoches), and southeast Texas (Huntsville) to southeast Georgia (Homerville, Chula, and Fargo).

Farm sizes varied from 5 to 50 acres of blueberries. Soils at all locations were typical of those used for blueberry planting, with a sandy loam texture, very low native fertility, and a strongly acidic soil pH (4.1 to 5.5).

Nitrogen (N) is readily leached, and frequent N fertilization is needed to ensure production. Average annual rainfall averaged 45 inches in Texas to 50 inches in Georgia, but rainfall patterns were erratic. A rainfall of only 0.5 to 2 inches per month was common during the east Texas summers.

The establishment of living mulch plots was frequently hampered by unseasonably wet or dry weather and severe winter freezes. Inadequate weed control was the only significant pest problem. Growers used conventional orchard floor management practices by applying combinations of preemergent (oryzalin [Surflan] and simazine [Princep]) and postemergent (glyphosate [Round-up] or paraquat) herbicides for control of weeds around the plants.

### Experimental Protocols

In Georgia, the following experimental protocol was followed. Living mulch crops were grown in the winter at three blueberry farms (Homerville, Fargo, and Chula). The Chula site had a moderate soil pH (5.5) and fertility, whereas the other sites were strongly acidic and less fertile. The crops that were evaluated are listed in Table 12-1. Seeding methods varied at each site, ranging from harrowing plus grain drill, grain drill into existing orchard floor vegetation, or broadcasting with or without harrowing, depending on the site. Plots were split, with one section receiving no fertilizer, and the other receiving up to 100 pounds of N per acre (lbs of N/ acre). There were two to four replications per treatment, depending on the site.

A somewhat different experimental protocol was followed in Texas. During the winter and summer, cover crops were planted at five grower locations and at the Overton experiment station farm. The crops that were evaluated are listed in Table 12-1. Plots were seeded with a small plot drill with six double-disk openers, spaced 9 inches apart in the middle of the blueberry rows. There were four replications of each crop per site. Plots were reseeded if the initial stands failed in response to environmental stresses. In general, nonlegume plots received 100 lbs of N/acre at planting, and all crops received phosphorus (P) and potassium (K) applications of 20 lbs/acre.

The effects of N on summer living mulches in Texas were evaluated on Tifleaf pearl millet, Headless Wonder sorghum, and Green Graze sorghum sudan. These crops were sown in May at 90 lbs/acre by using a broadcast seeder in the middle of the blueberry rows at the Nacogdoches site. Plots were fertilized with 0, 100, or 200 lbs of N/acre in mid-June. All plots received 36 lbs of K/acre and 22 lbs of magnesium (Mg)/acre. There were three plots per treatment, each of which was 300 feet long.

The capacity of winter legume to supply N for a succeeding summer grain cover crop was evaluated at the Overton site. Sorghum was grown

**TABLE 12-1**   Evaluation of Forage Cultivars for Blueberry Living Mulch Production in Texas and Georgia in 1989 and 1990

| Time Period and Forage | Location* | Overall Performance Rating |
|---|---|---|
| Winter 1988–1989 | | |
| Tifton 86 ryegrass | HO | Very poor to fair[†] |
| Gulf annual ryegrass | HO, FR, CH | Very poor to good[†] |
| Marshall ryegrass | HO, FR, CH, OV | Very poor to good[†] |
| GI-85 ryegrass | HO | Poor to fair[†] |
| Wrens abruzzi rye | HO | Very poor to good[†] |
| Elbon rye | OV, NC, HV, RU, TY | Poor to very good[†,‡] |
| Coker 227 oats | HO | Very poor[†] |
| Florida 302 wheat | HO, OV, NC, NC | Very poor[†] |
| Atlas 66 wheat | HO | Very poor[†] |
| Texas 182-85 wheat | OV | Very poor[†] |
| Beagle triticale | HO | Very poor[†] |
| B858 triticale | OV, NC | Very poor[†] |
| T20 triticale | OV, NC | Very poor[†] |
| Dixie crimson clover | CH, FR, OV, NC, HV, RU, TY | Poor to very good[†,‡] |
| Mt. Barker subterranean clover | OV, NC | Very poor to fair[†,‡] |
| Common hairy vetch | OV | Poor to good[†] |
| Summer 1989 | | |
| Tifleaf pearl millet | OV, NC, HV, TY, WB | Good to very good |
| Headless wonder sorghum | OV, NC, TY, WB | Poor to good |
| Green graze sorghum sudan | OV, NC, HV, WB | Fair to good |
| Iron and clay cowpeas | OV, NC, HU, TY, WB | Fair to very good[†,‡,§] |
| Sun hemp crotalaria | OV, NC, HV, TY, WB | Very poor to very good[†,§] |
| Everglades 41 Kenaf | OV | Fair to good |
| Sericae lespedeza | OV | Fair[†] |
| Winter 1989–1990 | | |
| Gulf ryegrass | HO, OV, NC, HV, TY, WB | Poor to good[†,‡,§] |
| Elbon rye | OV, NC, HV, TY, WB | Poor to very good[†,‡,§] |
| Dixie crimson clover | HO, OV, NC, HV, TY, WB | Poor to fair[†,‡,§,‖] |
| Common hairy vetch | OV, WB | Poor[†,§] |
| Yucchi arrowleaf clover | OV, WB, TY | Very poor to poor[†,§] |
| D-3 rose clover | OV | Poor to fair[†,§] |

*HO, Homerville, Georgia; FR, Fargo, Georgia; CH, Chula, Georgia; OV, Overton, Texas; NC, Nacogdoches, Texas; RU, Rusk, Texas; TY, Tyler, Texas; HV, Huntsville, Texas; WB, Winnsboro, Texas.

[†]Poor stand at several locations.

[‡]Damaged by record cold temperatures in December 1989.

[§]Damaged by deer foraging.

[‖]Drought during germination and/or growth severely reduced growth.

between blueberry rows in 1989 on soil treated with 0 or 50 lbs of N/acre or following a winter cover crop of crimson clover. The clover was roto-till incorporated into the soil in April, before seeding of the sorghum. Clover dry matter yield was 3,000 lbs/acre with a 1.2 percent leaf N concentration (36 lbs of N/acre in tops). The sorghum was sown in May with a precision seeder and was harvested in August.

An experiment was established at Overton in 1989 by using a split-plot design to evaluate the effects of irrigation and fertilization on mulch production of two cover crops. Whole plots had sprinkler irrigation or no irrigation, and subplots had factorial combinations of cover crops and fertilization. The cover crops were Sun Hemp crotalaria and Iron and Clay cowpeas. Subplots received 500 lbs of lime/acre, 50 lbs of P/acre, and 50 lbs of K/acre or were left unfertilized.

Weed suppression with cover crops was assessed by measuring weed densities in the field and allelopathic responses in the laboratory. Several crops were assessed for their allelopathic suppression of weed seed germination. Plant tops were air-dried (86°F), ground (40 mesh), and placed on the soil surface of a planting plug (2 by 2 inches) that was seeded with 25 seeds each of crabgrass, common bermudagrass, and pig weed. The percent germination was evaluated as a function of the cover crop and the amount of mulch applied.

## Results

The living mulch crops that were evaluated, the locations where these crops were planted, and a summary of their general performance are listed in Table 12-1. Living mulch systems employing rye, ryegrass, or crimson clover (winter) and pearl millet (summer) were rated the highest based on overall production, stand consistency, weed control, cost, and resistance to deer foraging and winter cold temperatures.

Living mulch yield performance in Georgia for 1989 was dependent on soil pH and N fertilization (Table 12-2). Marshall ryegrass and Wrens abruzzi rye were the two best cover crops evaluated. Sites with low soil pH (less than 5.0), low fertility, or with no or a low level of applied N failed to produce cover crop stands. The low-pH locations also failed to produce a significant amount of mulch in 1990 compared with that produced by sites with a higher pH (5.5). Nutrient analysis of the cover crops indicated that the mulch contained sufficient N and K for blueberry production (greater than 70 lbs/acre) and that the nutrients were released from the mulch within 4 months after cutting (data not shown). Ryegrass grown with 100 lbs of N/acre resulted in good overall mulch production.

In Texas, Elbon rye provided the greatest amount of winter mulch (Table 12-3). Wheat, triticale, ryegrass, and subterranean clover were less suitable

**TABLE 12-2**  Yields of Blueberry Living Mulches in Georgia

| Cover Crop | Treatment | Fresh Weight Yield (lbs/acre) Homerville[*] | Fargo[*] | Chula[†] |
|---|---|---|---|---|
| Gulf annual ryegrass | Fertilized | 2,310 | 1,182 | 16,299 |
|  | Unfertilized[‡] | 1,448 | 0 | 4,152 |
| Marshall ryegrass | Fertilized | 3,933 | 2,321 | 8,080 |
|  | Unfertilized | 352 | 0 | 543 |
| Wrens abruzzi rye | Fertilized | —[§] | 9,278 | — |
|  | Unfertilized | — | 586 | — |
| Crimson clover |  | — | — | 1,767 |

[*]Soil pH ≤ 4.4.
[†]Soil pH = 5.5.
[‡]Fertilized in spring only.
[§]— indicates that a crop was not planted.

**TABLE 12-3**  Yield of Blueberry Living Mulches for Different Locations in Texas

| Time Period and Crop | Dry Weight Yield (lbs/acre) Nacogdoches | Tyler | Huntsville | Winnsboro[*] | Overton |
|---|---|---|---|---|---|
| Winter 1988–1989 |  |  |  |  |  |
| Elbon rye | 2,948a[†] | 0 | 2,103a | [‡]— | 5,642a |
| Triticale | — | — | — | — | 3,799b |
| Marshall ryegrass | — | — | — | — | 1,963c |
| Wheat | — | — | 2,005a | — | 3,038bc |
| Crimson clover | 951b | 0 | 2,394a | — | 2,537bc |
| Subterranean clover | 587b | 0 | 0b | — | 1,991c |
| Hairy vetch | — | — | — | — | 4,224b |
| Summer 1989 |  |  |  |  |  |
| Pearl millet | 17,460a | 11,563a | 4,898a | 26,550a | 8,202b |
| Sorghum sudan | 16,942a | 1,410c | 4,269a | 21,895b | 7,715b |
| Sorghum | 20,803a | — | — | 21,044b | 7,056b |
| Cowpeas[§] | 0b | 5,009b | 6,151a | 12,901c | 7,558b |
| Crotalaria[§] | 0b | 0d | 1,251b | 14,463c | 17,910a |

[*]Overhead irrigation was used at this location.
[†]Different letters indicate separation of the means within columns by Duncan's test at the 0.05 percent level.
[‡]— indicates that a crop was not planted.
[§]Yields of cowpeas and crotalaria were very low at several locations because of deer foraging.

for a winter cover crop compared with rye. Several growers gave crimson clover high ratings as a cover crop. No applied N was required, yields were adequate, and it was visually attractive during bloom.

Establishment of good stands was hampered at several sites because of erosion and the subsequent washout of seeds or seedlings. Poor germination, soil infertility, or deer foraging also adversely affected the stands. The effects of an extremely dry fall and record low temperatures in the winter on stand establishment and growth of crimson clover during 1989–1990 were especially apparent. The weather also had an adverse effect on the other vegetations available for rabbit and deer populations in 1990. Consequently, mulch production at four of the five sites was markedly reduced by grazing from rabbits and deer. For the 1990 season, rye tolerated the adverse conditions better than did the other crops that were evaluated (data not shown).

Pearl millet was consistently the most productive living mulch crop in summer (Table 12-3). The other forage grains (sorghum and sorghum sudan) had similar levels of production, but they were not tolerant to frequent mowings to low heights. Yields of summer legumes (cowpeas and crotalaria) were erratic across all sites and were not tolerant to mowing. Legume mulches, if left unmowed, were undesirable because cowpeas grew into and up the blueberry plant and crotalaria grew too tall (greater than 12 feet).

Living mulch crops were highly dependent on fertilization and irrigation, with N limiting the mulch production of nonlegumes. At Nacogdoches, mean dry weight yields of sorghum, sorghum sudan, and pearl millet supplied with 0, 100, and 200 lbs of N/acre were 6,000, 15,500, and 30,000 lbs/acre, respectively. The Overton plot, where crimson clover was tilled into the row middles, produced a sorghum yield equivalent to that of plots supplied with 50 lbs of N/acre (data not shown). These plants, however, were still N deficient, with there being less than 0.7 percent leaf N. For summer legumes (cowpeas and crotalaria), fertilization with low levels of lime, P, K, and Mg increased the yield by 50 percent, while irrigation increased the yield by 150 percent (data not shown). The one site in Texas that used overhead irrigation—Winnsboro—had a significantly greater yield than other locations where the living mulch plots were not irrigated (Table 12-3).

The total cover crop yield multiplied by its N concentration indicated the theoretical amount of N available to the blueberry plants from the mulch system. In general, summer crops supplied three times the total N as that provided by winter crops (50 lbs of N/acre for rye and clover compared with 150 lbs of N/acre for pearl millet). This also showed the higher overall yield for summer cover crops and their greater demand on soil N. For example, when summer nonlegume cover crops were grown on a minimum N fertilizer program (50 lbs of N/acre), they exhibited severe N stress symptoms, and all had leaf N levels of less than 0.6 percent (data not shown). On a typical unamended blueberry soil, it was also apparent that a winter/summer rotation of legume/legume or legume/nonlegume would be only marginally acceptable with respect to N recycling and weed control. This was concluded because winter clover supplied only enough N for the blueberry plants (45 lbs of N/acre), and summer legumes stands were too inconsistent.

Weed control results were also evaluated. Pearl millet was the most effective cover crop for suppressing weeds at all locations (Table 12-4). The allelopathic effects of several mulches were tested in a greenhouse. As in the field evaluations, pearl millet usually suppressed germination more than did the other crops that were tested (Table 12-5).

## Discussion

Living mulches appear to be a practical and desirable cultural practice for controlling weeds and erosion and for reducing chemical fertilizer inputs for blueberries grown in the South. This system may also be appropriate for other perennial fruit crops. For this sustainable system to be viable, however, there are several criteria that must to be met.

Originally, the system was designed to operate with a winter legume/summer legume rotation. No summer legume was identified that was consistent across a variety of sites, tolerant to mowing, and easy and inexpensive to establish and that had allelopathic properties. Crimson clover met several of the criteria for a winter legume, but it did not grow when the soil pH was less than 5.0 and did not tolerate the extremes of fall drought, cold weather, or deer foraging. The poor crop stands of crimson clover in 1990

**TABLE 12-4** Weed Control Between Blueberry Rows as Affected by Living Mulch Cover Crops

| | Percentage of Ground Covered by Weeds* | | | |
|---|---|---|---|---|
| Crop | Nacogdoches Plot 1 | Nacogdoches Plot 2 | Winnsboro | Overton |
| Pearl millet | 15b[†] | 20c | 1a | 3a |
| Sorghum sudan | 52b | 31bc | 21ab | 26b |
| Sorghum | 39b | 36c | 9ab | 22ab |
| Cowpeas | 100c | 50b | 45bc | — |
| Crotalaria | 99c | —[‡] | 66c | — |
| Control | 100c | 94a | — | — |

*Weeds, mostly crabgrass.

[†]Different letters indicate separation of the means within columns by Duncan's test at the 0.05 percent level.

[‡]— indicates that a crop was not planted.

**TABLE 12-5**  Allelopathic Suppression of Weed Seed Germination Through the Use of Mulches of Different Cover Crops[*]

| | Percent Germination | | | | | |
| | Crabgrass | | Common Bermudagrass | | Pig Weed | |
| Cover Crop | 20 Days | 40 Days | 20 Days | 40 Days | 20 Days | 40 Days |
|---|---|---|---|---|---|---|
| Cowpeas | 13b[†] | 48b | 2a | 52d | 16b | 40b |
| Crimson clover | 15a | 53c | 1a | 36c | 10a | 30ab |
| Pearl millet | 13a | 37y | 1a | 28ab | 16b | 25a |
| Elbon rye | 22b | 52c | 2a | 22a | 21cc | 34ab |

[*]Stems and leaves of cover crops were air-dried at 25°C, ground to pass a 40-mesh screen, placed on top of soil that was planted with weed seeds, and misted with distilled water daily.

[†]Different letters indicate separation of the means within columns by Duncan's test at the 0.05 percent level.

were not limited to the blueberry plots but were also apparent under normal pasture situations. In most years, crimson clover could be expected to do well if soils received significant inputs of lime, P, K, and Mg. When native soil fertility was higher, a good stand could be achieved with only minor nutrient inputs. The visual attractiveness of clover during bloom and its reseeding ability, if not mowed too early, were other advantages of using clover.

Overall, the highest mulch production and year-round weed control were achieved with a winter Elbon rye/summer pearl millet rotation. This rotation required significant N fertilizer or manure inputs, because soils on which blueberries were grown were too infertile to support good cover crop growth.

Manure has potential as a nutrient source for living mulches. In 1990, manuring resulted in a higher yield of Elbon rye than did N fertilization (data not shown). A single application of manure per year, however, may not provide sufficient N to grow both a winter and summer living mulch crop, and two applications may be necessary. The high levels of manure required to supply the nutrient inputs for the cover crops may have an adverse effect on seed germination or, over the long term, may increase soil pH or P levels above the range desired for blueberry plants. At some sites, reduced cover crop germination was observed for seeds that were sown directly after a manure application. Future research will evaluate some of these problems encountered with the application of manure.

Yields from several winter/summer living mulch systems were adequate to provide 3 to 10 inches of mulch around blueberry plants. The amount of mulch produced increased in direct proportion to the increased width of the cover crop strip. For example, crimson clover or ryegrass, when sown row to row, resulted in a cutting area to mulched area ratio of 5:1, compared with a 2:1 ratio for a 5-foot band sown down the center of the row. Solid cover crop stands are more appropriate during the winter, when there is minimal competition between the cover crop and the blueberries. In the summer there should be at least an 8- to 20-inch buffer zone between the cover crop and the blueberry plant to avoid competition for water and nutrients.

The method of seeding determines the cover crop surface area and stand performance. The best stands were obtained with a grain drill that produced a 5-foot sward. For ryegrass and clover, good row-to-row stands at some sites were also obtained by broadcasting ryegrass or clover. Because many blueberry growers do not have access to a grain drill, an inexpensive hand broadcast seeder would be an appropriate low-cost alternative. Several growers in Georgia have adapted this technology and are broadcasting ryegrass in late fall. They report that this solid stand of ryegrass has suppressed winter and spring weeds, reduced erosion in the row middles, and helped to maintain the shape of the blueberry beds.

Cover crop harvesting was readily accomplished with a riding mower with a side-throw spout for crops with thin leaves (ryegrass and clover) and for crops that were not too tall. For thick-stemmed crops, like pearl millet, bush-hog mowers were necessary. Some summer crops, such as kenaf, cowpeas, and crotalaria, were not very tolerant of mowing and had little significant regrowth after they were mowed.

Living mulch systems may increase the potential for spring frost damage, as bare ground may provide increased temperatures (1° to 3°F) during a radiant frost over a vegetated orchard floor. Although this is a valid concern, earlier data on orchard floor management indicated that sodded row middles may reduce the radiant heat budget of the orchard floor, thereby delaying bloom and reducing the potential of frost damage (Patten et al., 1989).

The total cost of this system is a primary consideration for the blueberry grower. An enterprise budget detailing the cost of establishing and using a living mulch system is presented in Table 12-6. The estimated cost of growing and using cover crops for living mulch twice a year was $300/acre of blueberries. These costs represent approximately $80 more than those associated with fertilizers and herbicides in conventional blueberry production programs. Over time, soil fertility in the living mulch systems would increase and the cost of nutrient inputs would decrease. For blueberries cultivated on more fertile soils, a lower-cost living mulch system, such as hand broadcasting of crimson clover, could be cost-effective with respect to weed control and N inputs (less than $50/acre).

These small cost differences suggest that a living mulch agroecosystem could be advantageous to the grower in the long term. The sustainable

**TABLE 12-6** Farm Enterprise Budget for Blueberry Living Mulch Systems

| System | Cost ($) per Acre of Blueberries* |
|---|---|
| Living mulch cost† | |
| Seeding (seed + planting cost, regardless of crop) | 40 |
| Fertilizer | |
| Option 1: Chicken manure, delivered and spread at 12 tons/acre‡ | 200 |
| Option 2: 0.5 ton of NPK (13-13-13 fertilizer)/year + | |
| 5 N applications at 50 lbs/acre | 200 |
| Mowing four times per year | 60 |
| Total | 300 |
| Conventional cost§ | |
| Weed control | 80 |
| Fertilizer | 140 |
| Total | 220 |

*Cost per acre of blueberries is 65 percent of the actual cost per acre, since the living mulch plot only covers approximately 60 to 70 percent of the land on an acre of blueberries.

†Assumption included in this analysis are that the living mulch system has no effect on yield. Expenses are based on direct cost for supplies plus services as derived from the 1988 Texas Custom Rates Statistics Handbook.

‡The optimal rate and frequency of application for manure has yet to be determined.

§Values are estimates based on a budget for a 15-acre blueberry farm in Georgia.

agriculture system, however, is unlikely to be cost-effective during the transition time between systems. During this period, weed control and nutrient input from living mulches may not be completely satisfactory. A detailed, long-term economic analysis of the living mulch system will not be available until all of the horticultural variables have been evaluated over an extended period of time.

## OBJECTIVE TWO: SOLARIZATION FOR STRAWBERRY PRODUCTION

### General Farm Background

These experiments required closely monitored data collection that precluded the use of grower sites. Plots were established at experiment station farms at Lubbock, Texas; Overton, Texas; and Poplarville, Mississippi. Soils, planting methods, and pest control at all locations were typical of those usually recommended for strawberries. The soils were sandy loam to loamy sand, pHs were from 5.5 to 6.5, and native soil fertility was low. Nitrogen was the major limiting nutrient. All locations were watered by drip irrigation, and protection against frost was done by sprinkler irrigation. Record low temperatures in February 1989 caused some plant damage at all locations. The planting at Lubbock was killed in 1988–1989 because of freeze damage. The type and severity of indigenous soilborne pests depended on the site. Weeds were the major limiting soilborne pest at all sites. Research plots at Overton contained a high density of yellow nutsedge (*Cyperus esculentus* L.) tubers. Extremely heavy rains during harvest in 1989 (greater than 16 inches) caused 10 to 15 percent cullage loss because of fruit rots. Plastic mulch helped to minimized fruit rot and reduce the need for cover sprays. One early fungicide spray was used to control decay. In general, the levels of decay were not significant enough to make fungicide applications a necessity.

## Experimental Protocols

In the Overton experimental plots, the following experimental protocol was used. In 1987–1988, the soil treatments were 6 weeks of solarization, fumigation with 400 lbs of 98 percent methyl bromide and 2 percent chloropicrin per acre, or a control. Experimental units were 13-foot-long beds, 8 inches high and 28 inches wide, that were planted in double rows with 20 strawberry plants. Treatments were replicated 12 times. Solarization plots were covered with clear, 0.0015-inch-thick polyethylene for 6 weeks starting on August 10. Soil under the plastic was rewetted by drip irrigation three times during solarization. After 6 weeks, the plastic became too brittle for continued use. The old plastic was removed, and new clear 0.0015-inch-thick plastic was reapplied to provide plastic mulch for all plots. This plastic was then sprayed with a 7:1 ratio of water:white exterior latex paint to reduce soil heating and to allow strawberry planting without heat damage to the roots. Fumigation occurred 3 weeks before planting. Dormant Chandler strawberries were planted in double rows on September 22. Fertilizer was applied monthly with a trickle irrigation system in September, November, March, April, and May as N, P, and K at 36, 14, and 29 lbs/acre, respectively, in each application.

In 1988–1989, the experimental design was a split plot with six replications. Beds were varied at 8 or 13 inches in whole plots. Split plots (23 feet) were solarized, fumigated, or not treated. The solarization plastic (thickness, 0.0015 inch) was applied on July 29 and was removed October 14. All plots were drip irrigated three times during the solarization period. Fumigation of individual plots with 400 lbs of methyl bromide and chloropicrin per acre occurred on September 14. Beds were planted in double rows with 28 dormant Chandler strawberry plants on October 15. Preplant fertilizer (N, P, and K at 50, 65, and 75 lbs/acre, respectively) was tilled into the soil before bedding. Supplemental N was applied five times through a drip line (every 21 days from February to May) at 15 lbs of N/acre per application.

Weeds were controlled manually after ratings of weed density were taken. In the spring, soil samples were taken to evaluate nematode populations, and roots were examined for root-knot nematode infestations. Soil temperatures were monitored by placing thermocouples at soil depths of 4 and 8 inches.

For the 1989–1990 season, whole-plot soil treatments were expanded to include legume cover crops or manure combinations. Split plots were fumigated, solarized, or left untreated. Strawberries were planted in September. To date, only data on nematode populations and plant vigor ratings have been collected for these plots.

Because of differences in climate and growing conditions, a different experimental protocol was used in Lubbock. Plots were established in the fall of 1988 in west Texas by using a randomized complete block design with four replications. The experimental variables were factorial combinations of bed height (20 or 40 inches) and soil treatment (untreated, 1 month of solarization under clear plastic, metam sodium fumigation with water incorporation). Nematode populations were sampled before and after soil treatment. All plants were killed by cold temperatures in December.

In the Poplarville experiment, yet another protocol was required. Before the evaluation of solarization effects, Iron and Clay cowpeas were planted as a cover crop during the summer of 1988 and were tilled in the fall. Soil treatments were applied on September 1 and replicated six times. These treatments consisted of (1) clear plastic solarization for 2 months on raised beds, (2) clear plastic solarization on flat beds, (3) black plastic solarization on raised beds, (4) methyl bromide fumigation, or (5) an untreated control. On November 11, treatment plastic was removed, the plots were mulched with black plastic, and 12 Chandler strawberries were planted in each experimental unit. Strawberry yield and weed populations were evaluated in the spring.

The following protocol was used in Overton to study the impacts of various treatments on soil chemistry. To evaluate the effects of soil solarization and manures with variable carbon (C)/N ratios on nutrient availability, studies were initiated in the summer of 1989 on a fine sandy loam soil. The experimental design was a split plot, with whole-plot treatments being solarized or unsolarized soil. The split-plot treatments were factorial combinations of poultry manure applied at a rate to give 500 or 1,000 lbs of N/acre equivalents (5 or 10 tons/acre of manure) and fresh pine sawdust applied at 0, 40, or 80 tons/acre. Control plots that were either unamended or received fertilizer with N, P, and K (250, 40, and 75 lbs/acre, respectively) were included as split treatments. Split-plot treatments were applied in July and tilled into the top 10 inches of soil. A clear plastic tarp (thickness, 0.0015-inch) was applied from July 18 to September 26. Following solarization, soil samples were collected to a 6-inch depth. Suction lysimeters were placed at a depth of 16 inches in four of the split-plot treatments. Elbon rye was seeded as an indicator crop. Soil leachate was collected after each rainfall of greater than 2 inches. Chemical analyses of soil samples were conducted according to the testing procedures of Texas A&M University (College Station, Texas). Weed populations were evaluated during December.

## Results

### Soil Temperatures

The mean maximum soil temperatures at the 4-inch depth in the center of the 8-inch-high bed were 113° and 105°F for solarized and nonsolarized soils, respectively. The total time above 104°F in 1988 at 4- and 8-inch depths for the 8-inch solarized bed was approximately 590 and 300 hours, respectively. For nonsolarized soil, these values were 74 and 0 hours at 4- and 8-inch depths, respectively.

### Yield

At Overton, the total yield from solarized soil was greater than that from untreated soil in all years (Table 12-7). The yield was higher with fumigated soil than with solarized soil in 1988, but there was no difference between fumigated and solarized plots in 1989. Plants grown on 13-inch beds tended to have a greater total yield than did those grown on 8-inch

**TABLE 12-7**  Effect of Soil Treatment on Strawberry Yield and Weed Control

| Treatment | Total Strawberry Yield (lbs/acre) 1988 | 1989 | Surface Area Covered by Weeds (%) 5/30/88 Yellow Nutsedge | Annual Dicots* | Number of Annual Weeds/Plot 3/15/89 | Yellow Nutsedge Plants/Plot 6/24/89 |
|---|---|---|---|---|---|---|
| Fumigation | 17,329a[†] | 21,991a | 0.5a | 0.8a | 8a | 4a |
| Solarization | 15,526b | 20,628a | 10.5b | 2.6a | 11ab | 17b |
| Bare soil | 11,436c | 18,428b | 13.3b | 8.3b | 14b | 13b |

*Weeds included *Lamium amplexicaule* L., *Oenothera laciniata* Hill, and *Vicia dasycarpa*.
[†]Different letters indicate separation of means within columns by Duncan's test at the 0.05 percent level.

**TABLE 12-8**  Solarization and Manure Effect on Weed Populations at Overton in 1989

| Treatment | Number of Weed Seedlings/8 inches$^2$ | | | | |
| | Purple Nut Sedge | Dandelion | Total Dicots | Total Monocots | Total Number of Weeds |
| --- | --- | --- | --- | --- | --- |
| Solarization + manure | 3 | 1 | 3 | 4 | 7 |
| Solarization + no manure | 9 | 4 | 12 | 11 | 23 |
| No solarization + manure | 29 | 15 | 19 | 29 | 48 |
| No solarization + no manure | 42 | 29 | 39 | 42 | 81 |
| ANOVA* | | | | | |
| Solarization | 0.01$^†$ | 0.05 | 0.05 | 0.05 | 0.02 |
| Manure | 0.05 | 0.003 | 0.002 | 0.05 | 0.004 |
| Solarization by manure interaction | NS$^‡$ | 0.09 | NS | NS | NS |

*ANOVA, analysis of variance.
†Probablility of significance.
‡NS, not significant.

beds (data not shown).  Because of record freezing temperatures in February 1989, the entire crop at Lubbock was lost.  At Poplarville, soil treatments had no effect on the yields (data not shown).  This lack of an effect may have been the result of solarizing too late in the fall or the cold temperatures in February that killed flower buds and plants.

### Weed Control

Weed control in response to soil treatments varied by species.  There was no difference in the number of annual dicots per plot between fumigation and solarization plots in 1988, and both treatments had fewer annual weeds than the control plots did (Tables 12-7 and 12-8).  In 1989, fumigated soil had fewer annual weeds than did untreated soil, and plots treated by solarization had intermediate weed populations.  Fumigation was the only treatment that controlled yellow nutsedge (*Cyperus esculentus*) in both years.  No attempt was made to distinguish between *C. esculentus* seedlings and regrowth from tubers, but most plants resulted from established tubers, not seeds.  In other research plots, purple nutsedge (*Cyperus rotundus* L.) germinated from seeds and was controlled by solarization (Table 12-8).  Chicken manure also reduced the annual weed populations.  The combination of solarization and chicken manure resulted in the lowest weed density.  At Poplarville, weeds were not affected by solarization (data not shown).

## Nematode Control

Root-knot nematodes were not observed on plant roots in 1988. Highly variable nematode populations, with no significant differences across treatments, occurred in the spring of 1989. Fewer nematodes were observed in solarized soil than in the control plot in the fall of 1989 (Table 12-9). Compared with fumigated soil, solarized soil had similar levels of ring nematodes but more free-living nematodes. Soil incorporated with the sorghum cover crop had increased ring nematode levels compared with the untreated soil, while chicken manure reduced the levels of ring nematodes and increased the levels of free-living nematodes. Nematode populations (primarily *Pratyulenchus* spp.) at Lubbock in 1989 were reduced by 85 to 90 percent and 90 to 95 percent by solarization and chemical fumigation, respectively (data not shown). In general, solarization provided a level of temporary control over nematodes in the field that was less efficacious than that provided by fumigation.

## Soil Chemistry

Solarization increased soil pH and ammonium ($NH_4$) levels, while soil electrical conductivity and calcium, magnesium, and nitrate ($NO_3$) levels were reduced on solarized soil compared with those in untreated soil (Table 12-10). Manure applied before solarization markedly increased soil pH, electrical conductivity, potassium, calcium, magnesium, phosphorus, ammonium, zinc, manganese, and copper. The electrical conductivity and soil solution P, K, and calcium concentrations were also higher in leachate collected under soil treated with manure (data not shown).

**TABLE 12-9**   Solarization, Manure, and Cover Crop Effects on Ring and Free-Living Nematode Populations at Overton in 1989

|  | Number of Nematodes/500 ml soil | |
| --- | --- | --- |
| Treatment | Ring | Free-Living |
| Solarization | 38a[*] | 569b[*] |
| Fumigation | 0a | 331a |
| Control | 182b | 790c |
| Manure | 9a | 737c |
| Manure + sorghum | 174c | 651b |
| Control | 40b | 300a |

[*]Different letters indicate separation of the means within columns by Duncan's test at the 0.05 percent level.

**TABLE 12-10**    Effect of Solarization and Manure on Soil Fertility Parameters at Overton

| Treatment | pH | EC (dS/m) | Concentration (mg/kg) | | | | | |
|---|---|---|---|---|---|---|---|---|
| | | | P | $NO_3$ | $NH_4$ | K | Ca | Mg |
| Control | 4.8b* | 0.6a | 61 | 56a | 16b | 231 | 745a | 59a |
| Solarization | 5.3a | 0.3b | 49 | 42b | 46a | 182 | 559b | 45b |
| NPK[†] | 4.8b | 0.4b | 35b | 50ab | 34b | 153c | 546b | 34c |
| Manure (0 kg/ha)[‡] | 4.0b | 0.3c | 28b | 41b | 14c | 130c | 533b | 36c |
| Manure (550 kg/ha) | 5.2a | 0.4b | 68a | 51ab | 21c | 237b | 734ab | 61b |
| Manure (1,100 kg/ha) | 5.4a | 0.7a | 87a | 54a | 53a | 305a | 794a | 75a |

NOTE:  EC, electrical conductivity; P, phosphorous; $NO_3$, nitrate nitrogen; $NH_4$, ammonium nitrogen; K, potassium; Ca, calcium; Mg, magnesium.

*Different letters indicate separation of the means within columns by Duncan's test at the 0.05 percent level.

[†]NPK, 250 lbs/acre of nitrogen, 40 lbs/acre phosphorus, and 75 lbs/acre potassium.

[‡]Application rates were based on the total nitrogen content of the chicken manure.

## Discussion

Solarization resulted in strawberry yields comparable to those obtained with typical production systems located in areas with a hot summer climate. Solarization was not as effective as fumigation in certain areas, for example, in eradicating difficult to control perennial weeds (yellow nutsedge). The failure of solarization to eliminate perennials with an established deep root system, rhizomes, or tubers confirms results of previous studies (Pullman et al., 1984; Rubin and Benjamin, 1984). Poor control of perennial weeds was likely the result of a failure to achieve lethal heating below soil depths of 4 to 8 inches. *Cyperus* spp. can survive temperatures of greater than 140°F (Rubin and Benjamin, 1984). For most situations, however, control of annual weeds with solarization is sufficient to allow for production without herbicides or fumigation.

Obvious symptoms of plant diseases caused by soilborne pathogens were not indicated in these experiments. Pathogenic fungi or nematodes may not have limited yield. High populations of *C. esculentus* could account for the reduction in the yield with solarization compared with fumigation, but not for the increase in yield associated with solarization compared with the control. The reduction in annual weeds with solarization may explain some, but not all, of the increase in yield over that in untreated soil. In several other studies on solarization, increased plant growth response has also been reported in the absence of major soilborne pests (Katan, 1981; Stapleton and Devay, 1986). Modification of the soil pH or nutrient availability may also have been responsible for the increased plant growth response. The effect of solarization on nutrient availability in these experiments likely would have been attenuated by the fertilization program. Much remains to be discovered about the complex biological interactions of the rhizosphere, as discussed by R. James Cook in Chapter 3.

A concern that growers may have with the use of solarization is the selection of an appropriate type of plastic (ultraviolet stabilized). In the solarization system described here, the plastic should last for at least 1 year under field conditions. This allows for the plastic to be in the field for 2 months for the solarization period, sprayed white, planted with strawberries, and then replanted with melons the following summer. Several earlier attempts at solarization were not completely successful because the plastic's integrity did not last for longer than 4 to 5 weeks. In an initial evaluation of the effects of different types of polyethylene on soil heating, no major differences in film types were detected. One method that did enhance solar heating by several degrees was to decrease the soil albedo (darken) by spreading manure on the soil surface prior to solarization.

Typically, soils in the South have insufficient soil N because organic matter is low and $NO_3$ and $NH_4$ are readily leached out of the topsoil. The goal has been to attempt to build a supply of slowly released N and other nutrients in the soil, under the plastic, that could supply several sequential crops with adequate fertilization. Direct manure applications have resulted in the most vigorous plants at the lowest cost in comparison with a combination of winter and summer legume cover crops and manuring followed by a nonlegume cover crop (data not shown). The application of 20 tons of chicken manure per acre increased the soil availability of P and K levels by 100 and 260 lbs/acre in eastern Texas. Based on fertilizer rate-soil analysis correlation data, the application of 1,000 and 900 lbs of P and K per acre, respectively, from commercial fertilizer sources produced an equivalent increase in the levels of P and K in soil. The application of 20 tons of poultry manure ($240 delivered cost within 50 miles of the source) per acre would provide the equivalent of approximately $300 of commercial fertilizer if only half of the N were available (500 lbs/acre).

An economic evaluation of strawberry production with solarization compared favorably with fumigation and was advantageous over no treatment of the soil. Comparative enterprise budgets for conventional and solarization and manure systems are presented in Table 12-11. This analysis assumed very conservative production costs and fruit prices ($0.55/lb), and a 10 percent greater yield by fumigation over that by solarization. The solarization and manure system would cost $275 less per acre than fumigation and conventional fertilization, and $150 more than black plastic, conventional fertilization, and hand weeding. Assuming a 10 percent yield differential between the fumigation and solarization systems and assuming that the price received for fruit is of equal value, the conventional system

**TABLE 12-11**   Farm Enterprise Budget for Strawberry Production Using Different Cropping Systems

| System Analysis | Cost ($)/Acre of Strawberries |
|---|---|
| Nutrient inputs* | |
| Option 1: Chicken manure delivered and spread at 15 tons/acre† | 240 |
| Option 2: 600 pounds of 13-13-13 NPK fertilizer/year + 10 supplemental N applications at 20 lbs/acre through a drip line | 165 |
| Soilborne pest control‡ | |
| Option 1: Black plastic + hand weeding | 375 |
| Option 2: Solarization (cost of plastic laying, pigmenting, occasional hand weeding) | 450 |
| Option 3: Fumigation (custom application) | 800 |
| Yield comparison§ | |
| Average differential in total yield between conventional and low-input solarization systems (2 years of data) = 10 percent | |
| Average differential in total yield between solarization system and black plastic without fumigation (2 years of data) = 18 percent | |

| Sample Budget Analysis‖ | Cost ($) | | |
| | Conventional | Solarization | Black Plastic |
|---|---|---|---|
| Nutrient inputs | 165 | 240 | 165 |
| Soilborne pest control | 800 | 450 | 375 |
| Other production costs | 3,000 | 3,000 | 3,000 |
| Total cost | 3,965 | 3,690 | 3,540 |
| Marketable yields (lbs/acre)# | 12,000 | 10,800 | 8,856 |
| Net returns at $0.55/lb | 2,635 | 2,250 | 1,331 |
| Net returns at $0.60/lb | 3,235 | 2,790 | 1,774 |

*Costs are based on direct cost for supplies plus services cost as derived from the 1988 Texas Custom Rates Statistics Handbook.

†Cost of manure is based on rates in eastern Texas, which depend on distance from source ($8/ton delivered within 20 miles, $12/ton for 20 to 50 miles).

‡Assumes that cost for control of fruit decay is the same for all systems.

§Differentials are based on 2 years of data. Data comparing yield as a function of nutrient input sources is not yet available.

‖The estimates for these costs, yields, and returns are conservative.

#Assumes a conservative total yield for conventionally grown fruit of 16,000 lbs/acre with 25 percent loss because of cullage. Yield differential between systems are estimates from 2 years of data.

is 15 percent more profitable than the low-input system. With only a $0.05/ lb price differential between low-input and conventional fruit, however, the solarization system has a 5 percent greater return. The solarization system was more profitable than the system in which just black plastic and hand weeding were used. Solarization would also enable the grower to capitalize on the organic market niche. Additional advantages of solarization would be the avoidance of restricted-use pesticides and elimination of specialized fumigation equipment. Alternatively, solarization in combination with fumigation may substantially decrease the level of fumigant used (Stapleton and Devay, 1986).

## CONCLUSION

Results of this project have been presented at numerous growers meetings and field days. It is too early in the evaluation of these systems to estimate their potential significance and use by growers. Although these systems are effective replacements for conventional chemical input systems, the management intensity required for the successful use of the whole system may limit their use to only the most proficient growers. Individual components of these systems, however, could be adapted to other small fruit production and horticulture enterprises in the region or the nation.

## REFERENCES

Hartz, T. K., C. R. Bogle, and B. Villalon. 1985. Response of pepper and muskmelon to row solarization. Horticultural Science 20:699–701.

Katan, J. 1981. Solar heating of soil for control of soilborne pests. Annual Review of Phytopathology 19:211–236.

Patten, K. D., E. W. Nuenedorff, G. Nimr, S. C. Peters, and D. C. Cawthon. 1989. Growth and yield of rabbiteye blueberry as affected by orchard floor management practices and irrigation geometry. Journal of the American Society for Horticultural Science 114:728–732.

Pullman, G. S., J. E. De Vay, C. L. Elmore, and W. H Har. 1984. Soil Solarization—a Nonchemical Method for Controlling Disease and Pests. Leaflet 21377. Oakland: Cooperative Extension of the University of California.

Putnam, A. R. 1988. Allelochemicals from plants as herbicides. Weed Technology 2:510–518.

Rubin, B., and A. Benjamin. 1984. Solar heating of the soil: Involvement of environmental factors in the weed control process. Weed Science 32:138–142.

Stapleton, J., and J. E. Devay. 1986. Soil solarization: A non-chemical approach for management of plant pathogens and pests. Crop Protection 5:190–198.

Wagger, M. G. 1989. Time of desiccation effects on plant composition and subsequent nitrogen release from several winter annual cover crops. Agronomy Journal 81:236–241.

# 13

# Reactor's Comments

## Research and Education in the South

*Raymond E. Frisbie*

This reaction is to three sustainable agriculture projects in the southern region described in the following chapters: "Southeastern Apple Integrated Pest Management" by Dan L. Horton and colleagues, "Low-Input Crop and Livestock Systems in the Southeastern United States" by John M. Luna and colleagues, and "Solarization and Living Mulch to Optimize Low-Input Production System for Small Fruits" by Kim Patten and colleagues.

In general, the three projects described in those chapters have made very good progress in organizing and beginning their respective research and education projects. The projects dealt at the whole-farm level and involved interactions with related cropping or habitat systems. There was a definite sense of a farming systems approach. Although limited success was reported or expected because of the newness of each project, it was clear that the funding period for these types of research and education projects is far too short. Whole or mixed farm systems that consider multiple variables require several years of study to achieve reliable results. In this reactor's opinion, 2 to 3 years of funding is not adequate. Funding for 4 to 6 years is more realistic. For example, the STEEP program in Washington and surrounding states has been ongoing for several years with a singular focus on erosion control. The benefits of this program are only now being fully realized.

The Georgia low-input sustainable agriculture (LISA) apple production program described by Dan Horton and colleagues was well designed to consider horticultural, pest management, and erosion control factors. Market considerations regarding the timing of supplying fruit of high quality were evident in the program's design. The authors indicated they were accomplishing improved integrated pest management (IPM) of codling moth, red-banded leafroller, other insect pests, and plant pathogens through an integrated approach of pheromone trapping, phytosanitation, and habitat management. Intensive field sampling for insect pests and plant pathogens was used to time pesticide applications. Although a fairly comprehensive IPM system was discussed, the authors narrowly defined IPM as a "scout and spray program." This observation was pointed out.

The crop and livestock sustainable agriculture project at the Virginia Polytechnic Institute and State University described by John Luna and colleagues had, by far, the best system design in that it considered multiple production components and interactive linkages. It was clear that this interdisciplinary group had spent considerable time constructing a conceptual model on the operation of a fairly complex forage and livestock system. The group is to be complimented for using an innovative cover crop strategy that reduces erosion and provides supplemental nitrogen and for using a cultural management technique that controls armyworms and reduces insecticide use. Consideration of the impact of various forage production systems on calf weight gain and performance completes a very sophisticated system that should provide insight into the establishment of a sustainable system.

The project at Texas A&M University described by Kim Patten and colleagues that deals with strawberries and blueberries in eastern Texas is a good example of how to design a sustainable agriculture system in soils with extremely low levels of organic matter and that receive low levels of rainfall in summer. Although this research program has been under way for only a short time, it has shown that ground cover and other management strategies that may work in some areas of the country are not suitable everywhere. The group at Texas A&M University is to be congratulated for examining a series of innovative approaches, such as soil solarization for plant pathogen and nematode control. The fact that some of these techniques did not give positive results was not discouraging. It is as important to know what will not work in a sustainable agriculture system as to know what will work.

The research and extension faculty working on all three projects showed a very positive synergism in the development and conduct of their projects. They are to be congratulated.

# PART FOUR
# Research and Education in the North Central Region

# 14

# New Strategies for Reducing Insecticide Use in the Corn Belt

*Gerald R. Sutter and David R. Lance*

Corn rootworms (*Diabrotica* spp.) are the most serious pests of corn in North America; crop losses and control costs attributed to these pests are estimated to be near $1 billion annually (Metcalf, 1986). Within the Corn Belt, which encompasses 80 percent of the corn acreage in the United States, two species of rootworm, *D. virgifera virgifera* LeConte, the western corn rootworm, and *D. barberi* Smith and Lawrence, the northern corn rootworm, are the most important economic pests (Luckmann, 1978). Adults of these species lay their eggs in corn fields in late summer. The eggs hatch the following spring, and the larvae, the primary damaging stage, feed and develop almost exclusively on roots of corn (Branson and Ortman, 1971). When damage to roots is extensive, plant-water relationships are disrupted (Riedell, 1990) and the stability of the plant is reduced. If extensive root pruning coincides with heavy rains and strong winds, plants lodge (tip over), which hampers mechanical harvesting.

Although corn rootworms have been pests of corn for over a century (Forbes, 1886), several factors in crop production systems have elevated rootworms to the pest status they occupy today. Most *Diabrotica* spp. can be managed with proper crop rotation schemes; however, because of production needs, government farm programs, and other economic considerations, crop rotation has not always been a viable option for growers. In fields where corn is grown year after year, extensive use of soil insecticides has resulted because of such factors as (1) the introduction several decades ago of low-cost, presumptively effective soil insecticides, (2) difficulty in predicting damaging pest populations, (3) a prevalent philosophy among soil insect researchers that soil insecticides, like fertilizers, were essential inputs into corn production systems, and (4) promotion by agri-chemical industries that their products were the best crop insurance that corn farmers could buy (Turpin, 1977). The dependency on soil insecti-cides began in the 1950s, when growers began planting corn continuously throughout much of the Midwest and typically applied cyclodiene insecti-cides to control corn rootworms and other soil pests. Within a decade, resistance to cyclodienes became prevalent among corn rootworms (Ball and Weekmann, 1962). As a result, growers readily switched to organo-phosphate and carbamate insecticides as they became available for corn rootworm control. There is ample evidence that a high percentage of soil insecticide applications over the past four decades were applied unnecessar-ily and with limited or no knowledge of the potential of the pest population to produce economic damage (Stamm et al., 1985; Turpin and Maxwell, 1976). Peak usage of soil insecticides in corn production occurred in the late 1970s and early 1980s when 20 million to 30 million acres of corn were treated annually with 1 to 1.3 pounds of actual insecticide per acre (Sugui-yama and Carlson, 1985). In 1988, 35 percent of the corn acreage grown for grain was treated with soil insecticides, down from 41 percent in the 2 previous years (Delvo, 1989). Nevertheless, Delvo (1989) projected that of all insecticides used on row crops and small grains in the United States in 1989, 48 percent would be applied to corn, primarily for control of corn rootworm larvae. The insecticides cost up to $15 per acre, which did not include costs for labor or application equipment.

Most state extension personnel in the Corn Belt recommend crop rotation for optimal corn rootworm control. However, if growers intend to plant corn in the same field each year, they are encouraged to scout fields for beetles during August. If at any time growers find one or more beetles per plant, they are encouraged to either plant a nonhost crop or use a soil insecticide at planting time the following year.

If the trend toward rotating crops in the Corn Belt continues, it would appear that at least part of the corn rootworm problem will be solved. The literature shows, however, that corn rootworms were a problem long before classical insecticides became available and corn was grown in mono-cul-ture. Specifically, alternate-year rotation of corn with a nonhost crop some-times failed to control the northern corn rootworm (Bigger, 1932; Forbes, 1886). Researchers suspected that beetles migrated to nonhost fields, fed on vegetation, and oviposited, causing infestations the following year. How-ever, recent studies have shown that female northern corn rootworm beetles leave host fields to forage but typically return to corn fields to oviposit (Cinereski and Chiang, 1968; Gustin, 1984; Lance et al., 1989). A more feasible explanation for these infestations was advanced by Krysan et al. (1986). They found that 40 percent of northern corn rootworm eggs were capable of overwintering in the soil for two winters and that the trait was higher in populations from crop production areas that practiced crop rotation. What appears to be a genetic trait in this species places crop rotation as a pest management strategy in jeopardy, particularly when corn and a nonhost crop are planted in alternate years in a rigid pattern.

## EFFECT OF SOIL INSECTICIDES ON CORN ROOTWORM POPULATION DYNAMICS

Despite the popularity of soil insecticides and their extensive use over the past two to three decades, limited information is available on the effects of insecticides on corn rootworm population dynamics and corn production. Research on corn rootworms has been focused toward a crop protection mode rather than an offensive mode of pest population management. As a prime example, the most popular method to evaluate the efficacy of soil insecticides is to remove roots from plots just after larval feeding is completed, visually determine the amount of larval feeding damage, and assign numerical values between 1 and 6 that correspond to levels of root feeding and pruning by corn rootworm larvae (Hills and Peters, 1971). Insecticide efficacy is determined by comparing the numerical values of damaged roots from treated and untreated plots. These values may have little bearing on how an insecticide affects the pest population or protects yield loss (Sutter et al., 1990, in press). In a 4-year study, Sutter et al. (1989) infested field plots with known numbers of corn rootworm eggs per plant and found that the larval feeding damage inflicted at each pest density was consistent each year, but root protection by soil insecticides was highly variable from year to year and among the insecticides tested. Much of the variability was caused by edaphic and environmental conditions. Sutter et al. (1990) recorded consistent percentages of yield loss attributed to damage by corn rootworm larval feeding in untreated plots. Insecticides did not differ in their ability to protect the yield. More importantly, measurable yield protection by insecticides occurred only in plots infested with the higher egg densities; yields in plots infested with low to moderate levels of corn rootworm populations did not differ between treated and untreated plots (Figure 14-1). Correlations between root damage ratings and yields of untreated plants were highly significant, whereas root damage ratings were not significantly correlated to yield in treated plots. This suggests that root damage ratings should not be the only criteria for evaluating insecticide efficacy. These experiments did indicate that all insecticides applied at planting time did reduce root lodging at the high pest densities. Each year, in untreated plots, the amount of root lodging was extensive at the higher pest densities and, thus, would have interfered with mechanical harvesting.

Researchers rarely measure the effects of soil insecticides on survival of corn rootworms to the adult stage. As part of the previously mentioned

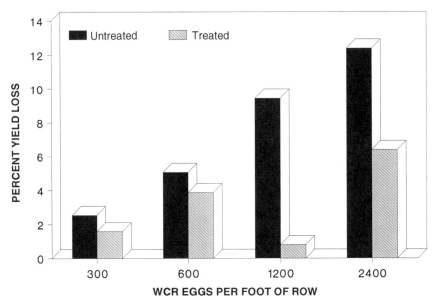

FIGURE 14-1    Ability of soil-applied insecticides to reduce yield loss in corn because of feeding by western corn rootworm (WCR) larvae at several population densities.  Source:  Data from G. R. Sutter, J. R. Fisher, N. C. Elliott, and T. F. Branson.  1990.  Effect of insecticide treatment on root lodging and yields of maize in controlled infestations of western corn rootworms (Coleoptera: Chrysomelidae). Journal of Economic Entomology 83:2414–2420.

study (Sutter et al., in press), it was found that reduction in beetle emergence from planting time applications of soil insecticides varied from 16.5 to 81.1 percent (Figure 14-2).  Rainfall, as it influences soil moisture, appears to affect pest survival and insecticide efficacy.  In 1981, the amount and distribution of rainfall were near normal.  Above-normal rainfall was recorded during the larval feeding period (June 10 to July 10) in 1982, and insecticides reduced survival rates in treated plots.  Rainfall was below normal in 1985, and insecticides had minimal effects on pest survival rates. Insecticides differed significantly in reducing beetle survival.  Water solubility of insecticides, which could affect the movement of the toxin into the soil profile, appeared to influence their effectiveness in reducing beetle survival more than did their inherent toxicity to the larval stage.

## FACTORS INFLUENCING INSECTICIDE USE IN CORN PRODUCTION SYSTEMS

Possibly, the greatest factor that promotes the extensive use of soil insecticides in corn production systems is the lack of reliable pest monitoring technology, more specifically, technology to predict accurately corn root-worm infestations that inflict measurable and significant crop losses that can be translated into an economic injury level (EIL) (Stern et al., 1959). Poston et al. (1983) suggested that EILs are the weakest links in most management programs because these values attempt to oversimplify very complex agroecosystems that may include several pests, variable environmental and agronomic conditions, and different host responses. This scenario typifies corn rootworm management in the Corn Belt. EILs for corn rootworms are based primarily on the amount of feeding damage larvae inflict on the root system; damage levels are then associated with yield differences between treated and untreated plots. The major flaw in this association is that all of the yield differences were assumed to be attributable to stress inflicted by feeding of corn rootworm larvae (Sutter et al., 1990). Research on corn rootworm thresholds has lagged behind similar research on other crops, in part because researchers have concentrated their efforts on insecticide-related questions rather than focusing on basic biological insect-related factors (Turpin, 1974).

Methods for sampling all life stages of *Diabrotica* spp. were recently reviewed in detail (Krysan and Miller, 1986). Most sampling methods are

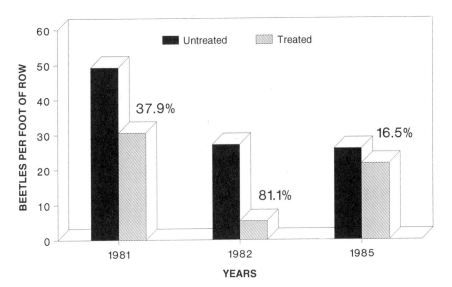

FIGURE 14-2 Emergence of beetles in field plots during three seasons. Numbers over the bars indicate the percent reduction in numbers of beetles in plots treated with soil insecticides. Source: G. R. Sutter, T. F. Branson, J. R. Fisher, and N. C. Elliott. In press. Effect of insecticides on survival, development, fecundity, and sex ratio in controlled infestations of western corn rootworms (Coleoptera: Chrysomelidae). Journal of Economic Entomology.

cumbersome and costly and have met with little success in corn rootworm pest management programs. Of the methods described, scouting of fields during August and visual counting of beetles on the plants has become the most accepted population-monitoring tool (Tollefson, 1986). However, Foster et al. (1986) concluded from an extensive study in Iowa that the value of sampling corn rootworm adults for predicting economic damage by corn rootworm larvae in the next growing season was low; the optimal strategy for managing corn rootworms in that state, they concluded, was not to sample for adults and always treat corn following corn with a soil insecticide at planting time.

## NEED FOR NEW CONTROL TECHNOLOGY FOR CORN ROOTWORMS IN THE CORN BELT

Reliance on the prophylactic application of insecticides for corn rootworm control has numerous problems that fall into the following broad interrelated categories.

• Use of soil insecticides in corn production systems can add up to $15 per acre in production costs, which may exceed the cost for energy used in corn production. During most years, a relatively low proportion of the fields in the Corn Belt harbor corn rootworm population densities that warrant treatment.

• Insecticides used routinely for corn rootworm control are among the most toxic pesticides on the market and carry a high risk of acute toxicity to growers and livestock (Metcalf, 1980). They have also been detected in groundwater and surface water (Williams et al., 1988) and have been implicated in numerous poisonings of wildlife and other nontarget organisms (National Research Council, 1989). In particular, birds are at extreme risk when they forage in fields that have been treated with carbofuran (Environmental Protection Agency, 1985). There is growing concern by farmers of the health risks involved with handling these compounds (McDonald, 1987). Furthermore, most soil insecticides are applied at a time when soils are vulnerable to erosion, particularly in conventional tillage systems, as well as during a season in which rainfall is typically prevalent.

• Because insecticides are placed in the soil at planting time, their persistence can be influenced by numerous edaphic factors such as soil moisture, degradation by microbial organisms, and differences in physical and chemical properties of soils. These factors cannot be regulated by the grower.

• Application of soil insecticides at planting time results in the highly inefficient use of resources. At planting time, the actual insecticide concentration in the upper soil profile (1 inch) is between 30 and 35 ppm, which is

100 to 200 times the average 50 percent lethal concentration ($LC_{50}$) for corn rootworm larvae (Sutter, 1982). Degradation of the parent compound begins almost immediately. By the time corn rootworm larvae are actively feeding on the plant's root system (6 to 8 weeks later), the amount of insecticide residue remaining can be reduced by over 100-fold and, depending on local conditions, may be well below the concentration needed to kill the larvae (Sutter et al., 1989).

## NEW MANAGEMENT STRATEGIES FOR CORN ROOTWORMS

New approaches to managing corn rootworms must be ecologically compatible with other corn pest management programs. If future corn production practices change from the conventional systems used for the past three to four decades to systems that require less input, other pest problems (weeds and insects) likely will emerge. At the same time, emphasis will shift toward the use of nonchemical pest management approaches. For growers to deploy biological control methods for pests other than corn rootworms successfully, they will no longer be able to apply chemical insecticides prophylactically for either larvae or adult corn rootworms without interrupting the delicate ecological balance needed to allow other management programs to function.

To reduce the level of chemical dependency that prevails at present in the Corn Belt, development of viable alternative management programs will be required for corn rootworms. It is unlikely that viable strategies can be developed for immature stages of corn rootworms since their habitat is in the soil and they are very inaccessible. Systems to accurately monitor and effectively control these stages have proven to be difficult. Populations of the adult stage, however, can be readily monitored (Tollefson, 1986). The concept of managing corn rootworm populations with adulticides has previously proved effective. Pruess et al. (1974) found that an ultra-low-volume application of malathion (9.7 ounces of active ingredient [AI] per acre) adequately suppressed beetle populations within a 16-square-mile management area and eliminated the need for planting time application of soil insecticides in the following growing season. They found, however, that pest populations rebounded after 1 year because of immigration of gravid females from surrounding areas. These data not only support the concept that corn rootworm can be managed through adult suppression but also indicate that management programs may be most successful if they are applied on an area-wide rather than an individual-field basis. Mayo (1976) applied carbaryl as an adulticide (1 pound of AI per acre) and suppressed beetle populations, on average, by 94.3 percent. Larval feeding damage to plants the following year did not differ from that to plants treated with a soil insecticide. Mayo did observe that carbaryl that was applied to plots often killed insect predators such as lady beetles and lacewings and created an environment for outbreaks of spider mites and other potential pests.

Recent advances in knowledge of the chemical ecology of *Diabrotica* beetles have opened new avenues for the development and deployment of effective management strategies for these pests (Lampman and Metcalf, 1988; Lance and Sutter, 1990). Specifically, attempts have been made to develop semiochemical-based technology for monitoring and, when necessary, suppressing populations of corn rootworm beetles. The latter management tactic involves enhancment of the efficiency of toxic baits by attracting beetles to particles of the bait and inducing them to feed.

Semiochemicals affecting *Diabrotica* beetles have been identified from two sources: the beetles themselves and members of the family Cucurbitaceae, which are ancestral host plants of diabroticites (Metcalf et al., 1980). Female western corn rootworm beetles produce a sex attractant pheromone (8$R$-methyl-2$R$-decylpropanoate) that lures males of both the northern and western species (Guss et al., 1984, 1985). The pheromone's usefulness for management programs is limited because it attracts only males and can elicit unusual responses from northern corn rootworms (Lance, 1988a,b). In contrast, compounds that attract beetles of both sexes (but that are more effective for females) have been discovered among squash blossom volatile and related compounds (e.g., see Table 14-1). Rootworm beetles also respond to cucurbitacins, which are tetracyclic triterpenoids that are found in most cucurbits. Cucurbitacins are not sufficiently volatile to act as attractants, but *Diabrotica* beetles stop and feed compulsively when they touch substrates that contain cucurbitacins. These compounds are very bitter and somewhat toxic to animals that are not adapted to feeding on them (Metcalf et al., 1980).

In the summer of 1989, studies were initiated to evaluate and develop the use of semiochemical attractants as tools to aid in monitoring populations of adult corn rootworms. Blocks of traps baited with various amounts of attractants for western ( $p$-methoxycinnamaldehyde) or northern (cinnamyl alcohol) corn rootworms were monitored throughout the season. The resulting data (not yet completely analyzed) will yield information on the relative precision of baited and unbaited traps, optimal levels of attractants for monitoring beetles, seasonal variations in the effectiveness of traps, and seasonal relationships between the number of beetles caught in traps (trap catch) and the deposition of eggs in fields. More comprehensive studies to relate trap catch to beetle population density will be conducted in 1990. Traps that kill beetles with a toxic bait (essentially modifications of the "vial" traps described by Shaw et al. [1984]) are currently being used. Compared with sticky traps, vial-type traps are less messy to handle and capture very few nontarget insects, which makes evaluation easier. Also,

**TABLE 14-1** Relative Attractancy of Selected Nonpheromonal Attractants for Western and Northern Corn Rootworm Beetles

| Compound | Attractancy of Rootworm Beetle Species[*] | | |
|---|---|---|---|
| | Western | Northern | Reference[†] |
| 1. 1,2,4-Trimethoxybenzene | 0 (?) | 0 (?) | 1 |
| 2. Indole | ++ | 0 | 2 |
| 3. *trans*-Cinnamaldehyde | + | 0 | 3,4 |
| TIC mixture[‡] | +++ | + | 1 |
| 4. Estragole | ++ | 0 | 3,5 |
| 5. *p*-Methoxycinnamaldehyde | +++ | 0 | 3,4 |
| 6. Eugenol | 0 | ++ | 1,6 |
| 7. Cinnamyl alcohol | 0 | ++ | 4,7 |

[*]0, not an attractant; +, ++, and +++, slight, moderate, and powerful attractancies, respectively. With sticky traps, powerful attractants often produce 100-fold increases in the numbers of rootworm beatles captured relative to those in unbaited traps.

[†]References: 1, R. L. Lampman and R. L. Metcalf. 1987. Multi-component kairomonal lures for southern and western corn rootworms (Coleoptera: Chrysomelidae: *Diabrotica* spp.). Journal of Economic Entomology 80:1137–1142. 2, J. F. Andersen and R. L. Metcalf. 1986. Identification of a volatile attractant for *Diabrotica* and *Acalymma* spp. from blossoms of *Cucurbita maxima* Duchesne. Journal of Chemical Ecology 12:687–699. 3, R. L. Metcalf and R. L. Lampman. 1989a. Estragole analogues as attractants for corn rootworms (Coleoptera: Chrysomelidae). Journal of Economic Entomology 82:123–129. 4, R. L. Metcalf and R. L. Lampman. 1989b. Cinnamyl alcohol and analogs as attractants for corn rootworms (Coleoptera: Chrysomelidae). Journal of Economic Entomology 82:1620–1625. 5, R. L. Lampman, R. L. Metcalf, and J. F. Andersen. 1987. Semiochemical attractants of *Diabrotica undecimpunctata howardi* Barber; southern corn rootworm, and *Diabrotica virgifera virgifera* LeConte, the western corn rootworm (Coleoptera: Chrysomelidae). Journal of Chemical Ecology 13:959–975. 6, T. L. Ladd, B. R. Stinner, and H. R. Kreuger. 1983. Find new attractant for corn rootworm. Ohio Report 68:67–69. 7, R. L. Lampman and R. L. Metcalf. 1988. The comparative response of *Diabrotica* species (Coleoptera: Chrysomelidae) to volatile attractants. Environmental Entomology 17:644–648.

[‡]TIC mixture = equal portions of 1,2,4-trimethoxybenzene, indole, and *trans*-cinnamaldehyde.

attractants can cause sticky traps to become loaded with insects in 24 hours or less, whereas vial-type traps can be designed with a sufficient capacity to be left in place for extended periods of time.

Optimal use of semiochemical-based technology for managing corn rootworm populations may require a shift in the size of the management unit.

Corn rootworm control decisions are currently made on a field-by-field basis (e.g., Foster et al., 1986; Stamm et al., 1985). To be effective, monitoring systems require accurate forecasts of future economic losses in individual fields. Unfortunately, even the most precise trapping data, as with data from visual counts, are probably not suitable for this purpose (Foster et al., 1986). Unpredictable variables such as weather and shifts in soil biota strongly affect crop losses by influencing, among other factors, survival of eggs over the winter, establishment of neonates in the spring, effects of feeding damage on yield, and when applicable, efficacy of soil insecticides (Gustin, 1981; Sutter et al., 1989). An alternative strategy is to use larger management units. Monitoring systems would be used to identify crop areas that produce corn rootworm population densities that, if not reduced, potentially could infest much broader areas. Control measures could then be directed at these "source" areas early in the season before beetles emigrate to surrounding fields. Although thresholds for suppression would remain somewhat arbitrary, an area-wide management system such as this could theoretically result in substantial reductions both in rootworm populations and in the amount of acreage treated for rootworms. Such a system, however, would require an effective and environmentally sound means of suppressing adult rootworms.

Semiochemical-based baits for suppressing rootworm beetles are currently being developed. The system that is envisioned will contain four categories of components. The first component is an agent to kill the beetles. Most studies to date have used carbamate insecticides such as methomyl or carbaryl, although other insecticides or other types of agents (e.g., growth regulators and pathogens) may eventually be used. With carbamates, 0.1 percent toxin in the bait produced substantial mortality of beetles (Lance and Sutter, 1990; Metcalf et al., 1987), but somewhat greater amounts (0.3 to 0.5 percent) are probably more practical if baits are to remain active for extended periods in the field (Lance and Sutter, 1990).

The second bait component is a feeding stimulant. Metcalf et al. (1987) clearly demonstrated that rootworm beetles readily feed upon and ingest a lethal dose of edible substrates that contain less than 0.1 percent cucurbitacins. In more recent studies, unpurified powdered root of the buffalo gourd, *Cucurbita foetidissima* H.B.K., has been added to bait at 5 percent of total dry weight. This powder contains about 0.5 percent cucurbitacins E, I, and E-glycoside (Metcalf et al., 1982), so again, the total bait is less than 0.1 percent cucurbitacins. Cucurbitacins are not only effective feeding stimulants, but because they are distasteful to nonadapted animals, they should tend to deter nontarget organisms from feeding on the bait.

Bait components in the third category are volatile attractants which, theoretically, will lure beetles to individual point sources of bait or from untreated portions of cornfields into treated portions (see Table 14-1).

Finally, the first three types of components are formulated into a carrier that can be consumed by beetles. Various carriers have been tested, including granules of bitter cucurbit fruit, corn grits (Metcalf et al., 1987), cereal-based flakes (D. R. Lance and G. R. Sutter, unpublished data), cereal-based pellets, and granules of cross-linked corn starch (Lance and Sutter, 1990). The starch granules have been studied most extensively; they are both palatable to beetles and effective at slowly releasing volatile attractants (Meinke et al., 1989).

Field tests of semiochemical-based toxic baits have been hampered somewhat by the mobility of the beetles. Movement of beetles between fields can mask the effects of baits in small experimental plots, and testing of baits on a large scale has not been feasible because of economic and legal considerations. To circumvent these problems, many aspects of bait performance were evaluated in walk-in cages (10 by 10 feet) containing corn plants that were treated with various bait formulations. Western corn rootworm beetles were released into the cages, and survivors were counted by using removal sampling after 24 to 72 hours, depending on the test.

In field cage tests in 1988 (Lance and Sutter, 1990), starch granule baits that were applied at a rate of 8 pounds per acre (equivalent to 0.6 ounces of carbaryl per acre) produced approximately 85 percent mortality of beetles in 24 hours (Figure 14-3). The cucurbitacins (feeding stimulants) appeared to be the key semiochemical component that enhanced the efficacy of baits in field cages. Baits that consisted of only the starch carrier plus carbaryl did not measurably affect survival, and within the limits of the field cage, addition of equal portions of 1,2,4-trimethoxybenzene, indole, and *trans*-cinnamaldehyde (TIC) as attractants (see Table 14-1) to baits did not improve the efficacy over that of granules containing only carbaryl and cucurbitacins. In 1989 field cage studies, the attractant, again, did not influence the efficacy of bait, even though a range of concentrations of attractants was tested and the bait particles were spaced farther apart than they were in the 1988 tests (unpublished data).

One of the initial concerns about bait performance was whether the particles could compete with naturally occurring sources of food. Specifically, an abundance of highly preferred food, such as green silk and fresh corn pollen, is available to beetles when corn is flowering; for many management purposes, bait should be applied during or soon after this period. The phenological effects on the efficacies of baits were tested by running field cage trials simultaneously in flowering corn and corn in the "dough" or "dent" stage (i.e., more mature) corn. In several such tests (e.g., Figure 14-3), the baits demonstrated their ability to compete favorably with highly preferred natural foods.

In 1989, several small (2- to 7-acre), partially isolated plots of field corn in eastern South Dakota were treated. The bait that was used was supplied

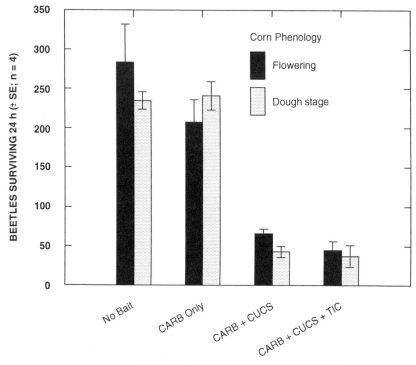

**CONTENTS OF THE BAIT FORMULATION**

FIGURE 14-3 Effects of bait components and corn phenology on efficacy of starch granule baits against western corn rootworm beetles in field cages. CARB = 0.5 percent carbaryl; CUCS = about 0.03 percent cucurbitacins; TIC = 1.5 percent TIC attractant (equal portions of 1,2,4-trimethoxybenzene, indole, and *trans*-cinnamaldehyde). Source: Data from D. R. Lance and G. R. Sutter. 1990. Field-cage and laboratory evaluations of semiochemical-based baits for managing western corn rootworm beetles (Coleoptera: Chrysomelidae). Journal of Economic Entomology 83:1085–1090.

by a private cooperator and contained 0.3 percent carbaryl, 5 percent powder of buffalo gourd root, and 0.5 percent of the TIC attractant in a flaky, cereal-based carrier. The bait was applied at about 8 pounds per acre; carbaryl, then, was applied at about 0.4 ounces per acre, which was a 98 percent reduction compared with conventional control procedures. In most cases, the bait produced substantial reductions in the populations of beetles in the plots (Figure 14-4). Unfortunately, the effectiveness of the bait was short-lived, and immigration into treated areas typically caused beetle populations to rebound within a week after treatment. This result was somewhat unexpected because bait that was aged in the field for a week showed a

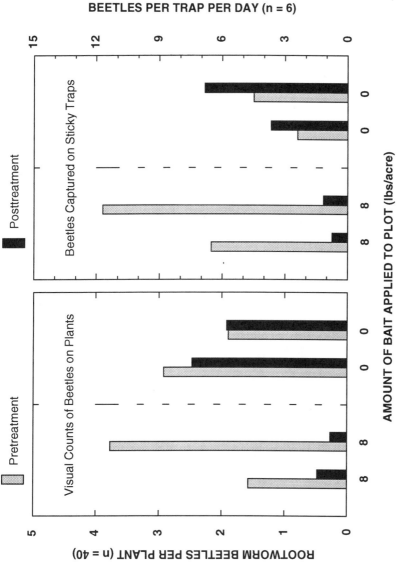

FIGURE 14-4 Visual counts of northern and western corn rootworm beetles on plants and capture of beetles on yellow, unbaited sticky traps in four plots of corn in Kingsbury County, South Dakota, during sampling periods before and after two of the plots were treated with semiochemical-based toxic bait. Source: G. R. Sutter and D. R. Lance, unpublished data.

high level of activity when it was assayed in the laboratory (D. R. Lance and G. R. Sutter, unpublished data). By 48 hours after treatment, however, almost all of the small (14-mesh or less), flaky particles of bait appeared to have been blown or washed from the corn plants, which is where the beetles spend most of their time. In a field cage study (D. R. Lance and G. R. Sutter, unpublished data), the bran-based bait produced over 90 percent mortality of beetles in 48 hours when it was sprinkled carefully on corn plants but did not produce any measurable mortality when it was broadcast on the ground. Thus, the position of the bait in the field appears to be very important. Currently, work in conjunction with several U.S. Department of Agriculture and private cooperators is being performed to develop a formulation that sticks to corn plants. Still, a different dry formulation (the starch granules) did not show such a rapid loss of efficacy in the field compared with that in laboratory assays (Figure 14-5). The starch granules

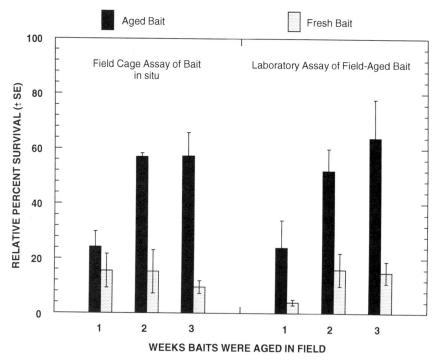

FIGURE 14-5 Effects of aging starch granule, semiochemical-based baits for 1, 2, or 3 weeks in the field on mortality of western corn rootworm beetles relative to mortality of beetles that were not exposed to bait. Source: Data from D. R. Lance and G. R. Sutter. 1990. Field-cage and laboratory evaluations of semiochemical-based baits for managing western corn rootworm beetles (Coleoptera: Chrysomelidae). Journal of Economic Entomology 83:1085–1090.

**TABLE 14-2** Mammalian Toxicity of Insecticides Used in Corn Rootworm Management

| Insecticide | $LD_{50}$ (mg/kg) (Rats)[*] | Toxicity Rating (EPA Signal Word and Category) | Comparative Toxic Exposure (Mammals) |
|---|---|---|---|
| Phorate | 2.6 | Danger (I) | 12,700 |
| Terbufos | 5.3 | Danger (I) | 6,240 |
| Carbuforan | 10.0 | Danger (I) | 3,300 |
| Fonofos | 12.7 | Danger (I) | 2,600 |
| Chlorpyrifos | 149 | Warning (II) | 222 |
| Carbaryl | 675 | Caution (III) | 1[†] |

[*]Where applicable, 50 percent lethal doses ($LD_{50}$) are means of values reported for male and female rats.

[†]In semiochemical-based toxic bait, based on a presumed application of 2 percent of currently recommended application rate of carbaryl for suppression of rootworm beetles.

SOURCE: Data from R. L. Caswell, K. J. DeBold, and L. S. Gilbert, eds. 1981. Pesticide Handbook (Entoma), 29th ed. College Park, Md.: Entomological Society of America.

were more dense than the cereal-based bait and tended to collect and remain in the leaf axils, silk, or other relatively protected areas on the plants (unpublished observations).

Another advantage of semiochemical baits that contain low concentrations of a toxin such as carbaryl is the relatively low mammalian toxicity of the insecticide compared with that of chemicals currently being used for managing diabroticites (Table 14-2). Furthermore, because the application rate of toxin is reduced in the baits by 98 percent, for example, the comparative toxic exposure to nontarget vertebrates would be greatly reduced.

There is currently a request, through a private cooperator, for an Experimental Use Permit from the U.S. Environmental Protection Agency (EPA). When the permit is granted, regional pilot-scale tests of bait formulations in production fields (80 to 160 acres in size) over several growing seasons will begin. The project will involve five states: Nebraska, Iowa, Illinois, Indiana, and South Dakota. Protocols for evaluating this management concept have been established to evaluate the system across these geographical areas uniformly. Factors to be investigated include overall efficacy of baits (reduction of beetle densities and oviposition and determination of how reductions affect plant protection the following growing season), optimal application patterns, influence of attractants on efficacy, effects of bait on nontarget organisms, and comparisons of bait technology with conventional methods of managing corn rootworm populations. A regional approach to field testing of this concept is essential. The baits must be tested under a variety of agronomic and environmental conditions; moreover, their effectiveness must be demonstrated locally before transfer of this technology to growers will be successful.

In view of the growing concern over the use of agrichemicals in the Corn Belt, this project can make a major contribution to the improvement of pest management programs. The concept, if proven successful, could be extended to other diabroticites that affect a variety of crops, primarily vegetables. The ultimate result of this innovation could be a very significant reduction in the amount of insecticide applied and the resultant environmental damage and human health risks.

## REFERENCES

Andersen, J. F., and R. L. Metcalf. 1986. Identification of a volatile attractant for *Diabrotica* and *Acalymma* spp. from blossoms of *Cucurbita maxima* Duchesne. Journal of Chemical Ecology 12:687–699.

Ball, H. J., and G. T. Weekmann. 1962. Insecticide resistance in the adult western corn rootworm in Nebraska. Journal of Economic Entomology 55:439–441.

Bigger, J. H. 1932. Short rotation fails to prevent attack of *Diabrotica longicornis* Say. Journal of Economic Entomology 25:196–199.

Branson, T. F., and E. E. Ortman. 1971. Host range of larvae of the northern corn rootworm: Further studies. Journal of the Kansas Entomological Society 44:50–52.

Caswell, R. L., K. J. DeBold, and L. S. Gilbert, eds. 1981. Pesticide Handbook (Entoma), 29th ed. College Park, Md.: Entomological Society of America.

Cinereski, J. E., and H. C. Chiang. 1968. The pattern of movements of adults of the northern corn rootworm inside and outside of corn fields. Journal of Economic Entomology 61:1531–1536.

Delvo, H. W., situation coordinator. 1989. Agricultural Resources Situation and Outlook Report. Washington, D.C.: Economic Research Service, U.S. Department of Agriculture.

Forbes, S. A. 1886. The Entomological Record for 1885. Pp. 1–25 in Miscellaneous Essays on Economic Entomology by the State Entomologist and His Entomological Assistants. Springfield, Ill.: H. W. Rokker.

Foster, R. E., J. J. Tollefson, J. P. Nyrop, and G. L. Hein. 1986. Value of adult corn rootworm (Coleoptera: Chrysomelidae) population estimates in pest management decision making. Journal of Economic Entomology 79:303–310.

Guss, P. L., P. E. Sonnet, R. L. Carney, T. F. Branson, and J. H. Tumlinson. 1984. Response of *Diabrotica virgifera virgifera, Diabrotica virgifera zeae,* and *Diabrotica porracea* to stereoisomers of 8-methyl-2-decyl propanoate. Journal of Chemical Ecology 10:1123–1131.

Guss, P. L., P. E. Sonnet, R. L. Carney, J. H. Tumlinson, and P. J. Wilkin. 1985. Response of northern corn rootworm, *Diabrotica barberi* Smith and Lawrence, to stereoisomers of 8-methyl-2-decyl propanoate. Journal of Chemical Ecology 11:21–26.

Gustin, R. D. 1981. Soil temperature environment of overwintering western corn rootworm eggs. Environmental Entomology 10:483–487.

Gustin, R. D. 1984. Effect of crop cover on oviposition of the Northern corn rootworm, *Diabrotica longicornis barberi* Smith and Lawrence. Journal of the Kansas Entomological Society 57:515–516.

Hills, T. M., and D. C. Peters. 1971. A method of evaluating postplanting insecticide treatments for control of Western corn rootworm larvae. Journal of Economic Entomology 64:764–765.

Krysan, J. L., and T. A. Miller, eds. 1986. Methods for the Study of Pest *Diabrotica*. New York: Springer-Verlag.

Krysan, J. L., D. E. Foster, T. F. Branson, K. R. Ostlie, and W. S. Cranshaw. 1986. Two years before the hatch: Rootworms adapt to crop rotation. Bulletin of the Entomological Society of America 32:250–253.

Ladd, T. L., B. R. Stinner, and H. R. Krueger. 1983. Find new attractant for corn rootworm. Ohio Report 68:67–69.

Lampman, R. L., and R. L. Metcalf. 1987. Multi-component kairomonal lures for southern and western corn rootworms (Coleoptera: Chrysomelidae: *Diabrotica* spp.). Journal of Economic Entomology 80:1137–1142.

Lampman, R. L., and R. L. Metcalf. 1988. The comparative response of *Diabrotica* species (Coleoptera: Chrysomelidae) to volatile attractants. Environmental Entomology 17:644–648.

Lampman, R. L., R. L. Metcalf, and J. F. Andersen. 1987. Semiochemical attractants of *Diabrotica undecimpunctata howardi* Barber, southern corn rootworm, and *Diabrotica virgifera virgifera* LeConte, the western corn rootworm (Coleoptera: Chrysomelidae). Journal of Chemical Ecology 13:959–975.

Lance, D. R. 1988a. Potential of 8-methyl-2-decyl propanoate and plant-derived volatiles for attracting corn rootworm beetles (Coleoptera: Chrysomelidae) to toxic bait. Journal of Economic Entomology 81:1359–1362.

Lance, D. R. 1988b. Responses of northern and western corn rootworms to semiochemical attractants in corn fields. Journal of Chemical Ecology 14:1177–1185.

Lance, D. R., and G. R. Sutter. 1990. Field-cage and laboratory evaluations of semiochemical-based baits for managing western corn rootworm beetles (Coleoptera: Chrysomelidae). Journal of Economic Entomology 83:1085–1090.

Lance, D. R., N. C. Elliott, and G. L. Hein. 1989. Flight activity of *Diabrotica* spp. at the borders of cornfields and its relation to ovarian stage in *D. barberi*. Entomologia Experimentalis et Applicata 50:61–67.

Luckmann, W. H. 1978. Insect control in corn—practices and prospects. Pp. 138–155 in Pest Control Strategies, E. H. Smith and D. Pimentel, eds. New York: Academic Press.

Mayo, Z. B. 1976. Aerial Suppression of Rootworm Adults for Larval Control. Agricultural Experiment Station Report. Lincoln: University of Nebraska.

McDonald, D. 1987. Chemicals and your health: What's the risk? Farm Journal 3(2):8–11.

Meinke, L. J., Z. B. Mayo, and T. J. Weissling. 1989. Pheromone delivery system: Western corn rootworm (Coleoptera: Chrysomelidae) pheromone encapsulation in a starch borate matrix. Journal of Economic Entomology 82:1830–1835.

Metcalf, R. L. 1980. Changing role of insecticides in crop protection. Annual Review of Entomology 25:219–256.

Metcalf, R. L. 1986. Foreword. Pp. 7–15 in Methods for the Study of Pest *Diabrotica*, J. L. Krysan and T. A. Miller, eds. New York: Springer-Verlag.

Metcalf, R. L., and R. L. Lampman. 1989a. Estragole analogues as attractants for corn rootworms (Coleoptera: Chrysomelidae). Journal of Economic Entomology 82:123–129.

Metcalf, R. L., and R. L. Lampman. 1989b. Cinnamyl alcohol and analogs as attractants for corn rootworms (Coleoptera: Chrysomelidae). Journal of Economic Entomology 82:1620–1625.

Metcalf, R. L., R. A. Metcalf, and A. M. Rhodes. 1980. Cucurbitacins as kairomones for diabroticite beetles. Proceedings of the National Academy of Sciences 77:3769–3772.

Metcalf, R. L., A. M. Rhodes, R. A. Metcalf, J. Ferguson, E. R. Metcalf, and P. Y. Lu. 1982. Cucurbitacin contents and diabroticite (Coleoptera: Chrysomelidae) feeding upon *Cucurbita* spp. Environmental Entomology 11:931–937.

Metcalf, R. L., J. E. Ferguson, R. Lampman, and J. F. Andersen. 1987. Dry cucurbitacin-containing baits for controlling diabroticite beetles (Coleoptera: Chrysomelidae). Journal of Economic Entomology 80:870–875.

National Research Council. 1989. Alternative Agriculture. Washington, D.C.: National Academy Press.

Poston, F. L., L. P. Pedigo, and S. M. Welch. 1983. Economic-injury levels: Reality and practicality. Bulletin of the Entomological Society of America 29:49–53.

Pruess, K. P., J. F. Witkowski, and E. S. Raun. 1974. Population suppression of western corn rootworm by adult control with ULV malathion. Journal of Economic Entomology 67:651–655.

Riedell, W. E. 1990. Western corn rootworm damage or mechanical root cutting: Effects on root morphology and water relations in maize. Crop Science 30:628–631.

Shaw, J. T., W. G. Ruesink, S. P. Briggs, and W. H. Luckmann. 1984. Monitoring populations of corn rootworm beetles (Coleoptera: Chrysomelidae) with a trap baited with cucurbitacins. Journal of Economic Entomology 77:1495–1499.

Stamm, D. E., Z B Mayo, J. B. Campbell, J. F. Witkowski, L. W. Anderson, and R. Kozub. 1985. Western corn rootworm (Coleoptera: Chrysomelidae) beetle counts as a means of making larval control recommendations in Nebraska. Journal of Economic Entomology 78:794–798.

Stern, V. M., R. F. Smith, R. Van Den Bosch, and K. S. Hagen. 1959. The integrated control concept. Hilgardia 29:81–101.

Suguiyama, L. F., and G. A. Carlson. 1985. Field Crop Pests: Farmers Report the Severity and Intensity. Agriculture Information Bulletin 487. Washington, D.C.: Economic Research Service, U.S. Department of Agriculture.

Sutter, G. R. 1982. Comparative toxicity of insecticides for corn rootworm (Coleoptera: Chrysomelidae) larvae in a soil bioassay. Journal of Economic Entomology 75:489–491.

Sutter, G. R., T. F. Branson, J. R. Fisher, N. C. Elliott, and J. J. Jackson. 1989. Effect of insecticide treatments on root damage ratings of maize in controlled infestations of western corn rootworms (Coleoptera: Chrysomelidae). Journal of Economic Entomology 82:1792–1798.

Sutter, G. R., J. R. Fisher, N. C. Elliott, and T. F. Branson. 1990. Effect of insecticide treatments on root lodging and yields of maize in controlled infestations of western corn rootworms (Coleoptera: Chrysomelidae). Journal of Economic Entomology 83:2414–2420.

Sutter, G. R., T. F. Branson, J. R. Fisher, and N. C. Elliott. In press. Effect of survival, development, fecundity, and sex ratio in controlled infestations of western corn rootworms (Coleoptera: Chrysomelidae). Journal of Economic Entomology.

Tollefson, J. J. 1986. Field sampling of adult populations. Pp. 123–146 in Methods for the Study of Pest *Diabrotica,* J. L. Krysan and T. A. Miller, eds. New York: Springer-Verlag.

Turpin, F. T. 1974. Threshold research on corn. Proceedings of the North Central Branch of the Entomological Society of America 29:61–65.

Turpin, F. T. 1977. Insect insurance: Potential management tool for corn insects. Bulletin of the Entomological Society of America 23:181–184.

Turpin, F. T., and J. D. Maxwell. 1976. Decision-making related to use of soil insecticides by Indiana corn farmers. Journal of Economic Entomology 69:359–362.

U. S. Environmental Protection Agency. 1985. Carbofuran; special review of certain pesticide products. Federal Register 50:41938–41943.

Williams, M. W., P. W. Holden, D. W. Parsons, and M. N. Lorber. 1988. Pesticides in ground water data base: 1988 interim report. Washington, D.C.: Office of Pesticide Programs, U.S. Environmental Protection Agency.

# 15

# On-Farm Research Comparing Conventional and Low-Input Sustainable Agriculture Systems in the Northern Great Plains

*Thomas L. Dobbs, James D. Smolik, and Clarence Mends*

The search for answers to questions about the economic, agronomic, and environmental sustainability of alternative agriculture systems requires multidisciplinary teamwork by groups of scientists, using a whole-farm perspective (Dobbs and Taylor, 1989; Madden and Dobbs, 1990). Farmers and public policymakers need much more information than is currently available on the agronomic performance, profitability, and riskiness of low-input sustainable (alternative) farming systems compared with those of more conventional systems. Information is needed on how systems compare under different agroclimatic and farm policy conditions. This information is best generated through two concurrent lines of inquiry: (1) studies involving operating farms and (2) studies based on research station trials. Neither type of study alone will suffice at the present stage of sustainable agriculture research.

Most economic studies of alternative farming systems based on research station trials are in the very early stages. An exception is that reported by Helmers et al. (1986), based, at the time, on 8 years of experimental results in Nebraska. Duffy and Nicholson (1987), Goldstein and Young (1987), and Dobbs et al. (1988) have reported economic results based on research station trials for shorter periods of time in Iowa and Pennsylvania, Washington, and South Dakota, respectively. South Dakota's experiment station research is now further along, and baseline economic results for a 5-year transition period (1985 to 1989) have been reported by Dobbs and Mends (1990); an article by J. D. Smolik and T. L. Dobbs recently submitted to the *Journal of Production Agriculture* contains detailed yield results for the transition period, as well as results of sensitivity analyses with economic variables.

Perhaps the most widely known studies of operating farms are the case studies in the National Research Council report *Alternative Agriculture* (National Research Council, 1989). A variety of other on-farm studies have recently been initiated with support from the U.S. Department of Agriculture's low-input sustainable agriculture (LISA) program and the Northwest Area Foundation.

Scientists at South Dakota State University initiated a study in 1984 in which they compared a set of conventional and alternative farms in east-central South Dakota. This study has provided the agronomic and economic data needed to systematically compare two operating farms using contrasting systems over the past 5 years (1985 to 1989). These two farms are located in the transition zone between the corn-soybean growing region of the Midwest and the small-grain growing region of the northern Great Plains. Thus, a wide variety of foodgrain, feedgrain, and oilseed crops are grown in this region. None of the case studies presented in the National Research Council (1989) report represented agricultural practices in the northern Great Plains, and there is a need for information that sheds light on the comparative productivity and profitability of conventional and alternative farms in that region. Policymakers, in particular, are unsure how certain proposed farm program revisions might affect the profitability of, and therefore the adoption of, alternative agriculture systems in that region. Results of this ongoing case study, which is being conducted by a multidisciplinary team of plant scientists and agricultural economists, are therefore timely.

Through a series of surveys, case studies, and policy analyses supported by the Northwest Area Foundation, researchers at South Dakota State University are gaining a detailed understanding of sustainable agriculture practices in different crop regions of South Dakota (Dobbs et al., 1989; Taylor et al., 1989a,b). Detailed analysis of yields and economic performance for the pair of farms reported in this chapter was supported by grants from the U.S. Department of Agriculture (USDA) LISA program. Analysis of these two farms has interpretive value in its own right; moreover, the analysis has enabled the development of methodology that is being adapted and used in other South Dakota case studies. The present analysis of two case study farms will be placed in the context of a larger set of case studies in South Dakota later in 1990.

## THE TWO CASE STUDY FARMS

Agronomic data were collected over a 5-year period for two farms, one that uses an alternative agriculture system and one that uses a conventional agriculture system, in Lake County in east-central South Dakota. The topography in the study area is gently rolling. The climate is continental, with a 7-month growing season (April to October), and the long-term average growing season precipitation is 19.68 inches.

Crop acreage on the conventional farm ranged from 317 to 998 acres over the 5-year study period (1985 to 1989). The farm is basically a corn-soybean operation in which these two crops are alternated every other year on most fields (except for farm program set-aside acres). The farm also has a small cow-calf operation, in which the calves are finished on the farm, and a hog-finishing operation, from which approximately 1,000 hogs are marketed each year. Commercial fertilizers (primarily nitrogen and phosphorus) and herbicides are used each year on the conventional farm, and corn is cultivated at least once each year. In recent years, soybeans have been drilled in narrow rows and, therefore, have not been cultivated. The conventional farmer has not been using a moldboard plow. An all-in-one soil finisher (involving disk, field cultivator, and harrow) is the conventional farmer's principal implement for soil preparation before planting.

The amount of land cropped on the alternative farm ranged from 537 to 900 acres over the 5-year period studied. The principal rotation is a 4-year system consisting of a small grain overseeded with alfalfa-alfalfa-soybeans-corn, in that order. Small-grain crops normally consist of some combination of barley, oats, and spring wheat. Alfalfa is harvested as hay only 1 year before it is chiseled. On average, during the study period, 50 percent of the crop acreage was in alfalfa and small grains (including farm program set-aside acres, generally with a small-grain cover crop), and the other 50 percent was split roughly equally between corn and soybeans. At the time that agronomic data collection started (1985), the alternative farm had not applied any commercial fertilizers or herbicides for the previous 8 years. Land continued to be farmed that way during the study period, but moderate amounts of commercial fertilizers and herbicides were used on some newly rented land coming into the alternative farm's system. Most of this farm's cropland qualifies as organic under the certification standards being used in South Dakota and nearby states.

The alternative farm has a farrow-to-finish hog operation involving 40 to 50 sows. The operator of this farm also finishes a few cattle and is in the process of rebuilding the cattle portion of his operation. A small portion of some of the alternative farm's fields receive light applications of composted manure from the livestock operations; however, over the 5-year study period, less than 5 percent of the fields were treated with manure. The alternative system relies primarily on legumes (alfalfa and soybeans) for nitrogen.

The inclusion of alfalfa in the rotation also helps control weeds. Rotary hoeing, harrowing, hand weeding, and cultivating are also important weed control measures in row crops (corn and soybeans) on the alternative farm. The moldboard plow is not being used in the alternative system.

## PRODUCTION PERFORMANCE

Yield data and soil samples were collected from areas within fields with Egan soil associations (fine-silty, mixed, mesic Udic Haplustolls; slopes, 0 to 6 percent). Egan soils are deep and well drained and have medium to high fertility. Row crop yields were estimated by hand harvesting 10 randomly selected 3-foot lengths of row. Soil samples were collected in late September-early October, and 8 to 10 subsamples were collected from each plot area. Samples for the alternative system were obtained from fields that had not received commercial fertilizer or herbicides for at least the previous 8 years. Yield and soil test data were statistically analyzed by using years as replications.

Corn and soybean yields varied considerably from year to year (Table 15-1). Although the alternative system had higher average yields for corn and the conventional system had higher average yields for soybeans, the differences between the alternative and conventional systems were not significant ($p > 0.05$). An earlier study (Lockeretz et al., 1978) conducted in the Midwest also found no significant differences in corn and soybean yields between organic and conventional farms.

The alternative farmer was asked in early 1989 to recall his small-grain and alfalfa yields for the years 1985 through 1988, and he was asked to estimate his 1989 yields of those crops following harvest that fall. These estimates were needed for the whole-farm economic analyses. At the same

**TABLE 15-1** Corn and Soybean Yields for Alternative and Conventional Farming Systems, Lake County, South Dakota, 1985 to 1989

| Year | Yield (bu/acre) | | | |
| | Corn | | Soybeans | |
| | Alternative | Conventional | Alternative | Conventional |
| --- | --- | --- | --- | --- |
| 1985 | 88.1 | 110.6 | 23.1 | 30.5 |
| 1986 | 115.3 | 107.0 | 36.3 | 38.4 |
| 1987 | 136.6 | 134.7 | 25.0 | 39.1 |
| 1988 | 130.7 | 79.0 | 38.7 | 39.0 |
| 1989 | 128.7 | 128.5 | 31.4 | 36.1 |
| Average* | 119.9 | 112.0 | 30.9 | 36.6 |

*Averages are over 5 years for hand-harvested yield estimates with Egan soil associations. Yield differences between alternative and conventional farming systems for corn and soybeans were not statistically significant ($p > 0.05$).

times, both the alternative and conventional farmers were also asked for their estimates of whole-farm average yields for corn and soybeans in each of the years 1985 through 1989. These yield estimates, which were not limited to the Egan soil portions of the two farms, were as follows when averaged over the 5-year period: alternative corn, 82.6 bushels/acre; conventional corn, 97.2 bushels/acre; alternative soybeans, 25.4 bushels/acre; and conventional soybeans, 36.0 bushels/acre. These farmer-estimated whole-farm average yields were all lower than the hand-harvested average yield estimates for Egan soils on the two farms. Moreover, the farmer-estimated corn yields on the alternative farm were lower than the farmer-estimated corn yields on the conventional farm, whereas the alternative system corn yields were higher than those of the conventional system when estimates were made by hand harvesting. The corn yield differences were not statistically significantly ($p > 0.05$) in either case, however. Farmer-estimated soybean yield estimates for the alternative and conventional farms did differ significantly ($p < 0.05$).

Three factors might explain why the farmers' estimates of yields were less than the estimates obtained by hand harvesting: (1) the greater efficiency of hand harvesting, (2) the exclusion of lower-yielding knolls and low-lying areas from areas that were hand harvested (these landscape positions were excluded because they did not contain Egan soil associations), and (3) the farmers may not have accurately recalled all of their yields and may have tended to be conservative in their estimates.

For purposes of comparing conventional and alternative farming systems, the use of yield estimates obtained by hand harvesting (rather than estimates made by the farmers) for the row crops on the better (Egan) soil areas has the effect of eliminating a possible source of bias caused by different soil properties on the two farms that are unrelated to the choice of a conventional or an alternative management system. Therefore, the economic analysis reported in this chapter is based on the yields (for row crops) obtained by hand harvesting.

Soil nitrate and potassium levels during the fall did not differ significantly ($p > 0.05$) between the two systems (Table 15-2). The lack of differences in nitrate levels between the two systems suggests that the rotation pattern in the alternative system resulted in adequate levels of nitrogen fixation. Soils in the study area are naturally high in potassium. Phosphorus levels were significantly higher in the conventional system; however, no symptoms of phosphorus deficiency were observed in the alternative system. Organic matter was significantly higher in the alternative system (Table 15-2). Higher levels of organic matter in low-input sustainable agriculture systems were also reported in previous studies (e.g., Lockeretz et al., 1978; Sahs and Lesoing, 1985).

**TABLE 15-2** Soil Test Results for Alternative and Conventional Farming Systems, Egan Soil Associations, Lake County, South Dakota

| Matter Tested* | Alternative System | Conventional System | Statistical Analysis |
|---|---|---|---|
| Nitrate ($NO_3$) nitrogen (0–2 feet) (lbs/acre) | 41.1 | 39.0 | NS |
| Phosphorus (0–6 inches) (lbs/acre) | 10.8 | 19.0 | $F = 9.5$[†] |
| Potassium (0–6 inches) (lbs/acre) | 574 | 568 | NS |
| Organic matter (0–6 inches) (%) | 4.4 | 3.6 | $F = 6.7$[†] |

NOTE: System averages are for 1986 to 1989. Values are averages over 4 years. Samples were obtained in the fall of each year. NS, not significant.

*Values in parentheses are soil depths at which tests were taken.
[†]The differences were significant at $p < 0.05$.

## ECONOMIC PERFORMANCE

Crop enterprise budgets were developed for the two case study farms, and the budgets were aggregated to whole-farm bases, taking into account available acreage and farm program provisions for each year of the study. Thus, the acreage devoted to each crop and to set-aside acres varied from year to year, depending on farm program provisions and each farmer's cropping decisions. Sometimes, one or both farmers participated in optional paid diversion programs for corn. In the baseline analysis, premium prices for organically certified crops were ignored. Crops were valued each year by using the higher of the applicable (1) average market year prices estimated for South Dakota or (2) federal loan rates for eastern South Dakota; applicable federal deficiency payments for each qualifying crop were then added. Thus, year-to-year variations in federal farm program provisions and market prices, in addition to variations in weather and yields, affected whole-farm gross returns. For purposes of valuation, it was assumed that all hay was sold; hay was valued on the basis of estimated market prices in South Dakota for each year. Thus, at this stage of the analysis, livestock have not been integrated into the economic analysis. Explicit integration of livestock into the whole-farm analysis will take place in late 1990. Input prices were held constant over the study period in the analysis.

### Baseline Results, Assuming No Premium Prices for Organic Crops

Five-year average results of the baseline economic analysis for the alternative and conventional farming systems are given in Table 15-3. The first two columns of data show various cost and return measures on a per acre basis. Gross income calculations were based on the corn and soybean yields reported in Table 15-1 and on yields for other crops (mainly the alfalfa and small grains of the alternative farmer) reported by the farmers themselves. The same cost and return measures are shown on a whole-farm basis in the fourth and fifth columns of data, assuming that there were 700 tillable acres on each farm. Each farm actually had an average of slightly over 700 tillable acres (i.e., acres available for crops, including hay, and farm program set-asides) over the 5-year period. The first whole-farm net income figure (return over all costs except land, labor, and management) is one indication of the earnings available for family living expenses, taxes, and savings for a family-owned and -operated farm in which all farm labor is provided by the family. In arriving at the third net income figure (return over all costs except management), all labor, including family labor, was calculated at the going local wage rate, and a rental value or opportunity cost of the land was also subtracted from the gross income. Any additional net income attributable to livestock operations (to be calculated in the next stage of analysis at South Dakota State University) could be added to family income, as could off-farm income earned by family members.

Direct costs (sometimes referred to as operating or cash costs) were $39/acre higher on the conventional farm than they were on the alternative farm ($85/acre compared with $46/acre) (Table 15-3). This difference was largely due to the fertilizer and herbicide inputs on the conventional farm. Gross income on the conventional farm ($216/acre) was $50/acre higher than that on the alternative farm ($166/acre). Thus, the higher gross income on the conventional farm more than offset that farm's higher direct costs. Although average corn yields were slightly higher on the alternative farm (Table 15-1), the heavy concentration of corn and soybeans in the conventional farm's rotation, together with higher soybean yields on that farm, caused the conventional farm's gross income to be higher.

Before accounting for land, labor, and management costs, the conventional system showed net income that averaged $15/acre more than that of the alternative system ($103/acre compared with $88/acre). Because the alternative farm required 58 percent more labor than that required on the conventional farm, when all labor was included in the costs, the difference increased from $15 to $19/acre (see net income over all costs except land and management in Table 15-3).

Both systems provided positive net returns to management (see net income over all costs except management in Table 15-3), on average, and in

**TABLE 15-3** Five-Year (1985 to 1989) Average Baseline Economic Results for East-Central South Dakota Case Study Farms

| Cost or Income Measure | Costs and Returns ($/acre) | | | Whole-Farm Costs and Returns ($) | | |
|---|---|---|---|---|---|---|
| | Alternative System | Conventional System | Difference (conventional minus alternative) | Alternative System* | Conventional System* | Difference (conventional minus alternative)* |
| Direct costs other than labor | 46 | 85 | 39 | 31,920 | 59,360 | 27,440 |
| Baseline gross income | 166 | 216 | 50 | 116,200 | 151,480 | 35,280 |
| Baseline net income over All costs except land, labor, and management | 88 | 103 | 15 | 61,460 | 72,100 | 10,640 |
| All costs except land and management | 76 | 95 | 19 | 53,340 | 66,780 | 13,440 |
| All costs except management | 41 | 60 | 19 | 28,700 | 41,720 | 13,020 |

*For a farm with 700 tillable acres.

each of the 5 years studied (Table 15-3 and Figure 15-1). Only in one year (1988) did the alternative farm have higher net returns than the conventional farm.

When each farmer's own whole-farm yield estimates, rather than the estimates obtained by hand harvesting, for corn and soybeans were used in the analyses, differences in net income in favor of the conventional farm were increased substantially. For example, the difference of net income over all costs except management was increased from $19 to $29/acre. Using farmer-estimated yields, the alternative farm experienced net losses (negative net returns to management) in 2 of the 5 years, whereas the conventional farm experienced no such losses.

### Implications of Premium Prices for Organic Crops

The crop return calculations underlying the results for the alternative farm shown in Table 15-3 were based on the assumption that all production was marketed conventionally. In actual practice, most of the alternative farm is operated without purchased chemical inputs, and a portion of the production (mainly soybeans) normally goes into markets for organically produced products that bring premium prices ("organic premiums").

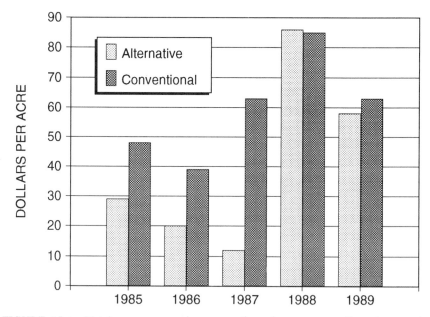

FIGURE 15-1   Net income comparison over time: income over all costs except management, 1985 to 1989.

The impact of organic premiums on the net returns of the alternative farm was analyzed by using four different combinations of assumptions about the portion of the soybean crop that went into organic markets and the magnitude of the price premium for the organically marketed portion. These assumptions were derived from experiences reported by organic farmers in South Dakota (Taylor et al., 1989a). The combinations of assumptions about soybeans, ranging from conservative to optimistic, were as follows: (1) option B, 50 percent marketed organically at a 25 percent price premium over the price for conventionally produced soybeans; (2) option C, 70 percent marketed organically at a 25 percent price premium over the price for conventionally produced soybeans; (3) option D, 50 percent marketed organically at a 40 percent price premium over the price for conventionally produced soybeans; and (4) option E, 70 percent marketed organically at a 40 percent price premium over the price for conventionally produced soybeans. Additional cleaning is required for a crop that is marketed organically, so it was assumed that each harvested bushel of crop results in 0.90 bushel on an organically cleaned basis. Calculations were carried out on a whole-farm basis, recognizing that soybeans constitute just one part of the alternative system rotation. Whole-farm analyses were carried out for each year of the 5-year study by using the combinations of assumptions listed above.

Results are shown in Figure 15-2 in terms of 5-year averages for net income over all costs except management per acre. The baseline (no organic marketing) case for the alternative farm ($41/acre) is given on the left, and the conventional farm comparison ($60/acre) is given on the right. Each of the organic marketing sensitivity analyses added $1 or $2/acre, on a whole-farm basis, to the net income for the alternative farm. The most optimistic option tested for all 5 years (option E) left the average net income for the alternative system ($48/acre) lower than that for the conventional system, however.

Another analysis (not shown in Figure 15-2) was carried out for the alternative farm in 1989 only, because more detailed information was available for the alternatives farm's organic marketing that year. In that year, 42 percent of the soybean crop was actually marketed at a 77 percent premium over the conventional price, and 29 percent of the oats crop was actually marketed at a 190 percent premium over the conventional price. Incorporation of these organic amounts and premiums resulted in net income over all costs except management (on a whole-farm basis) of $75/acre that year (Table 15-4) compared with $58/acre when the premiums were ignored (baseline; see Figure 15-1). Thus, in some years, the actual organic premiums can result in higher income for alternative than for conventional farms under the conditions of this study. Whether organic premium prices can remain as high as those observed in 1989, however, remains to be seen.

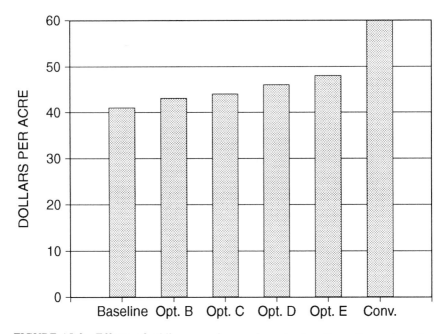

FIGURE 15-2   Effects of adding organic premiums in the alternative system: income over all costs except management, 5-year average.

## Policy Analyses

Extensive policy analyses have begun under South Dakota State University's sustainable agriculture research grant from the Northwest Area Foundation. However, results of only two types of these analyses are reported here.

The first policy analysis presented represents a 25 percent increase in purchased chemical fertilizer and herbicide input prices. This could result from a significant tax on these inputs, which conceivably could be implemented to discourage their use by internalizing the possible external costs associated with chemical input use. Thus far, special state taxes on chemical fertilizers and pesticides generally have been set at rates that are quite low. These state taxes raise some revenues for research and education on input use, but the tax rates are probably too low to directly discourage the use of inputs. It is possible that stiffer taxes will be used in the future to discourage chemical input use, however. Developments in world petroleum markets also could push chemical input prices up significantly in the future. This would have the effect of increasing fuel costs as well; however, fuel costs were not increased in the analysis reported here.

When chemical input prices were increased by 25 percent, the 5-year

**TABLE 15-4** Comparison of Conventional and Organic Farms When Actual Organic Price Premiums Are Included in Calculations, 1989 Crop Year

| Cost or Income Measure | Costs and Returns ($/acre) | | | Whole-Farm Costs and Returns ($) | | |
|---|---|---|---|---|---|---|
| | Alternative System | Conventional System | Difference (conventional minus alternative) | Alternative System* | Conventional System* | Difference (conventional minus alternative)* |
| Direct costs other than labor | 48 | 87 | 39 | 33,600 | 60,900 | 27,300 |
| Baseline gross income | 205 | 223 | 18 | 143,500 | 156,100 | 12,600 |
| Baseline net income over All costs except land, labor, and management | 123 | 107 | −16 | 86,100 | 74,900 | −11,200 |
| All costs except land and management | 111 | 99 | −12 | 77,700 | 69,300 | −8,400 |
| All costs except management | 75 | 63 | −12 | 52,500 | 44,100 | −8,400 |

*For a farm with 700 tillable acres.

average of net income over all costs except management decreased by only $1/acre (from $41 to $40/acre) for the alternative farm, because chemicals (in limited quantities) were used on only a portion of that farm. On the conventional farm, however, average net income for the 5-year period decreased by $9/acre (from $60 to $51/acre). Chemical input price increases of this magnitude do not appear to be sufficient, by themselves, to equalize the net returns for the two types of farming systems. However, the higher chemical input prices, together with organic premiums for some of the products of the alternative farm, could be sufficient to bring net returns of alternative systems close to or higher than those of conventional systems.

The second policy analysis presented here consists of a 25 percent reduction in federal farm program target prices for corn and small grains in the years 1985 through 1989. Since there has been much discussion of possible further reductions in target prices, the effects of hypothetically lower target prices during the past 5-year period were calculated. Starting with the baseline figures (no organic premiums), net income over all costs except management decreased by $17/acre (from $41 to $24/acre) on the alternative farm and by $19/acre (from $60 to $41/acre) on the conventional farm as a result of setting target prices 25 percent lower than their actual levels in each of the past 5 years (Figure 15-3). The adverse effect was greater, in

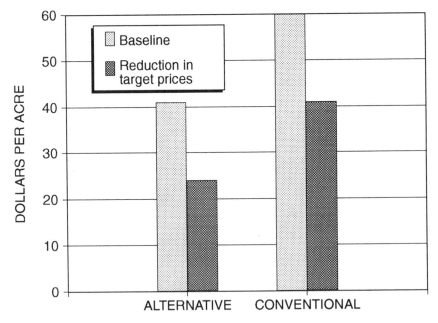

FIGURE 15-3   Effects of reducing federal target prices by 25 percent: income over all costs except management, 5-year average.

absolute terms, on the conventional farm, as expected, but by only a slight amount ($2/acre). As a percentage of net income, the effect was greater on the alternative farm, however; there was a 41 percent decline on the alternative farm and a 32 percent decline on the conventional farm.

The policy analyses and interpretations for these two case farms were still in progress at the time of preparation of this chapter. Hence, caution must be used in drawing conclusions at this point. Nevertheless, it is clear that, at present, both alternative and conventional farms in the northern Great Plains depend a great deal on government programs for their economic sustainability. Both farms benefited from government payments in such forms as deficiency payments, payments for optional paid acreage reductions (including participation in the "0-92" program), and amounts by which government commodity loan levels exceeded market prices in some years. These payments averaged $27 and $33/acre over 5 years for the alternative and conventional farms, respectively. On a 700-acre whole-farm basis, the government payments averaged $18,900 for the alternative farm and $23,100 for the conventional farm. The payments were 16 percent of the average gross income and 66 percent of the average net income for the alternative farm, and they were 15 percent of the average gross income and 55 percent of the average net income for the conventional farm.

## DISCUSSION

This on-farm research case study found conventional farming to be more profitable, on average, than alternative (low-input sustainable) farming in an area of east-central South Dakota when price premiums for organic products were ignored. In this paired comparison, corn and soybeans averaged 83 percent of the conventional farm's crop land acreage compared with 49 percent of the alternative farm's acreage. For the alternative farm, an average of 20 percent of the acreage was in harvested small grains (barley, oats, and wheat) and 16 percent was in alfalfa. (The remaining cropland of each farm was in various forms of paid and unpaid acreage set-asides.) Given the combination of crop prices and federal farm program provisions that were in effect from 1985 through 1989, it appears to have been difficult for rotation systems containing substantial small-grain and forage legume acreage to have been fully competitive with systems that were heavily dominated by corn and soybeans. Organic crop price premiums, when available to qualifying alternative farms, could have reduced and perhaps, in some cases, eliminated the net return difference, however.

Case studies now under way for farms in other crop regions in South Dakota may give different results about comparative profitability. A whole-farm economic analysis based on experiment station trials at South Dakota State University's Northeast Experiment Station (Dobbs and Mends, 1990) showed that alternative systems are more economically competitive with conventional systems than did the case farm analysis reported here. The Northeast Experiment Station, where those trials were conducted, is in an area with somewhat lower average rainfall and a shorter growing season than in the east-central county, where the case study farms discussed here are located. Consequently, small grains constitute a greater portion of the conventional rotations on farms near the Northeast Experiment Station than was the case for the conventional case study farm reported here. It appears that the greater the role of small grains in conventional crop rotation systems, the more likely are alternative (low-input sustainable) systems to be economically competitive with those conventional systems, at least with the current federal farm program.

There is a need for more extensive analyses of various farm program options on the relative economic competitiveness of conventional and alternative farming systems. Some such analyses are currently under way by South Dakota State University and Washington State University agricultural economists.

Matters other than on-farm profitability are also relevant to public policy regarding alternative agriculture. For example, the implications of different types of farming systems for the economic health of rural communities and regions are important; these implications will receive attention over the coming year in studies at South Dakota State University. Environmental externalities related to the chemicals that are used in different types of systems also need greater research attention to provide bases for the development of public policy. When contamination of groundwater may be related to a particular farming system, for example, public policies may need to either internalize such costs or provide positive incentives to farmers to shift to more environmentally benign systems. Farming systems that erode the soil resource base, as evidenced by the reduced organic matter content and the need to purchase nutrients in conventional systems, sometimes may be more profitable in the short term but very well may not be sustainable from a long-term environmental perspective. Even when water quality and other environmental concerns weigh heavily on the development of policies toward alternative agriculture, it is crucial for empirical information to be available regarding the impacts of policy options on the economic profitability (and, therefore, the economic sustainability) of farms that use conventional versus alternative farming systems in different agroclimatic regions. Experiment station and on-farm studies such as the one reported here constitute the principal source of such information.

## ACKNOWLEDGMENTS

Research on which this paper is based received support from the South Dakota State University Agricultural Experiment Station, U.S. Department of Agriculture Low-Input Sustainable Agriculture grants LI-88-12 and LI-89-12, and Northwest Area Foundation grant 88-56.

## REFERENCES

Dobbs, T. L., and C. Mends. 1990. Profitability of Alternative Farming Systems at South Dakota State University's Northeast Research Station: 1989 Compared to Previous Transition Years. Economics Research Report No. 90-1. Brookings: South Dakota State University.

Dobbs, T. L., and D. C. Taylor. 1989. Economic considerations in evaluating alternative agricultural practices. Pp. 109–131 in Proceedings of Great Plains Agricultural Council Annual Meeting. Fort Collins, Colo.: Great Plains Agricultural Council.

Dobbs, T. L., M. G. Leddy, and J. D. Smolik. 1988. Factors influencing the economic potential for alternative farming systems: Case analyses in South Dakota. American Journal of Alternative Agriculture 3:26–34.

Dobbs, T. L., D. L. Becker, and D. C. Taylor. 1989. Farm Program Participation and Policy Perspectives of Sustainable Farmers in South Dakota. Economics Staff Paper No. 89-7. Brookings: South Dakota State University.

Duffy, M., and S. Nicholson. 1987. Two of Iowa's low input agriculture research programs. Presented at Symposium on Current Economic Prospects for Sustainable Farming Systems, American Agricultural Economics Association Annual Meeting. American Journal of Agricultural Economics 69:1067 (Abstract).

Goldstein, W. A., and D. A. Young. 1987. An agronomic and economic comparison of a conventional and a low-input cropping system in the Palouse. American Journal of Alternative Agriculture 2:51–63.

Helmers, G. A., M. R. Langemeier, and J. Atwood. 1986 An economic analysis of alternative cropping systems for east-central Nebraska. American Journal of Alternative Agriculture 1:153–158.

Lockeretz, W., G. Shearer, R. Klepper, and S. Sweeney. 1978. Field crop production on organic farms in the Midwest. Journal of Soil and Water Conservation 33:130–134.

Madden, P., and T. L. Dobbs. 1990. The role of economics in achieving low-input farming systems. Pp. 459–477 in Sustainable Agricultural Systems, C. A. Edwards, R. Lal, P. Madden, R. H. Miller, and G. House, eds. Ankeny, Iowa: Soil and Water Conservation Society.

National Research Council. 1989. Alternative Agriculture. Washington, D.C.: National Academy Press.

Sahs, W. W., and G. Lesoing. 1985. Crop rotations and manures versus agricultural chemicals in dryland grain production. Journal of Soil and Water Conservation 40:511–516.

Taylor, D. C., T. L. Dobbs, and J. D. Smolik. 1989a. Sustainable Agriculture in South Dakota. Economics Research Report No. 89-1. Brookings: South Dakota State University.

Taylor, D. C., T. L. Dobbs, D. L. Becker, and J. D. Smolik. 1989b. Crop and Livestock Enterprises, Risk Evaluation, and Management Strategies on South Dakota Sustainable Farms. Economics Research Report No. 89-5. Brookings: South Dakota State University.

# 16

# Low-Input, High-Forage Beef Production

*Terry Klopfenstein*

The trend in the beef industry has been to raise large cattle with faster body weight gains. This method requires high-grain feeding (including corn silage, which is half grain)—a type of cattle production that is incompatible with the forage resources available on many farms in the Corn Belt.

The low-input sustainable agriculture (LISA) project on which this chapter is based is a study of low-input systems of beef production. These systems emphasize the use of animal-harvested (grazed) forages (pasture, range, cornstalks, and other residues consumed in the field) in place of supplemental feeds (such as soybean meal, grain, and other mechanically harvested or purchased feeds) to reduce costs and increase profits. The purpose of this study is to examine ways to optimize the use of crop residues produced in cropping sequences and forages, especially legumes, which supply nitrogen to the soil at a lower cost and minimize soil erosion.

## PRINCIPLES OF BEEF PRODUCTION IN THE CORN BELT

This study is based on the principles that use of forages and manure can reduce input costs in agricultural production systems and that forage production in cropping systems can greatly reduce soil erosion. Because of their ability to utilize forages, beef cattle fit into farming systems that are based on forage production and manure utilization.

Much of the land in southern Iowa, northern Missouri, and eastern Nebraska is subject to erosion. However, a large percentage of the land in this region can be used for crop production (with appropriate tillage and management practices), and the resulting crop residues can be partially removed without causing excessive soil erosion (Lindstrom et al., 1979). In addition, most farms have some land that should not be tilled; it is suitable only for pasture. The challenge of this study is to integrate economical beef production systems with the available forage resources.

Analysis of data collected from beef producers in Iowa and Nebraska shows large differences in feed costs among producers. Producers who rely more heavily on grazed forages and, consequently, who incur reduced expenditures for off-farm purchases of supplements had significantly higher profits. Better management of forages and careful attention to reducing input costs will be necessary in the future to keep production costs down and to ensure the profitability of systems that reduce soil erosion and groundwater contamination.

Just as the combinations of forage and grain that can be used to feed growing-finishing cattle are numerous, variations in cattle types have also increased dramatically in the past few years. Extremes have been observed, from heifers that weigh from 800 to 1,000 pounds (lbs) at low Choice grade to steers that weigh from 1,300 to 1,400 lbs at the same grade and degree of fatness. Recognizing the growing consumer preference for leaner meats, the packing industry has moved rapidly to provide boxed beef at the low Choice grade with an acceptable carcass range of 600 to 900 lbs.

The feeding system interacts with cattle type to produce various carcass weights at the low Choice grade. Cattle with similar growth potentials that are grown on roughages before they are finished on grain are older when they are sent to market (low Choice grade) and have heavier carcasses. This is because they have developed farther along their growth curve and because they have attained more skeletal and muscle growth before they are fattened.

While the sizes of mature beef cattle have increased, the feed efficiencies of cattle taken to the same degree of fatness probably have not increased (Smith et al., 1976). Efficiencies of feed conversion are primarily affected by the composition of the weight gain rather than by mature weight. Therefore, the current trend to try to produce "uniform" cattle is not really necessary. It is important, however, to match the feeding system to the cattle type, to produce cattle with acceptable weights at the low Choice grade. As a generalization, for animals with smaller frames (mature size), more roughage should be fed to avoid overfinishing (too much fat) at an acceptable carcass weight.

It is more difficult to meet the market requirement of low Choice grade combined with yield grade 1 or 2. The correlation of external fat to marbling (intramuscular fat) is quite high. Grain is fed to cattle to fatten them. The desirable fat is marbling, and outside fat is wasteful. Producers can make cattle as lean as they desire without the use of growth promoters or repartitioning agents simply by feeding them more roughage and less grain.

However, no sure method currently exists to produce low Choice yield grade 1 or 2 cattle consistently.

The beef industry is also facing a critical economic challenge. One problem is competition from lower-cost meats, primarily poultry and pork. Although promotion may sell more beef, especially in the short run, it seems that cost of production must be reduced if the beef industry is to remain competitive and profitable. The ultimate goal of beef production systems is to produce a product that is suitable to meet market demand while using the available resources in an environmentally sound way and at a price sufficient to encourage further production and consumption.

Because of their ability as ruminants to utilize large quantities of fiber, cattle are competitive with other meat-producing species only when they are fed forage. The trend of the beef industry has been in the opposite direction in the past 30 or more years, with more grain being fed (including that in corn silage). It seems that the beef industry can go in two primary directions: high-grain or high-forage systems of production.

In the high-grain system, calves would be placed on high-grain rations after an adjustment period of approximately 30 days postweaning. This system best fits large-frame (large mature size) steers, bulls, or both. It is important to note here that these cattle will reach the necessary fatness for the Choice grade with body weights that are 50 to 200 lbs lighter than those of the same cattle grown in a high-forage system. The smaller size is an advantage for the high-grain system, because the carcass weights will not exceed the standards of the beef packers, and therefore, the price will not be discounted. A disadvantage is that less beef is marketed per cow, and therefore, the basic cost of raising the animal must be covered with fewer pounds of beef. The primary advantage of the high-grain system is the rapid and efficient rate of weight gain, which reduces interest on invested input costs and the daily feedlot yardage costs. The use of some corn silage in place of grain in this system does not reduce feed costs because the price of silage is based on the price of the grain in the silage.

The following are some principles for designing beef production systems:

• Animal harvesting of feeds by grazing is economical. If the costs of fuel, equipment, and labor increase in the future, this will be even more economical.

• Crop residues are less expensive to produce than are conventional forages because the cost of production (land, fertilizer, etc.) is charged against the grain. Admittedly, harvesting costs may be high for crop residues, but conventional forages must be harvested as well, and the cost may be nearly as great.

• Grasses should primarily be grazed, not harvested. Some harvesting may be needed to provide winter forage and supplemental needs.

• There will be a premium on lean growth in beef cattle, but there is no premium for fat.

• Beef cattle should be finished on grain so that they have an acceptable amount of fat to meet current grading standards and market demands and, therefore, so that a profitable market price can be obtained.

• Cattle raised in high-forage systems make excellent compensatory weight gains during the early stages of finishing.

Because the market requires that beef carcasses be in an acceptable weight range, cattle used in high-forage systems should be heifers of large- or small-framed breeds or steers of smaller-framed (British) breeds. About 40 percent of the cattle fed in feedlots are heifers. In addition, many British breed steers are produced from young (first and second calf) cows, even if exotic terminal cross (large-framed) sires are to be used in later parturitions. Therefore, over half of the beef animals produced in the future would be expected to have frame sizes that are compatible with high-forage systems.

## DESIGN OF THE STUDY

The team conducting this study includes researchers and extension specialists from Nebraska, Missouri, and Iowa. The disciplines represented on the study team include beef nutritionists, agronomists, and agricultural economists. Cattle producers from the three states serve as consultants on the project.

Eight extension scientists are members of the project team. In addition to their input into the research, their most significant contributions will be in information dissemination. The research findings will be reported in cattle reports that are published by each state, and at least one field day will be held each year in each state in conjunction with the ongoing research. In addition, a three-state symposium was held on June 13 and 14, 1990, for producers and extension personnel in the three-state area. A publication will be prepared in the future for agent training and producer meetings. The primary target audiences are agricultural extension agents, U.S. Soil Conservation Service personnel, and farmers-cattle producers, particularly those who have marginally productive, highly erodible soils. Many of these producers will earn increased profits from these forage-based beef production systems. An additional expected benefit is greatly reduced water pollution and soil erosion.

In Missouri, the first objective will be to determine whether summer annuals can be planted into fall fescue grass sod by the no-till method. This information is needed because of the need to convert fungus-infested tall fescue pastures to fungus-free cultivars. The second objective will be to compare winter grazing of stockpiled tall fescue grass (allowed to grow tall without grazing or mowing) versus drylot feeding of hay as they relate to subsequent weight gains by steers. To optimize the utilization of forages, low-cost alternatives to winter feeding, such as grazing of forage that is stockpiled during the fall, should be associated with improved management of cattle in the pastures in summer. Cattle often do not gain weight at high levels during winter grazing of relatively low-quality forage, but they grow rapidly (known as compensatory growth) during the spring, when high-quality forage becomes available. Therefore, a period of spring grazing should be used in a beef production system so that the low-cost gain as a result of compensatory growth can be captured by the beef producer. The study will determine the efficacy of growing calves on stockpiled forage and the use of high-quality wheat pastures to accomplish compensatory growth in calves during the spring.

In Iowa, the project team is developing cow-calf production systems that maintain beef cow reproductive efficiency with minimal use of hydrocarbon fuels. Different systems of grazing of summer pastures composed of grass or grass-legume mixtures are being studied. Fall and winter feeding systems will compare grazing of crop residue and minimal hay supplementation with feeding of hay ad libitum (allowing the animals to eat as much as they want) to cows on drylots.

## EARLY FINDINGS

The following findings are based on the research that has been conducted and completed in Nebraska.

An accounting model was developed to aid in understanding biological and economic relationships and to study the impact of variations in resource costs on net returns earned through different beef production systems. The model compared production costs and break-even prices of cattle placed on high-grain finishing diets immediately after weaning with those of cattle grown on forage diets prior to finishing. Two types of experiments were conducted to establish a biological basis for the model. The first type of experiment compared high-grain versus high-forage growing-finishing systems. The second evaluated the effect of the rate of weight gain in winter on total system performance.

### High-Grain Versus High-Forage Experiments

Each year for 3 years, calves from 136 British breed cows and Charolais bulls born in the spring were weaned at an average age of 187 days and were used to evaluate the two systems. After an initial 30-day period to allow adjustment to weaning, the cattle were allotted randomly to either a high-grain system, in which they were placed directly into the feedlot for finishing on a 90 percent grain diet, or to a high-forage system, in which they were grown on forage diets prior to finishing (Table 16-1). Cattle in the high-grain system were adjusted to the finishing diet in 21 days and were then fed for an additional 185 days. Cattle in the high-forage system were wintered on crop residues (165 days), grazed on summer pasture (115 days), and then finished in the feedlot (122 days) in the same manner as cattle in the high-grain system were.

## Wintering Systems Experiments

Eighty mixed British breed steers (20 head per system per year) averaging 520 lbs were used each year for 2 years to evaluate wintering systems. These trials had two objectives: (1) to determine what level of performance could be expected with different wintering systems and (2) to establish three different levels of weight gain during winter to evaluate the effect of the rate of weight gain during winter on subsequent performance. Three rates of weight gain during winter were used: 0.62, 0.84, and 1.1 lbs/day. Averages of 2 years of data were used. After wintering (106 days), cattle were grazed on summer pasture (116 days) and then were finished in the feedlot (122 days).

Across both years, six different wintering systems were evaluated. These wintering systems used harvested crop residues supplemented with different levels of supplemental protein and alfalfa hay, as well as cornstalk grazing supplemented with harvested crop residues and protein supplement or alfalfa hay (Table 16-2).

**TABLE 16-1**  Finishing Performance Input of High-Forage and High-Grain Systems

| Item | High-Forage* | High-Grain[†] |
|------|--------------|---------------|
| Initial weight[‡] (lbs) | 521 | 521 |
| Final weight (lbs) | 1,162 | 1,058 |
| Days on feed | 112 | 189 |
| Daily gain[§] (lbs) | 3.72 | 2.84 |
| Daily feed (lbs) | 26.6 | 18.0 |
| Gain/feed | 0.140 | 0.158 |

*Grown on high-forage diets prior to finishing.
[†]Finished immediately after weaning.
[‡]Weighed at 207 days of age, i.e., the start of the experiment.
[§]Gain only in the feedlot: 112 days for forage cattle, 189 days for grain cattle.

**TABLE 16-2** Compositions of Diets Used in Wintering Systems

| Ingredient (lbs/head/day) | System* | | | | | |
| --- | --- | --- | --- | --- | --- | --- |
| | 1 | 2 | 3 | 4 | 5 | 6 |
| Husklage† | 9.7 | 7.8 | 6.78 | 3.21 | | |
| Alfalfa hay | | 2.2 | 6.78 | | 1.0 | 4.20 |
| Urea | 0.24 | 0.09 | 0.20 | | | |
| Corn | 0.59 | 0.15 | | 0.40 | | |
| Soybean meal | | | | | 0.95 | |
| Blood meal | | 0.18 | | 0.11 | | |
| Corn gluten meal | | 0.24 | | 0.15 | | |
| Molasses | | | | | 0.09 | |
| Vitamin and mineral premix‡ | 0.15 | 0.15 | 0.15 | 0.15 | 0.15 | |
| Cornstalks (acres/head) | | | | 1.0 | 1.0 | 1.0 |

*Cattle in systems 1, 2, and 3 were fed in drylot (106 days). Cattle in systems 4, 5, and 6 grazed cornstalks. Cattle in system 4 received husklage from a silo-press bag while grazing stalks.

†Husklage is ensiled corn residue that was passed through the combine at the time of harvest.

‡Vitamins: 100 international units (IU) of vitamin A, 20 IU of vitamin D, and 0.02 IU of vitamin E per gram of premix. Minerals: 55 percent dicalcium phosphate, 21 percent limestone, 19 percent salt, 0.33 percent magnesium, 0.20 percent zinc, 0.15 percent iron, 0.07 percent manganese, 0.017 percent copper, 0.01 percent iodine, and 0.002 percent cobalt.

## Model Development

The rate of winter weight gain used in the model was dependent on the particular wintering system selected. The research findings indicate that the rate of weight gain during winter negatively affected the subsequent rate of weight gain when the cattle grazed on pasture. The prices of corn and alfalfa hay are important elements of the model. The cost of supplemental protein is linked to the price of corn. Specifically, the price of soybean meal was calculated as 2.4 times the price of corn on a weight basis for the 10-year average relationship. The prices of all other protein sources were based on the price of soybean meal on an equal cost per unit of protein. The cost of corn silage was based on the price of corn.

## BREAK-EVEN BEEF PRICES OF ALTERNATIVE SYSTEMS

### Wintering Systems

Systems 1 and 2 were designed to produce two levels of weight gain by increasing the level of supplemental protein. Cattle wintered through system 2 gained 0.46 lb/day more than those in system 1, but their break-even

**TABLE 16-3** Effect of Wintering System and Level of Inputs on Break-Even Beef Price[*]

| | Gain (lbs/day) | | Break-Even Price |
|---|---|---|---|
| System | Winter | Summer | ($/100 lbs) |
| 1.  Husklage, drylot low input[†] | 0.57 | 1.45 | 67.10 |
| 2.  Husklage, drylot high input[†] | 1.03 | 1.08 | 67.75 |
| 3.  Husklage, drylot alfalfa[†] | 1.03 | 1.08 | 68.07 |
| 4.  Cornstalks, silo-bag[‡] | 0.73 | 1.32 | 66.26 |
| 5.  Cornstalks, protein supplement[‡] | 0.90 | 1.19 | 66.67 |
| 6.  Cornstalks, alfalfa[‡] | 0.73 | 1.32 | 66.88 |
| High grain | | | 68.37 |

[*]Medium input prices except high feeder cattle price.
[†]Yardage was $0.25/day.
[‡]Yardage was $0.10/day.

costs were higher by $0.65/100 lbs of gain (Table 16-3). This result occurred because of the cost of the increased weight gain during winter and the associated reduction in compensatory weight gain made on grass in system 2 compared with that in system 1.

The systems that used cornstalk grazing (systems 4, 5, and 6) were the most feasible, as long as they did not require large amounts of supplemental protein. This was due to the lower cost of cornstalk grazing compared with that of feeding harvested feedstuffs. Wintering feedlot yardage fees were included to cover the costs of facilities, labor, and management. These costs were $0.15 lower for cattle wintered on cornstalk fields compared with those for cattle wintered in drylots, thereby giving an additional advantage to the cornstalk grazing systems. Alfalfa proved to be an excellent protein and energy supplement for cattle wintered on cornstalk fields (system 6).

Total weight gain in the forage phase of the system was affected only slightly by the wintering phase. Therefore, expenditures for inputs to increase performance during winter were not profitable. Reductions in wintering yardage costs or other fixed costs were advantageous, while any added costs for extra performance during the winter were not.

### High-Grain Versus High-Forage Systems

The break-even beef price was lower for the high-forage system than for the high-grain system, except when the price of corn was very low relative to the prices of the other inputs. The extended feeding period required by the high-grain system (189 versus 112 days) implies that twice as much corn (or other grain) was required. However, because of the increase in size and compensatory growth potential, cattle in the high-forage system consumed more feed per day when they were placed on a finishing ration. They gained weight faster, but were less efficient in terms of weight gain per pound of grain (Table 16-1). Nonetheless, the price of corn had only a small effect on the comparative economic efficiency (break-even price) of the high-grain system versus that of the high-forage system (Figure 16-1).

Increasing interest rates increased the break-even price faster for the forage system than it did for the high-grain system (Figure 16-2). This was due to the need for a greater total investment and the longer time period required for the high-forage system. This effect was not large, however. For example, an increase in the interest rate of 5 percent with the high-input system increased the break-even price by only $2.49/100 lbs. Likewise, the differences in break-even beef prices because of the purchase price of feeder cattle were small and favored the high-forage system.

In general, cattle on feeding systems that increased weight gain during winter through higher inputs (between 0.62 and 1.1 lbs/day) were found to have decreased weight gains during summer. Since weight gain during winter required higher costs than did weight gain during summer, an increase in wintering costs decreased the economic feasibility of the system,

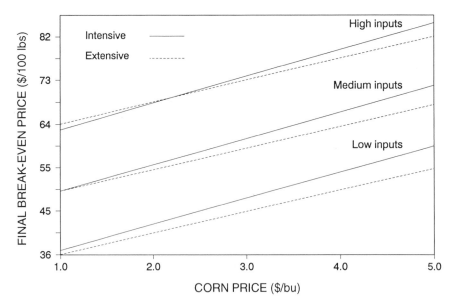

FIGURE 16-1   Effect of price of corn and finishing systems on final break-even price.

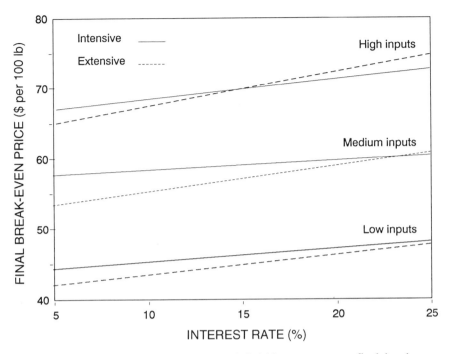

FIGURE 16-2   Effect of interest rate and finishing systems on final break-even price.

as reflected in the higher break-even price of beef.  Changes in corn prices and interest rates had relatively small effects on the comparison of high-grain versus high-forage feeding systems, even though total interest costs were higher for the high-forage system.  The greatest benefit of high-forage systems is increased total production per animal unit.  This increased product diminishes the average fixed cost of the feeder calf and, along with other efficiencies, yields a lower break-even price than that obtained with high-grain systems. Cattle finished after being grown on high-forage diets gained weight much faster, but they consumed almost as much total grain as those that were finished immediately after weaning.  Forage systems can produce more total beef at a lower cost per unit of product, except in times of very low grain costs relative to the costs of other inputs (interest, feeder cattle, etc.) or high wintering costs.

## Increased Grazing of Forages

Since it was observed that the cattle fed in the high-forage system made about 57 percent of their weight during high-grain finishing, the study was expanded to include the objective of increasing the amount of weight gain made while the cattle were on pasture. Ninety-six steer calves were used in this research, which was initiated on October 19, 1988. The calves grazed cornstalks from October 19, 1988, through March 1, 1989. They received different protein supplements, thus providing the opportunity to evaluate low-input supplementation with alfalfa or inexpensive by-products. The calves were then fed ammoniated wheat straw until May 3, 1989, when they were moved to brome pastures. Methods of increasing weight gain were (1) use of Sudan grass for summer pasture, (2) supplementation with bypass protein (protein that bypasses digestion in the rumen but is digested in the small intestine) and an ionophore (a feed additive that enhances the efficiency of fermentation in the rumen), and (3) extension of the grazing period until November 20, 1989.

Calves gained 0.95 lb/day while grazing cornstalks and 0.55 lb/day when they were given the ammoniated wheat straw. They gained 1.85 lbs/day when they were given brome until June 27 and gained 0.6 lb/day more if they were fed the supplement. Because of the drought during the period of this research, the cattle required more acres of pasture than normal, and they had better than normal weight gains on the brome from June 27 to September 4, 1989 (2.16 lbs/day). Cattle on the Sudan grass gained 2.35 lbs/day during that period, and none of these cattle responded to the supplement (data not shown). Overall, weight gains during summer were increased by the supplement (Table 16-4). The cattle that remained on the brome regrowth gained 2.38 lbs/day from September 4 to November 20, 1989.

On September 4, 1989, two-thirds of the cattle entered the feedlot and were finished in 101 days. They gained 4.08 lbs/day, and feed conversion was 6.4 (Table 16-5). The remaining one-third of the cattle that entered the feedlot on November 20, 1989, gained 3.40 lbs/day for 94 days at a feed conversion of 8.7. Break-even prices of these cattle were about $6.00 to $13.00/100 lbs better than those of the cattle in the high-grain system, that is, those that were finished on grain immediately after weaning (Table 16-5).

**TABLE 16-4**  Weight Gains During Summer Grazing

| System | Weight Gain (lbs) | |
| | Summer | Daily |
| --- | --- | --- |
| Continuous brome | 240 | 1.95 |
| Brome and Sudan grass | 248 | 2.02 |
| Brome and supplement | 275 | 2.24 |

**TABLE 16-5** Finishing Performance and Break-Even Prices Under Various Systems

| | High Grain* | Continuous Brome† | Brome and Sudan Grass† | Brome and Supplement† | Brome Fall Regrowth* |
|---|---|---|---|---|---|
| Daily gain (lbs)‡ | 2.73 | 4.00 | 4.07 | 4.18 | 3.40 |
| No. of days on system‡ | 207 | 101 | 101 | 101 | 94 |
| Lbs of feed/lbs of gain‡ | 6.19 | 6.16 | 7.02 | 5.98 | 8.70 |
| Final weight (lbs) | 1,075 | 1,276 | 1,291 | 1,337 | 1,389 |
| Break-even price ($)§ | 74.15 | 67.02 | 71.45 | 65.77 | 65.25 |
| Lbs of grain/lbs of gain‖ | 4.89 | 2.39 | 2.72 | 2.27 | 2.31 |

*Cattle were placed on the finishing diet at the time of weaning.

†Cattle were wintered on cornstalks and ammoniated straw. Summer grazing was continuous brome from May 3 to September 4, 1989, brome from May 3 to June 27, 1989, and Sudan grass from June 27 to September 4, 1989; continuous brome plus a protein and ionophore supplement and continous brome were given from May 3 to November 20, 1989.

‡For the feedlot finishing period only.

§Final break-even price for the cattle.

‖Pounds of grain fed per pound of total weight gain achieved from the time of weaning to the time of market.

## CONCLUSIONS AND FUTURE CHALLENGES

Corn Belt farmers can produce cattle competitively by using low-input, soil-conserving, high-forage systems. The management needed in this system is very intensive, however. It is relatively easy to put cattle in a feedlot and feed them corn. It is much more difficult to manage cattle that are grazing cornstalks when fields are muddy or covered with snow. It is a major challenge, even to the most capable livestock manager, to maximize gains on summer pasture while working around droughts and other forces that constantly change the amount of feed (grass) available. It can also be done economically, however, and in a manner that conserves soil, enhances the environment, and is sustainable.

The trend in the beef cattle industry has been toward high-grain feeding. The preliminary findings of this LISA project are contrary to that trend. However, several constraints limit adoption of high-forage systems:

1. *Cattle producer attitudes.* Compared with high-grain systems, forage utilization has the image of being old-fashioned and not progressive. University researchers and extension personnel frequently share this attitude; they are partially responsible for the attitudes of producers that use of grain is the only way to feed cattle. Major changes in attitudes are required to increase forage utilization.

2. *Government policies favor grain production.* It is amazing that gov-

ernment programs are primarily developed for grain producers and that the consequences on livestock production are generally ignored. Grain subsidies encourage feeding of grain at the expense of forage utilization. This is not consistent with the goals of sustainability. For example, alfalfa is an excellent crop for use in rotations with corn and soybeans because it adds nitrogen, reduces erosion, and improves soil tilth. Alfalfa production is much less profitable than subsidized grain production is, however.

3. *Applied research.* A mixture of basic and well-designed applied research is urgently needed to solve today's problems, but it is not being adequately funded. Funding is now out of balance. With the exception of the LISA program, most research funding for forage and beef research is for very basic research (primarily biotechnology). There is a widely held perception that biotechnology will provide a quick fix for all of agriculture's problems through bioengineering of animals or production of a magic drug. The major limiting factor is consistent, good-quality, long-term applied research that will help producers today and that will move the industry toward sustainability.

## REFERENCES

Lindstrom, M. J., S. C. Gupta, C. A. Onstad, W. E. Larson, and R. F. Holt. 1979. Tillage and crop residue effects on soil erosion in the Corn Belt. Journal of Soil and Water Conservation 34:80–82.

Smith, G. M., D. B. Laster, L. V. Cundiff, and K. E. Gregory. 1976. Characterization of biological types of beef cattle. II. Post weaning growth and feed efficiency of steers. Journal of Animal Science 43:37.

# 17

# Reactor's Comments

## Low-Input Sustainable Agriculture Projects, *Alternative Agriculture,* and Related Issues

*Harold F. Reetz, Jr.*

This review first comments on the three low-input sustainable agriculture (LISA) projects presented in this section of the volume and then provides some general comments on the workshop on which this volume is based and the report *Alternative Agriculture* (National Research Council, 1989).

### THREE NORTH CENTRAL LISA PROJECTS

#### New Strategies for Reducing Insecticide Use in the Corn Belt

This review of new strategies for reducing insecticide use by Gerald R. Sutter and David R. Lance emphasized the need to control the corn rootworm, which causes over $1 billion in crop losses annually. They discussed several pest management strategies that offer the potential for reliable control of corn rootworm, while at the same time reducing the use of chemical insecticides.

Much of this project is targeted at the management of adult beetle populations as opposed to the common practice of controlling the larvae by applying insecticides to the soil.

Crop rotation has been a common method of adult beetle population management. For most Corn Belt farmers, rotation has been the main defense against corn rootworms. By eliminating the host plant from the field where the eggs are laid, the life cycle is broken. Recent studies, however, have shown that up to 40 percent of the rootworm eggs may overwinter for 2 to 5 years, so rotation is not always successful.

The use of various attractants to concentrate adult populations, so that insecticides can be applied to a smaller percentage of the total corn acreage, has the potential of greatly reducing the amount of insecticide used. This could greatly reduce insecticide costs for continuous corn systems, and could reduce the potential for environmental contamination from insecticides.

New bait systems are under development in which plant-applied insecticides are used at much lower rates compared with soil-applied materials. Many details must be worked out, but the environmental and economic benefits look promising.

These studies must be expanded to large field-scale tests to study the true effects on rootworm population dynamics. This will be an interesting project to watch in the coming years. This work is an example of the development of innovations from conventional systems that are helping to make U.S. agriculture more competitive and more sustainable.

## On-Farm Research Comparing Conventional and Low-Input Sustainable Agriculture Systems

This project described in the chapter by Thomas L. Dobbs and colleagues involves an economic analysis of whole-farm crop systems. While the low-input system's livestock component was not included in the analysis, the value of the crop products used in the livestock systems was included in the enterprise analysis.

An important philosophy discussed in the review of this project was that whole-farm systems are analyzed to identify specific components that can be studied in greater detail in traditional research projects. These specific studies determine the best management practices for the soil-climate-crop system and the goals and abilities of the manager. The practices thus identified are then applied to the whole-farm system, and the impact on the system is then measured.

This approach is important to the development of site-specific management recommendations, which will become increasingly important in crop management in coming years.

One problem identified in the analysis was the difficulty in dealing with changes in the crop rotation during the course of the project. This is a common problem in conducting on-farm research, but in a way, it is more reflective of real-world conditions.

The low-input case study suffers from low-return crops in the rotation. This is one of the main reasons that such rotations have been abandoned in conventional farming systems. It is probably even more dramatic in the central and eastern Corn Belt, where corn and soybean yields are 50 to 100 percent greater than those obtained in this study conducted in South Dakota. When one adds to that the comparative competitive advantage of producers in the central and eastern Corn Belt for corn and soybeans because of their proximities to markets (domestic and foreign), input sources, and transportation arteries, as well as more favorable climate and soil conditions, it becomes very difficult to see an opportunity for low-input rotation systems to compete with the existing intensive corn and soybean management systems used in the Midwest.

The researchers concluded that the low-input system could be made more competitive with conventional systems by adding a premium price to organically grown crops and imposing a 25 percent "tax" on fertilizer and chemical inputs. There are two problems with this approach:

1. Organic premiums would not exist if a large number of farmers adopted the same management system and organically grown crops were readily available to all consumers.

2. Imposition of such a tax on inputs simply for the reason of reducing their use is agronomically unsound and economically unacceptable. Such a policy is tantamount to legislating production practices.

It is inconceivable that the political system at the federal or state level can develop an equitable system for determining the appropriate cultural practices (such as fertilizer and chemical application rates) to be used on the wide range of crops-soil-environment-management systems across the country. In fact, more success will be found in working toward the site-specific management approach mentioned above, where fertilizer use is based on detailed soil samples and realistic yield goals, and pesticide use is based on an integrated pest management system, including regular field scouting and combinations of appropriate chemical, cultural, and biological control methods.

Overall, the first 5 years of this comparison have provided some interesting data. Better control over the rotation plan will perhaps help improve the comparison in the future. The addition of more farms to the study could be helpful as well. Management system comparisons require many years of study to provide sufficient data to establish the trends and allow for the selection, study, and reintroduction of specific management components.

Projects such as this one are a critical part of the research needed to determine which practices can fit into profitable management systems. New practices must first be tested in more intensive research projects, but they should eventually be incorporated into management system comparisons such as the one described by Dobbs and colleagues, so that the overall impact on a farming system can be evaluated.

## Low-Input, High-Forage Beef Production

This low-input, high-forage beef production project described by Terry Klopfenstein is a comparison of extensive management versus intensive management systems for beef production. It involves a search for ways to use crop residues as a main component of the beef production system. Grain finishing is still used, but for a shorter period of time. The overall feeding cycle is lengthened, but differences in feeding costs help offset the time efficiency factor.

The crop residues that are used are less expensive than conventional forage crops. Corn stalk grazing provides the main roughage component, with alfalfa hay used as a protein supplement. Animal harvesting is used when possible to reduce harvesting costs and labor.

This project has the potential to help keep small-scale beef feeding competitive as conventional feedlots are challenged by the swine and poultry industries for grain feeding efficiency.

This project is a good example of a cooperative effort between research and extension programs, which are essential elements of maintaining a viable research and extension system.

*Future Challenges*

The report concludes that the trend toward high-grain feeding must be reversed if the LISA concepts of this project are to be implemented. This comment presupposes that these concepts should be implemented. That conclusion, like other management decisions, should be made on the basis of research data. The data from this project provide some of that support, but more data are needed. If the high-forage system does prove to be more profitable, it will be adopted. The trend toward high-grain systems has resulted from the fact that they were more efficient.

It is not fair to attack university research and extension personnel or cattle producers for their attitude that forage utilization is old fashioned. I have worked with a number of people who consider forages an important part of livestock production and who have developed high-yielding and profitable forage and livestock systems. Grain feeding is also an important part of the crop and livestock management system.

The authors state that government policies favor grain production. While this is partly true, there have also been cost-sharing programs that have provided some support for forage production. Government programs for lime, rock phosphate, conservation plans, and several similar government programs have tended to support forage production and make it more profitable. Most of these programs have been terminated, but a few still exist.

Alfalfa is an excellent crop for rotation with corn and soybeans, as stated above, and it provides help in controlling erosion, improving tilth, and supplying nitrogen. It is not fair, however, to conclude that alfalfa production would be dramatically increased if grain subsidies were reduced. Alfalfa production must be tied to demand for alfalfa for this system to be viable. A substantial increase in the acreage of alfalfa would destroy the market for those who now depend on alfalfa as a cash crop. There is much more than grain price support programs involved in the overall balance of grains and forages.

Regarding applied research, I share the concern that funding for applied research is not in balance with that for basic research. This is also a complicated issue, however. Politically, it is easier to get funding for programs that have quick turnaround times. The issue of accountability and the need to get quick results is a major driving force. Reinforcement of the formula funding program for research and extension would help provide the long-term funding needed for applied research.

Perhaps a more critical problem is the system that is in place for evaluating research and extension performance. Scientific publications are the main "measuring stick" for professional accomplishments. Applied research is slow to accomplish and difficult to publish. This makes it unattractive to young scientists who are trying to build a career. Until the recognition of applied research is improved and publication of applied research data is acceptable to the scientific community, this imbalance will continue. Progress is being made, but more is needed.

Establishment of a separate research and extension system for sustainable agriculture is not the answer. That would merely establish another bureaucracy that would skim off already limited resources. The change should be made within the existing U.S. Department of Agriculture (USDA) and land-grant research and extension system.

Biotechnology is not a quick fix, but there are important gains to be made from biotechnology. The basic versus applied pendulum will continue to swing. Attempts must be made to keep a balance between the two extremes. Both are needed to sustain agriculture as a viable industry.

## GENERAL COMMENTS

I appreciate very much the opportunity to represent the agribusiness community as a participant in this workshop and the opportunity to comment on these projects and the LISA program in general. My training is in crop ecology, and I have always been concerned about protecting the soil and water resources on which the agricultural production system depend. Continued communication and dialogue are essential to help avoid misinterpretation and misunderstanding on all sides of the issues raised by the LISA program.

The fertilizer and chemical industries have sometimes overreacted to the ideas presented in relation to the LISA program. The industry has been put in a defensive position, however, by many of the LISA-related policy suggestions. Policies proposed under the 1989 Fowler Bill (U.S. Senate, The Farm Conservation and Water Protection Act of 1989), for example, which recommended a 40 percent reduction in pesticide and fertilizer use, are an overreaction to the perceived potential for problems associated with the use of these materials.

A whole series of suggestions for the 1990 farm bill proposed by a coalition of environmental advocacy organizations fly in the face of scientific reality. They are based on emotion and philosophy more than on scientific facts.

The USDA (including the various research and educational agencies within it) and the National Research Council cannot afford to abandon their long-standing insistence on sound scientific methods in research and sound research evidence upon which to base extension and other education programs.

Production practices cannot be effectively legislated. True progress toward a more environmentally responsible, economically sustainable agriculture system will be made only through more site-specific, intensive management systems that attempt to identify and systematically eliminate limiting factors that are holding down productivity or creating potential environmental hazards. These recommendations must be based on solid research information and local experience for the given soil-plant-climate-environment system and for the experience and management ability of the individual farmer and the team of advisers (extension adviser, crop consultant, Soil Conservation Service conservationists, dealers, etc.) who provide technical support. Farmers are good stewards. They depend on their soils and water supplies for their businesses and their families. They depend on their management systems to sustain their businesses. They will not intentionally destroy the resources on which that business is built.

Farmers do not intentionally buy excess pesticides and fertilizers. In fact, farmers would prefer to not buy any pesticides and fertilizers. Farmers buy pesticides because they want dead weeds and insects—pests that rob them of their narrow profit margins. When they buy fertilizers, they are buying increased yield potential—increased profitability and quality for the crops they produce. Farmers buy these inputs to the extent—and only to the extent—that they can expect to improve the profitability of their farming operations.

Terms such as *satanic pesticides* and *synthetic chemical fertilizers* have been used by some of the speakers at this symposium, but they are inaccurate and cause the uninformed listener to develop misleading impressions of the pesticides and fertilizers used by farmers and of the industries that produce them. Increasingly, they cause consumers to fear for the safety of the food supply, when, in fact, the United States has the safest, most highly regulated, and lowest-cost food supply in the world.

The fertilizers commonly used on U.S. farms are naturally occurring nutrients that undergo a minimum amount of processing to make them more easily handled for uniform application and more readily soluble for most efficient use by growing crops. Fertilizers merely supplement the natural supplies of the same nutrients in the soil and replace the nutrients that are removed by the harvested crop.

## COMPETITIVE GRANTS

Charles Hess and others have expressed support for expanding the funding for competitive grants under the 1990 farm bill. Such grants give an opportunity to fund projects that will provide quick turnaround, provide information to supply the need for accountability in the political arena, and support the pressures for scientific publications within university and USDA promotion systems. These are all important goals. A strong level of support for the long-term research programs on traditional subject areas such as soil management, plant breeding, and crop nutrition must be maintained, however.

In these subject areas, answers come slowly because researchers are dealing with climatic variabilities that can mask real treatment differences, or because several generations of materials must be evaluated to make progress.

 These processes take time. They do not produce instantaneous or exciting results. They do provide, however, the basic information and technological developments that must continue to serve as the framework on which the new technologies of genetic engineering, biotechnology, computer simulation, and other high-technology projects can be tested and implemented.

Since the mid-1970s, I worked closely with the late Herman Warsaw, the world-record corn producer (370 bushels/acre in 1985), as he developed a crop management system that not only broke yield records but also revitalized a badly eroded, low-productivity farm. By rebuilding high fertility levels, Warsaw was able to grow increasingly higher yields, which returned increasingly large amounts of crop residues to the soil. (University of Illinois and Purdue University research has shown that for each additional pound of grain produced, a corn crop produces an additional 1 pound of aboveground stover and up to 0.75 pound of roots. This ratio holds fairly constant throughout a yield range from 100 bushels/acre to over 300 bushels/acre.) Thus, when Warsaw chisel-plowed his corn residue, leaving one-third of it on the surface, he was leaving the equivalent of the residue from an *average* corn yield on the surface to control erosion, while turning under twice that amount of plant material to help build soil structure, organic matter, and overall tilth.

People like Herman Warsaw challenge their fellow farmers and researchers to look for the limiting factors in their own production systems. Small-plot research and demonstration studies are needed to test new ideas. Then, these ideas must be incorporated into mainstream farming systems to measure their impact and potential for widespread adoption. Whether high yield or low input is the goal, the research procedures are the same. A low-input research and extension program is not needed. Continued support for existing research and extension programs is needed so that they can test the low-input alternatives along with conventional practices. Then, the alternative practices that prove to have merit can readily be moved into mainstream production channels.

For the past 25 years, the Potash & Phosphate Institute has promoted maximum economic yield (MEY) crop production. During the 1980s this was a major thrust of its research and educational programs. Determination of the maximum potential yield for a given site is an important first step. This is done with small-plot research. Then, economic analysis is used on small-plot and field-scale tests to determine the MEY level. The goal is to improve profits, but to do so in a way that is agronomically sound, economically efficient, and environmentally responsible. MEY is low-input per unit of output. MEY is efficient, profitable, and sustainable. In the end, MEY production systems and sustainable agriculture production systems may not be much different. The goal should be to determine, on the basis of scientific research and on-farm evaluation, what are the best management practices for a given soil-plant-environment system.

## ALTERNATIVE AGRICULTURE

The report *Alternative Agriculture* (National Research Council, 1989) has stimulated much discussion and attracted much attention since it was released in the fall of 1989. The report contains a great deal of information about farming systems and their potential impact on the environment and the economic viability of U.S. agriculture. The report has also raised some serious concerns that must be addressed:

1. Major changes in production systems must be based on science. The report states that research is lacking in many of the subjects discussed. The LISA projects described in this volume are a step in the right direction, but too much of the report is based on emotion and philosophy and not enough is based on research.

2. The environmental problems mentioned in the report are not well documented. Most are actually isolated cases of accidents or mismanagement.

3. Reading of the report by the public and news media interpretations of it have caused undue concern among the general public about the safety of the U.S. food supply. As stated earlier, U.S. consumers enjoy the safest, most abundant food supply in the world—and at very low cost.

4. The report implies that manure and legumes are environmentally benign sources of nitrogen (N). Researchers throughout the Midwest have shown that legumes and manure may actually be more likely to release N at a time when it is highly susceptible to leaching, denitrification losses, or both. N fertilizers can be more readily controlled for rate and timing of application. This does not mean that legumes and manures are not a valuable source of nutrients, but careful attention to management details is required for their most efficient use.

5. Erosion is presented in the report as being a serious threat, which is true. Replacement of chemical weed control with mechanical cultivation can lead to increased erosion, however. Well-fertilized crops develop better root systems and more total dry matter, which help improve the permeability and water-holding capacity of the soil. Below-maintenance fertilizer applications result in reduced biomass production and lead to a greater potential of erosion. Robert Klicker's discussion (this volume) on the importance of high fertility for maintaining erosion control and profitability in wheat production in the Palouse area of the Northwest is a good example, as is the system of high-yield corn production developed by Herman Warsaw in Illinois. Low-input systems generally reduce root development and total dry matter production. This is not sustainable.

6. As a scientist and former extension specialist, I am concerned about the weak scientific basis for the *Alternative Agriculture* report. The prestige and integrity of the National Academy of Sciences, National Research Council, and USDA are threatened by some of the conclusions presented without sound scientific basis. As an extension specialist—and as an industry agronomist—I have always insisted on the use of sound research as the basis for promoting changes in production practices. I cannot maintain my integrity as a scientist unless I demand the same from the LISA program.

## INTERGRATED CROP MANAGEMENT PROGRAM

Charles Hess reported (this volume) on the initiation of a special program of the USDA Agricultural Stabilization and Conservation Service, known as the SP53 Integrated Crop Management Program. This pilot program was designed to determine the effects of reducing pesticides and fertilizers by 20 percent on a group of 100 farms in each state. The concept of a categorical reduction of inputs by 20 percent without a scientific basis (pest scouting, soil tests) is philosophically wrong. In fact, that approach is an insult to the research and extension programs and agribusiness efforts that have been directed toward the development of integrated crop management systems.

As stated earlier, production practices cannot be effectively legislated. A better approach would be to use consultants as provided in the protocol, but use them to evaluate a farm's needs and make sound recommendations specifically designed for a given farm. Fertilizer applications and pest management decisions must be more site specific and must be based on detailed local field data from soil survey, soil testing and plant analysis, and pest scouting.

It is also unlikely that the 3-year duration of the program is sufficient to evaluate the effects of reducing inputs. Such short-term projects may be sufficient to evaluate a component, but as has been shown in the South Dakota project reported in this volume (see the chapter by Thomas Dobbs and colleagues), farming systems projects are long-term studies. The intent of the SP53 Integrated Crop Management Program may be different from the initial program of implementation. There is still time to make it workable.

## ARE THE VARIOUS GROUPS REALLY AT ODDS?

The fertilizer and chemical industries depend on a viable, profitable, sustainable agriculture system. I am not at odds with the LISA program's overall goals of sustainable production systems. The problem lies in the ideas on how to reach those goals and how we define sustainable. In fact, the problem really lies in the ability and willingness of the various groups to communicate with each other. Agriculture has a poor image among the general public—either as a wasteful, irresponsible, environmentally destructive industry or as the farm couple in Grant Wood's portrait, *American Gothic*. Neither of these is an accurate portrayal of the technically advanced, business-oriented, environmentally conscious farmer of today who will be the farmer of the twenty-first century as well.

I do not apologize for my involvement in the high-yield research and education programs in universities and industry that have helped make the United States the low-cost producer of high-quality food, fiber, and energy products that it is today. I do not ask environmentalists to apologize for their concern that society should strive to protect soil, air, and water resources. Our goals are compatible. Let's all work together to keep U.S. agriculture number one in the world—sustainable agronomically, economically, and environmentally.

## REFERENCE

National Research Council. 1989. Alternative Agriculture. Washington, D.C.: National Academy Press.

# PART FIVE
# Research and Education in the Northeastern Region

# 18

# Long-Term, Low-Input Cropping Systems Research

*Rhonda R. Janke, Jane Mt. Pleasant, Steven E. Peters, and Mark Böhlke*

One of the most challenging areas of research within the mandate of low-input sustainable agriculture (LISA) involves long-term cropping system studies. Agricultural scientists are continually faced with a dilemma when implementing research of this nature. The problem is how to conduct component research that tests a specific hypothesis about one or two factors (e.g., nitrogen fertility or weed control) while maintaining the realism and complexity of the cropping systems described in whole-farm comparisons and case studies. One of the myths that dominated the research community in 1981 was that LISA-type cropping systems were nothing more than conventional systems without the use of chemicals (Harwood, 1984). As a result, component research at that time "proved" that these systems were inferior to the conventional management approach. Experiments designed with this primary assumption (bias) failed to take into account the fact that chemical usage is only one of many factors involved in designing the efficient, well-structured, integrated, and biologically stable systems that have been found on many commercial organic farms (National Research Council, 1989; U.S. Department of Agriculture, 1980). The interactions among the components of an efficient farm enterprise, such as the use of cover crops and animal manures, biological nitrogen fixation, nutrient uptake and release rates, weed dynamics, disease and insect suppression, and the rotation effect, are poorly understood (Harwood, 1985). Replicated and controlled settings are needed to explore, evaluate, and comprehend these vital processes. In addition to answering questions about basic biological processes, well-researched model cropping systems provide valuable agronomic and economic information for farmers, extension agents, and agribusinesses.

This chapter describes three studies of long-term cropping systems currently being conducted in the United States. Several other studies are ongoing at other locations (California, Nebraska, Ohio, and Michigan, to name a few). The three studies described in this chapter are located in New York and Pennsylvania and received funding from the LISA program of the U.S. Department of Agriculture (USDA) for the 1988 and 1989 growing seasons. Some of the agronomic and economic results from these studies are highlighted, although two of the three studies were established in 1988, and from a pragmatic point of view, these data should be considered preliminary results.

This raises a question of concern for researchers involved with long-term studies: How are sustainable funding sources for low-input sustainable agriculture research obtained? If it takes between 5 and 10 years to begin to get useful data from experiments (posttransition period), who should be asked to bear the cost of supporting this research? Private industries, by nature, tend to support research that is usually geared only toward product development and testing. Public research funds are mandated for research that benefits the public at large. Low-input sustainable agriculture now has the support of the public because of the increasing awareness of the negative environmental effects of some conventional agricultural practices, including groundwater and surface water pollution and soil erosion. The public has also become more concerned about food quality and safety. It is clear that long-term research is needed to develop and improve upon options available to farmers. However, the public funds allocated to sustainable agriculture (the LISA program of USDA) continues to flow in small, 1-year grants. This is a contradiction that must be resolved.

## THE RODALE FARMING SYSTEMS TRIAL

A long-term study was initiated in 1981 at the Rodale Research Center in southeastern Pennsylvania to examine the process of converting from a conventional to a low-input/organic cropping system. Three representative cropping systems were designed:

1. A low-input (i.e., low purchased input) system with animals (LIP-A) that simulates a crop and livestock farm (typical of farms in Pennsylvania and several other regions of the country) uses a 5-year rotation that includes corn grain, soybeans, small grains, legume hay, and corn silage. Animal manure is applied prior to each crop of corn to supplement the legume nitrogen from the previous hay or soybeans (Table 18-1).

2. The low-input cash grain (LIP-CG) system is based on the assumption that a cash crop is needed each year for cash flow and that no animal manure is available.

**TABLE 18-1** Five-Year Rotations for the Farming Systems Trial, Rodale Research Center

| System | Crops | | | | |
| | Year 1 | Year 2 | Year 3 | Year 4 | Year 5 |
|---|---|---|---|---|---|
| Low-input with animals | Wheat/ alfalfa + red clover* | Alfalfa + red clover | Corn | Soybeans | Corn silage |
| Low-input cash grain | Barley/ soybeans† | Wheat/ red clover* | Corn | Barley/ soybeans† | Wheat/ red clover* |
| Conventional cash grain | Corn | Soybeans | Corn | Soybeans | Corn |

*Alfalfa and/or red clover was frost seeded (broadcast seeded in March) into wheat.
†Soybeans were relay planted (drilled into a small grain) into spring barley.

3. The conventional (CONV) rotation of corn and soybeans is grown with purchased fertilizer, herbicides, and insecticides applied at rates as provided in guidelines of Pennsylvania State University (University Park, Pennsylvania).

The low-input (LIP) rotations rely on crop rotation, cover crops, relay cropping, and mechanical cultivation for weed control and only on green manures and animal manures as nitrogen sources. Each of the three crop rotations was started at three different points in the rotation, for a total of nine treatments. These treatments were replicated eight times in a split-plot, randomized complete block design. Whole plots are 60 × 300 feet, and subplots are 20 × 300 feet. Cropping systems (LIP-A, LIP-CG, or CONV) were assigned to whole plots, which were split by rotation entry point. More detailed descriptions of and results from this experiment can be found in the literature (Andrews et al., 1990; Culik, 1983; Harwood, 1984, 1985; Liebhardt et al., 1989; Peters et al., 1988; Radke et al., 1988).

During the biological transition period (1981 to 1984), that is, during the process of converting from CONV to LIP methods, a change in the equilibrium between soil processes and plant growth was anticipated and observed. Evidence that supports this includes the fact that both LIP systems (animal and cash grain) had lower corn yields than the CONV cropping system did in 1981 to 1984 (Figure 18-1), whereas all three systems have had similar corn yields since then. This is attributed to a lack of nitrogen in the soil that was available to plants in the LIP systems during the transition years, as reflected in corn ear leaf nitrogen analysis (Figure 18-2), but there has been an adequate supply of nitrogen from 1985 to the present. Excessive weed growth in some years may have also limited crop yield and may have been compounded by the lack of nitrogen, although no consistent pattern of weed increases, decreases, or species shifts was apparent from the transition period data. Annual weeds dominate in the LIP rotations, and perennials dominate in the CONV system, partly because of herbicide selectivity.

Soybean yields were similar in all three systems during the transition years (Liebhardt et al., 1989), and small-grain and hay yields compared favorably to county averages (Peters et al., 1988). Thus, corn was the only crop that appeared to be limited during this biological transition period. Conclusions about the best way to go through the transition agronomically (begin the rotation with a small-grain and cover crop or soybeans) were verified in an economic analysis (Duffy et al., 1989) of the transition years. When the rotation begins with a small grain and legume or soybeans, the result is a greater return to land and management than that from "cold turkey" corn in the LIP systems. If higher crop prices because of federal government price support programs are not factored in, the order of average returns from each system (combining all rotation entry points) was as follows: LIP-A > CONV > LIP-CG. The CONV system was the most profitable if the government price support program for corn was included. However, because rotation entry point was a significant factor, the most profitable LIP-CG treatment resulted in a higher average annual net return than that from the least profitable CONV treatment.

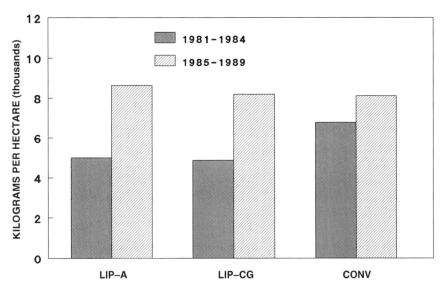

FIGURE 18-1   Average corn yields from the farming systems trial from 1981 to 1984 (during conversion) versus those from 1985 to 1988 (postconversion).

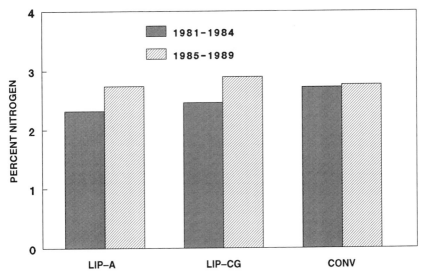

FIGURE 18-2   Average tissue nitrogen content of corn ear leaves at the time of silking from the farming systems trial for 1981 to 1984 (during conversion) versus that for 1985 to 1988 (postconversion).

Initially called the conversion experiment, the study was renamed the farming systems trial in 1986. Emphasis shifted toward an assessment of the long-term reliability of LIP practices. The four major objectives are (1) to compare crop performance in the three cropping systems; (2) to compare the economic viability of these rotations and input regimes; (3) to test the sustainability, regenerative capabilities, and environmental impact of the LIP approaches by monitoring soil chemical and physical properties, weed pressure, and nitrogen cycling processes over time; and (4) to continue to encourage active collaboration with the agricultural research community for increasing understanding of the mechanisms of soil and plant processes in a biologically complex environment.

Total biomass production and grain yields have essentially been the same in all systems since 1985. Nitrogen does not currently limit corn yield in any system, as determined by the ear leaf tissue test (Table 18-2) and by extensive testing of soil nitrate nitrogen levels during the growing season (Figure 18-3A and B). Similar levels of nitrogen are supplied to corn crops in each rotation, amounting to 130 pounds (lbs) of nitrogen per acre as fertilizer in the CONV rotation, an average of 75 lbs of nitrogen per acre in the top growth of the clover before plowing in early May in the LIP-CG rotation, and an average of 205 lbs of nitrogen in beef manure per acre applied to the LIP-A system from 1986 to 1990. Most agronomic literature advises that only 50 percent of the nitrogen in the beef manure is available

**TABLE 18-2**    Corn Yield, Tissue Nitrogen Content of Ear Leaves at Silking, and Weed Biomass for 1988 and 1989, Farming Systems Trial, Rodale Research Center

| Treatment | Corn Yield (bu/acre) | Tissue Nitrogen (% N) | Weed Biomass (lbs/acre) |
|---|---|---|---|
| 1988 | | | |
| LIP-A* | 110a† | 2.86a | 900a‡ |
| LIP-CG | 109a | 2.96a | 449b |
| CONV (soybeans) | 104a | 2.79a | 88c |
| CONV (corn) | 85b | 2.74a | 250bc |
| 1989 | | | |
| LIP-A | 124ab | 2.64a | 1,343a |
| LIP-CG | 111c | 2.72a | 100c |
| CONV (soybeans) | 130a | 2.87a | 251bc |
| CONV (corn) | 116bc | 2.86a | 486b |

NOTE: LIP, low input; A, animals ; CG, cash grain; CONV, conventional.

*Previous crop for LIP-A treatment was 2-year-old red clover-alfalfa mixture; for LIP-CG was a 1-year-old red clover stand; and CONV treatments followed either soybeans or corn.

†Letters designate statistical differences at the $p < 0.05$ level by using analysis of variance (performed with Statistical Analysis System software) and Duncan's multiple range test. The Duncan letters should be read within a year, within a column only.

‡A plus/minus weed study was superimposed on the experiment to determine whether ambient weed levels caused yield suppression. Only the LIP-A treatment in 1988 showed statistically significant yield reduction in the "plus" weed subplot (yield was about 80 percent that of the hand-weeded control plot).

to crops the year of application, with a fraction of that amount of nitrogen being available in subsequent years. The literature also indicates that the legume nitrogen is only available to corn in the long term and not the short term, but data from these studies indicate that no supplemental nitrogen fertilizer is currently needed to meet the nitrogen needs of either the LIP-A or the LIP-CG systems.

Weeds are often more abundant in the LIP systems than they are in the CONV systems, but weeds have reduced the corn yield in only two systems since 1986. Corn in the LIP-CG system in 1986 (data not shown) and the corn in the LIP-A system in 1988 showed statistically significant yield reductions (Table 18-2) in unweeded versus hand-weeded subplots. The highest yields in 1988 were from corn in the LIP-A system, despite the weeds. The critical weed threshold levels change from year to year, depending on weather and growing conditions. Over 1,300 lbs of weeds (dry weight) per acre did not decrease the corn yield in the LIP-A system in 1989, probably because rainfall was plentiful and timely and moisture was not limiting (it was limiting in 1988).

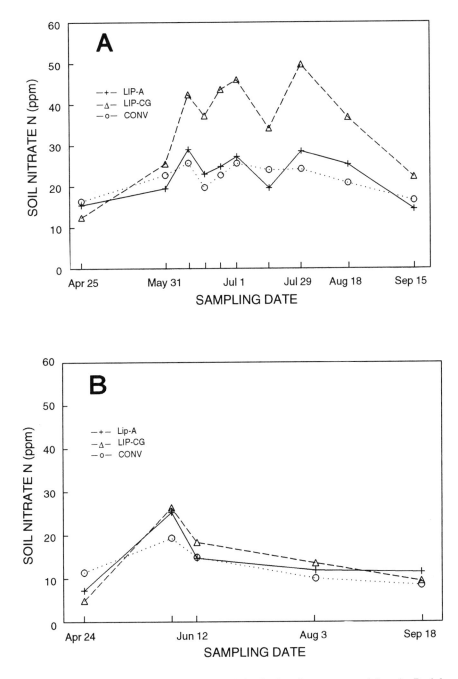

FIGURE 18-3   Soil nitrate nitrogen levels in the farming systems trial at the Rodale Research Center in (A) 1988 and (B) 1989.

Soil biology data (Doran et al., 1987; Fraser, 1984; Werner, 1988) indicate that the LIP systems have greater microbiological activity and a greater abundance of many microarthropods. The microbial activity is attributed more to the diversity of crops in the rotation, especially the legume cover crops and hay, and to the application of animal manure in the LIP-A system than it is to the absence of pesticides and chemical fertilizers. This increased biological activity may partially account for preliminary results from $^{15}N$ studies (Harris et al., 1989), which indicate that less nitrogen is lost from the LIP-CG system than is lost from the CONV system and that more nitrogen is retained in the soil in the LIP-CG system. Preliminary work by a graduate student at Ohio State University (Columbus) will help to determine the role of various fractions of organic matter in holding onto the nitrogen, and sampling of vesicular-arbuscular mycorrhizae by a researcher of the Agricultural Research Service of USDA in Wyndmoor, Pennsylvania, will help to determine the role of these organisms in nutrient cycling and availability of nutrients of plants.

An approximation of a whole-farm analysis with the yields from 1981 to 1989 was made by using FINPACK farm management software to simulate a 750-acre Maryland farm (Hanson et al., 1990). In that analysis only the CONV and the LIP-CG systems were compared, and similar average annual profits over the 9-year period were found without the government price support program ($29,891 for the CONV system versus $27,614 for the LIP-CG system). However, with the government price support program for corn (requiring base acres and set-asides) the CONV system averaged $39,193 per year. The farmer of the LIP-CG system would have averaged $32,464 if the same set of price supports was used for corn, wheat, and barley, but this is a fictitious scenario because this farmer would not have been in compliance (would not have met the base acre cross-compliance requirements for wheat and barley). Another interesting result of the 1981 to 1989 economic comparison was that the farmer of the LIP-CG system experienced less fluctuation in annual income. The standard deviation in annual income over the 9-year period was $16,985 compared with that of the CONV farmer who did not participate in the government price support program (standard deviation, $37,811) or the CONV farmer who did participate in that program (standard deviation, $26,416).

None of these results would be possible if this were a 2-year or even a 5-year trial. In this trial, biological processes became most interesting after the initial 5-year transition period, and economic analyses of long-term performances and variabilities in income are possible now that the trial is entering its tenth cropping season. Ironically, this experiment was supported by the private sector (Rodale Press) for the first 7 years, and USDA funding from the LISA program of USDA during 1988 and 1989 has supplied only a fraction of the total cost of conducting the experiment, analyzing the data, supporting the work of collaborators, and presenting the results. Funding from the LISA program of USDA has, however, allowed the investigators involved in the study described here to strengthen collaborative relationships with researchers at Cornell University (Ithaca, New York), and has facilitated the initiation of the Cornell cropping systems experiment. Soil scientists from Cornell University have conducted a detailed taxonomic description of the soils at the Rodale site (Waltman and Scott, 1989) and are involved in measuring the physical properties of the soil (such as hydraulic conductivity, by Harold van Es, Soil Science Department, Cornell University, Ithaca, New York) and characterizing roots (Rich Zobel, USDA, Ithaca, New York).

## THE CORNELL CROPPING SYSTEMS EXPERIMENT

A long-term experiment was initiated at Cornell University to address concerns specific to farmers in that climatic region. Dairy farming is the dominant agricultural enterprise in much of New York State and other states in the Northeast. The growing season is shorter and soils are colder than soils in Pennsylvania. Corn silage is more common than corn grain in New York, and farm rotations often include alfalfa. The Cornell experiment was designed to compare standard practices with alternative strategies that reduce the use of agrichemicals for corn silage production. Alternative practices include (1) ridge tillage, (2) manure substitution for inorganic nitrogen, (3) interseedings of cover crops, and (4) band application of herbicide or cultivation for weed control. A total of 10 cropping systems (Table 18-3) are being compared, with three weed control regimes imposed on each cropping system. The experiment is being conducted at two sites (Aurora and Mt. Pleasant research farms), and there are five replications at each site.

The entire project includes more than 30 people, including eight farmers, six extension agents, and six faculty members of Cornell University with primary extension responsibilities (see Mt. Pleasant [1990] for a list of participants). In addition to the long-term trials initiated on the research farms, field-scale trials are being conducted on six New York dairy and cash-grain farms. These trials are comparing several practices that reduce fertilizer or pesticide use in corn to conventional farming practices. Faculty members from several disciplines, including soil and crop science, entomology, plant pathology, and plant breeding, contribute significant time and expertise to the project.

Although 1989 was the second year of funding from the LISA program for this project, it was the first year that all treatments were established at both sites. The 1988 growing season was used to prepare the sites, perform the ridging operations, and plant cover crops. Corn silage yields for the

**TABLE 18-3** Cropping Systems Treatments, Mt. Pleasant and Aurora Research Farms, Cornell University

| Treatment | Cropping System |
| --- | --- |
| 1 | Corn-corn-alfalfa-alfalfa[*,†] |
| 2 | Corn-oats/alfalfa-alfalfa-alfalfa[*] |
| 3 | Continuous corn |
| 4 | Continuous corn + manure[‡] |
| 5 | Continuous corn (ridge till) |
| 6 | Continuous corn (ridge till) + manure[‡] |
| 7 | Continuous corn (ridge till) + interseeded red clover |
| 8 | Continuous corn + interseeded red clover |
| 9 | Continuous corn (ridge till) + interseeded ryegrass |
| 10 | Continuous corn + interseeded ryegrass |

[*]All four rotation sequence entry points are represented in the experiment.

[†]All treatments were subjected to three weed control regimens in a split-plot design: (a) cultivation only, (b) banded herbicide (10-inch over row), and (c) broadcast herbicide. For rates and compounds, see J. Mt. Pleasant. 1990. Alternative Cropping Systems for Low-Input Agriculture in the Northeast. Lisa Program Progress Report. Ithaca, N.Y.: Cornell University. Mimeograph.

[‡]Manure supplied the major portion of the recommended nitrogen. For all other treatments, inorganic fertilizers were used.

1989 growing season showed the effectiveness of several practices for reducing inputs while maintaining yields and protecting soil and water resources (Table 18-4). For example, banded application of herbicides combined with one cultivation reduced herbicide use by more than 60 percent compared with broadcast application. Weed levels in the banded herbicide application plus cultivation plots were the same or slightly higher than those in the broadcast application plots (Table 18-5), but corn yields were equal to or higher than those from the broadcast herbicide plots (Table 18-4). The plots that received no herbicides had higher weed levels than the band or broadcast applications did, and the yields from both sites were significantly lower than those from plots that were treated with herbicide. The yield reduction, however, was less than 10 percent, and it remains to be seen whether that reduction is economically significant (Mt. Pleasant, 1990). Because of the wet weather conditions during the 1989 growing season, only one cultivation was performed on the plots that were not treated with herbicides. Greater efficacy of the cultivation treatment is anticipated in future years as operator experience with the cultivator improves and if more favorable weather conditions allow more timely operations.

The use of red clover and ryegrass interseeding was successfully demonstrated on several field-scale, on-farm trials in New York. Ground cover from the interseedings ranged from 11 to 37 percent in mid- to late summer, with no effect (negative or positive) on corn yields (Mt. Pleasant, 1990). Interseedings could potentially reduce nitrate leaching in the fall, provide soil cover through the winter, and contribute substantial amounts of organic matter and nitrogen to the cropping system the following year. There is also some indication that interseeding may promote weed suppression (i.e., ridge-till plots at the Mt. Pleasant site in 1989; see Mt. Pleasant [1990]).

**TABLE 18-4** Effects of Cropping System and Weed Control Treatments on Corn Silage Yields and Weed Levels, Mt. Pleasant Research Farm, Cornell University, 1989

| | Weed Control | | | |
|---|---|---|---|---|
| Cropping System | Cultivated | Band | Broadcast | Mean |
| *Corn Silage Yield (tons/acre at 70% moisture)* | | | | |
| Conventional till | 11.92 | 13.73 | 12.22 | 12.62 |
| Conventional till + manure | 8.15 | 10.36 | 10.73 | 9.75 |
| Ridge till | 6.63 | 9.35 | 9.63 | 8.53 |
| Ridge till + manure | 4.15 | 5.10 | 8.27 | 5.84 |
| Ridge till + RC | 13.85 | 12.94 | 10.47 | 12.42 |
| Conventional till + RC | 12.98 | 14.87 | 11.95 | 13.27 |
| Ridge till + RG | 12.01 | 12.18 | 10.55 | 11.58 |
| Conventional till + RG | 12.53 | 14.41 | 13.20 | 13.38 |
| Mean | 10.83 | 11.83 | 11.22 | |
| *Weed Cover (%)* | | | | |
| Conventional till | 71.8 | 21.2 | 0.0 | 31.0 |
| Conventional till + manure | 78.6 | 23.4 | 0.4 | 34.1 |
| Ridge till | 120.0 | 73.2 | 6.2 | 66.5 |
| Ridge till + manure | 73.0 | 67.0 | 2.4 | 47.5 |
| Ridge till + RC | 47.6 | 23.0 | 6.2 | 25.6 |
| Conventional till + RC | 73.6 | 20.2 | 0.8 | 31.5 |
| Ridge till + RG | 62.2 | 42.0 | 15.2 | 39.8 |
| Conventional till + RG | 71.0 | 21.0 | 0.4 | 30.8 |
| Mean | 73.3 | 31.0 | 2.9 | |

NOTE: RC, red clover interseeding; RG, ryegrass interseeding. Analysis of variance values for corn silage yields: system, $p = 0.0001$; weed control, $p = 0.0079$; system $\times$ weed control interaction, $p = 0.0005$. Analysis of variance values for weed levels: system, $p = 0.0001$; weed control, $p = 0.0001$; system $\times$ weed control interaction, $p = 0.0001$.

**TABLE 18-5**  Effects of Cropping System and Weed Control Treatments on Corn Silage Yields and Weed Levels, Aurora Research Farm, Cornell University, 1989

| Cropping System | Weed Control | | | |
| --- | --- | --- | --- | --- |
| | Cultivated | Band | Broadcast | Mean |
| *Corn Silage Yield (tons/acre at 70% moisture)* | | | | |
| Conventional till | 23.40 | 24.65 | 25.15 | 24.40 |
| Conventional till + manure | 20.14 | 20.85 | 20.36 | 20.45 |
| Ridge till | 22.47 | 24.10 | 24.14 | 23.57 |
| Ridge till + manure | 21.75 | 23.86 | 22.62 | 22.74 |
| Ridge till + RC | 23.06 | 24.09 | 24.33 | 23.83 |
| Conventional till + RC | 24.71 | 25.31 | 22.88 | 24.30 |
| Ridge till + RG | 22.66 | 24.54 | 23.61 | 23.60 |
| Conventional till + RG | 21.37 | 23.28 | 22.72 | 22.46 |
| Mean | 22.62 | 23.92 | 23.59 | |
| *Weed Cover (%)* | | | | |
| Conventional till | 8.6 | 5.4 | 3.3 | 5.8 |
| Conventional till + manure | 20.1 | 4.3 | 0.6 | 8.2 |
| Ridge till | 8.8 | 3.1 | 0.2 | 4.0 |
| Ridge till + manure | 4.4 | 1.2 | 0.8 | 2.1 |
| Ridge till + RC | 7.9 | 5.0 | 0.6 | 4.5 |
| Conventional till + RC | 11.0 | 5.1 | 4.5 | 6.9 |
| Ridge till + RG | 8.6 | 7.2 | 1.8 | 5.9 |
| Conventional till + RG | 10.4 | 4.0 | 10.1 | 8.2 |
| Mean | 9.9 | 4.1 | 2.8 | |

NOTE: RC, red clover interseeding; RG, ryegrass interseeding. Analysis of variance values for corn silage yields: system, $p = 0.1220$; weed control, $p = 0.0002$; system $\times$ weed control interaction, $p = 0.5796$. Analysis of variance values for weed levels: system, $p = 0.1559$; weed control, $p = 0.0001$; system $\times$ weed control interaction, $p = 0.0072$.

This result was also observed in some early studies on cover crop inter-seeding conducted by the Rodale Research Center (Palada et al., 1982). Corn also appears to benefit from cover crop interseeding established in the previous year. In 1989, corn at the Mt. Pleasant site showed a significant positive response to cover crops that were established in 1988. This response was not observed in a previous study of interseeded systems in New York (Scott et al., 1987), because interseeding was not examined under different tillage practices or under various levels of weed control. It is hypothesized that changes in soil structure or effects on weed infestation were responsible for the corn yield response in the long-term study.

Other results from 1989 indicate that corn yields were lower in plots that received manure than they were in plots that received sidedressed inorganic nitrogen. In 1989, corn response to manure was not consistent with the large body of data from previous research on crop response to manure at Cornell University. Apparently, the very wet spring in the Northeast was largely responsible for the reduced nitrogen availability from manure in 1989. The yield for ridge-tilled corn was less than that for conventionally tilled corn when manure was the primary nitrogen source, but yields were the same for both tillage systems that received inorganic nitrogen. Ridge tillage is believed to be a promising reduced-tillage system for New York farmers, because soil in the ridge may warm up faster than soil in a high-residue, flat, no-till system.

Other interesting results from the Cornell study show that there are significant interaction effects of tillage system and cover crops with corn pests. For example, corn eyespot disease was significantly higher in ridge-till treatments compared with those in conventional-till treatments (17.5 versus 0.7 infected plants per 200 square feet). Slug feeding was higher in plots with interseeded cover crops, with an average of 15 plants per 200 square feet showing damage compared with 9 plants per 200 square feet on land without cover crops. However, in 1989 neither of these pests was above the economic threshold level. The interactions of pests with various components of cropping systems are best understood and described in long-term experiments of cropping systems rather than in shorter-term trials, in which these interactions may not be exhibited.

The net value of this experiment as a long-term cropping systems trial has already become apparent during the first year after its establishment. For example, the effects of tillage and interseedings in corn at some sites were contrary to the effects found in past research. This result suggests that examination of cultural practices as part of a cropping system may yield different conclusions than those obtained from single-factor experiments. In addition, organic sources of nitrogen (animal manure) did not always perform as predicted from past research and experience. Nitrogen mineralization from organic nitrogen sources is highly dependent on soil and climatic conditions. The ability to use organic sources of nitrogen efficiently is limited because the processes that control nitrogen mineralization and the cycling of nitrogen within the soil-crop system are not fully understood, especially in new or novel tillage systems. Also, changes in weed control practices affect other pests (disease and insect) as well as the physical properties of the soils. The research at Cornell University has begun to identify and record the complex web of interactions that occurs within a cropping system. It is the first step toward managing that cropping system with greater efficiency.

## THE RODALE LOW-INPUT, REDUCED-TILLAGE EXPERIMENT

This long-term experiment was initiated in 1988 at the Rodale Research Center to determine the feasibility of reduced-tillage alternatives to the moldboard plow. Reduced-tillage options include chisel plow/disk (some residue left on the surface after tillage), ridge-tillage (no primary tillage but ridge tops cleared of residue at the time of planting), and no-tillage (no primary tillage and all residue left on the soil surface) (Figure 18-4). A mixed-tillage regime is also included in the design. In this treatment the primary tillage method is determined on a year-to-year basis, depending on the presence of weeds and the crop to be planted. The study emphasizes the energy and economic implications of various cropping systems as well as changes in the chemical and physical properties of the soil, weed species diversity and population shifts, and crop growth and yield (Vargas et al., 1989). This is one of the few experiments in the United States that is evaluating the feasibility of no-till crop production without the use of herbicides.

Three cropping systems are compared in Table 18-6. A CONV (recommended rates of fertilizers and pesticides) rotation of corn and soybeans is compared with two different LIP rotations. Both LIP rotations rely primarily on cover crops for nitrogen supply and weed control in the no-till system and do not include pesticides. One LIP rotation (LIP-1) is being run parallel to the CONV corn-soybean rotation and includes cover crops prior to corn and soybeans that are relay cropped into small grains. The relay crop consists of spring barley drilled in early spring, followed by soybeans planted into the tillering barley in May. The barley is harvested for grain in July, and the soybean seed is harvested in the fall at the same time that the CONV soybeans are harvested. The second LIP rotation (LIP-2) is a 4-year rotation with corn and soybeans as the principal crops, but 1 year of a small grain-legume mixture is added to diversify the rotation.

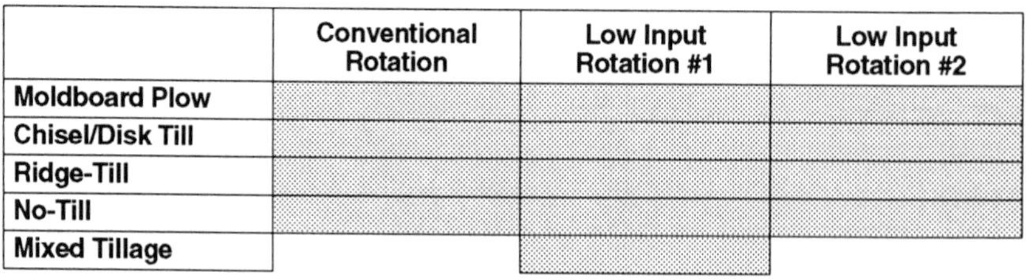

| | Conventional Rotation | Low Input Rotation #1 | Low Input Rotation #2 |
|---|---|---|---|
| **Moldboard Plow** | | | |
| **Chisel/Disk Till** | | | |
| **Ridge-Till** | | | |
| **No-Till** | | | |
| **Mixed Tillage** | | | |

FIGURE 18-4  Experimental design for the low-input (LIP), reduced-tillage experiment at the Rodale Research Center. Four tillage levels × three cropping system regimens resulted in 12 treatments. A thirteenth treatment examined a mixed tillage regimen that was run parallel to LIP-1.

**TABLE 18-6** Crop Rotation Sequence for the Low-Input, Reduced-Tillage Experiment, Rodale Research Center

| Treatment | Year 0, 1987 | Year 1, 1988 | Year 2, 1989 | Year 3, 1990 | Year 4, 1991 |
|---|---|---|---|---|---|
| CONV | Corn | Soybeans | Corn | Soybeans | Corn |
| LIP-1 | Corn | Barley/ soybeans | Hairy vetch/corn | Barley/ soybeans | Hairy vetch/corn |
| LIP-2 | Corn | Hairy vetch + rye/corn | Wheat/ soybeans | Oats/ Berseem clover | Hairy vetch/ corn |

NOTE: CONV, conventional; LIP, low input.

The field plots are 40 feet wide and 100 feet long, and only the center section (20 × 80 feet) is used for data collection. This center area is subdivided into two areas; half is designated for within-season sampling, and half is designated for final yield determinations, to minimize the impact of sampling on yield determinations and to plan space for trials superimposed by collaborators over the life of the experiment. There is potential for some plot-to-plot interaction effects from some pests (insects, diseases, weeds) but not others, and a combination of sampling only the plot centers and knowledge of the pest species' ecology will minimize errors in pest data interpretation. The 13 treatments are replicated six times in a randomized complete block design. Tillage regimes within each cropping system can be compared every year. In all years, corn and soybeans can be compared between the CONV and the LIP-1 cropping systems, and in years 4, 8, 12, and so on, corn will be present in all three cropping systems for across-system comparisons.

Abnormally dry weather in 1988 greatly reduced the yields, regardless of the cropping system or tillage regime that was used. The moldboard plow regime achieved the highest yields, the chisel/disk and ridge-till regimes were intermediate, and the no-till regime, regardless of cropping system, had the lowest yields (Peters et al., in preparation). In the CONV systems, soybean yields averaged 36 bushels/acre (bu/acre) and ranged from 47 (moldboard plow) to 20 (no-till) bu/acre. In the LIP-1 cropping system, spring barley yields averaged 42 bu/acre, but the relay-cropped soybeans failed to produce any yield because of the drought. The corn in the LIP-2 cropping system yielded between 44 (moldboard plow) and 0 (no-till) bu/acre. The failure of the hairy vetch and competition for moisture from the rye contributed to the agronomic and economic failure of this system in 1988.

The most noteworthy result from 1989 was that corn that was no-till planted in the mixed-till regime yielded as much as corn in the moldboard plow regime did and was not significantly different from the yield in the best-yielding CONV corn treatments (Table 18-7). Cover crop establishment in the strict no-till system was difficult in 1988, resulting in a poor corn crop in 1989. However, in the mixed-till system, in which plowing was used to establish hairy vetch, it was possible to plant corn by the no-till method without herbicides, and it was found to be the most profitable treatment in 1989 (Wes Musser, Agricultural Economics Department, The Pennsylvania State University, University Park, personal communication of preliminary analysis, 1990). Corn in the strict no-till system (i.e., the LIP-1 no-till regime) was deficient in nitrogen, and the poor cover crop stand in the no-till treatment did not provide adequate weed suppression. Weed control in the other LIP-1 systems was adequate and was not statistically different from the weed control achieved in the CONV cropping system with herbicides (Table 18-7). The lesson in all of this is that good cover crop establishment is essential for reduced-tillage and especially for no-till, LIP rotations to work. It was also found that in a strict, multiyear, no-till system, the establishment of cover crops is difficult. This was not known before this experiment was conducted,

**TABLE 18-7** Corn Grain Yield, Percent Nitrogen in Ear Leaf, and Weed Biomass from the Low-Input, Reduced-Tillage Experiment, Rodale Research Center, 1989

| Treatment | Grain Yield (bu/acre) | Ear Leaf (% N) | Weed Biomass (lbs/acre) |
|---|---|---|---|
| CONV | | | |
| Plow | 150a* | 3.16ab | 173de |
| Chisel | 112c | 3.24a | 372bcde |
| Ridge till | 131abc | 3.20a | 194de |
| No-till | 147a | 3.20a | 352bcde |
| LIP-1 | | | |
| Plow | 129abc | 3.09abc | 144de |
| Chisel | 130abc | 2.86cd | 17e |
| Ridge till | 117bc | 2.81d | 121de |
| No-till | 63d | 2.12e | 1,318a |
| Mixed till | 140ab | 2.91bcd | 234cde |

NOTE: CONV, conventional; LIP, low input.

*Duncan's multiple range test was performed by analysis of variance using Statistical Analysis System software; comparisons are within columns.

because hairy vetch was used only in short-term experiments to determine, for example, the optimal date of corn planting and equipment needs.

A more detailed look at the hairy vetch showed that hairy vetch germination appeared to be adequate in all treatments, but subsequent slug feeding in the high-residue plots (chisel/disk, ridge-till, and no-till) resulted in diminished hairy vetch stands. Large differences were measured in the spring nitrogen contributions from the aboveground biomass of hairy vetch; 260 lbs/acre in the moldboard plow and mixed-till system, 132 lbs/acre in the chisel/disk system, 89 lbs/acre in the ridge-till system, and 49 lbs/acre in the no-till system. In the spring of 1989, the mixed-till regime and the no-till system were both no-till planted to corn, with the planter set up to plant the seeds in a narrow slot, minimizing soil disturbance. Hairy vetch that was not killed by the planting operation was mown. Interesting differences between the CONV and LIP-1 cropping systems were noted in nitrate nitrogen levels in the soil throughout the growing season (Figures 18-5 and 18-6) and among the tillage regimes within each cropping system.

The LIP-2 rotation was planted with wheat and soybeans in 1989. Wheat yields ranged from 43 to 23 bu/acre, and soybean yields ranged from 28 to 21 bu/acre, with the higher yields being in the moldboard plow tillage regimes and the lowest being in the no-till system (Table 18-8). Expenses are minimized in this system because no herbicides are used and the soybeans are drilled into the established grain crop. However, the high cost of harvesting both crops and the relatively low yield of soybeans in 1989 compared with the soybean yield in the CONV system in 1988 indicate that for the moldboard plow, chisel/disk, and ridge-till systems, monocultures of soybeans are more profitable, while in the no-till system the LIP relay crop is more profitable than the CONV monoculture crop (Wes Musser, Agricultural Economics Department, The Pennsylvania State University, University Park, personal communication of preliminary results, 1990).

More time is needed to determine whether the yields in the reduced-tillage plots go up over time relative to those in the moldboard plow regime. Reports in the literature indicate that 10 years or more is sometimes needed before no-till systems come to equilibrium. The value of the conversion experiment was not fully appreciated until after the fourth and fifth growing seasons, and it is anticipated that this trial will also increase in value, in terms of information generated, over time. It has already provided a useful comparison of energy use in LIP versus CONV cropping systems by using various tillage regimes for 2 years, and the comparison will be more representative after 4 and 8 years of data are collected (the second and third rotation cycles). Shifts in weed populations and abundance can be examined and long-term economic performance assessed as the trial matures.

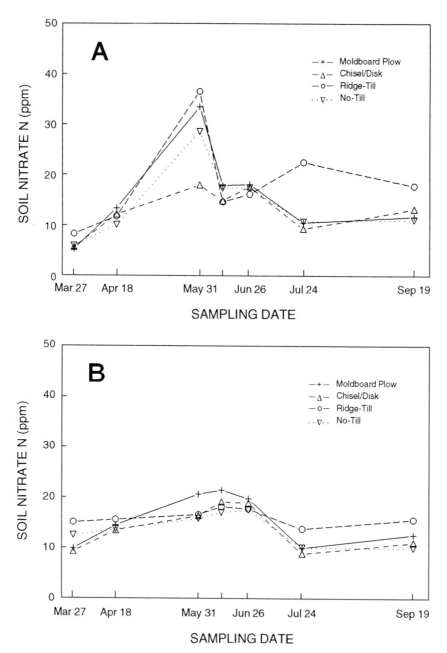

FIGURE 18-5　Soil nitrate nitrogen levels in the conventional cropping system (30 lbs/acre starter and 100 lbs/acre sidedress nitrogen applied as ammonium nitrate) for the (A) 0- to 2-inch and (B) 2- to 8-inch soil layers in the low-input, reduced-tillage cropping system experiment, Rodale Research Center, 1989.

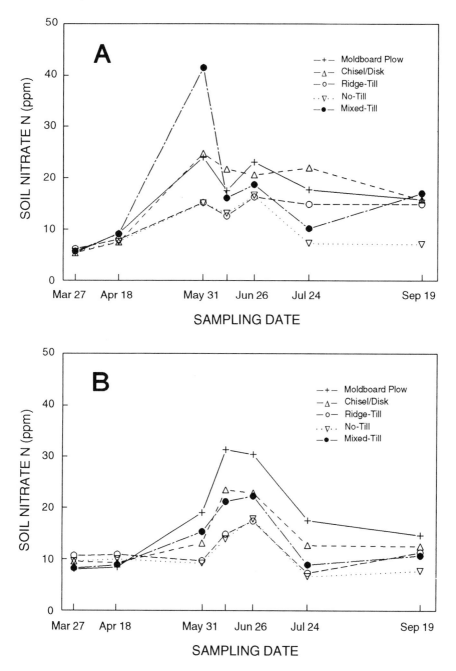

FIGURE 18-6   Soil nitrate nitrogen levels in the low-input (LIP-1) cropping system (nitrogen from hairy vetch cover crop) for the (A) 0- to 2-inch and (B) 2- to 8-inch soil layers in the low-input, reduced-tillage cropping system experiment, Rodale Research Center, 1989.

**TABLE 18-8**   Wheat Yield, Soybean Yield, and Weed Biomass in the Low-Input, Reduced-Tillage Experiment, Rodale Research Center, 1989

| Treatment (LIP-2) | Wheat Yield (bu/acre) | Soybean Yield (bu/acre) | Weed Biomass (lbs/acre) |
| --- | --- | --- | --- |
| Plow | 43a[*] | 28a | 330bcde |
| Chisel | 33b | 28a | 599bc |
| Ridge till | 30b | 24ab | 668b |
| No-till | 23c | 21b | 490bcd |

NOTE: LIP, low input.

[*]Duncan's multiple range test was performed by analysis of variance using Statistical Analysis System software; comparisons are within columns.

## LONG-TERM CROPPING SYSTEMS COMPARISONS

In addition to the three experiments highlighted in this chapter, several other experiments with similar objectives have begun or are being initiated at various locations around the country.   Recently, a working conference was held at the Rodale Research Center (in preparation) to bring together researchers working on long-term cropping systems studies, to discuss common concerns, and to search for ways to strengthen existing collaborations and develop new working relationships among the various institutions represented.   Ongoing trials represented at the conference (besides those at the Rodale Research Center and Cornell University) were those at Michigan State University (East Lansing), the University of Georgia (Athens), North Dakota State University (Fargo) (Gardner, 1989), Ohio State University (Columbus) (Edwards and Creamer 1989), North Carolina State University (Raleigh) (King 1990), and Clemson University (Clemson, South Carolina). Trials are also ongoing at the University of Nebraska (Lincoln) (Helmers et al., 1986; Sahs and Lesoing, 1985,), South Dakota State University (Brookings), and the University of California (Davis). Experiments are in the planning stages or are about to be initiated at Rutgers University (New Brunswick, New Jersey), the University of Wisconsin (Madison), the University of Minnesota (St. Paul) (Crookston and Nelson, 1989), and possibly other locations as well.   All these trials have the common theme of a long-term planning horizon, and each is looking for LIP sustainable cropping systems as alternatives to the CONV systems for their region. Some trials include a tillage component as well as different levels and different types of inputs; in addition, they compare various crop rotations. Several experiments have been conducted to examine the crop rotation effect and

An aerial view of Pennsylvania countryside illustrates the relationship of farms and communities to water resources. The Susquehanna River is pictured. (See Chapter 19.) Credit: Steven Williams.

Reduced tillage alternatives to the moldboard plow (top) include chisel plow tillage whereby some residue is left on the surface after tillage (bottom), ridge tillage in which no primary tillage occurs but ridge tops are cleared of residue at the time of planting (top right), and no tillage in which all residue (here, hairy vetch) remains on the soil surface when crops are planted (bottom right). (See Chapter 18.) Credit: Rhonda R. Janke and Steven E. Peters.

The parasitic wasp, *Lemaphagus curtus,* lays eggs in a larvae of the cereal leaf beetle, *Oulema melanopus,* which was introduced into the United States from Europe in the 1970s. (See Chapter 20.) Credit: Raymond I. Carruthers.

The cluster of dead grasshoppers pictured above were infected with *Entomophaga grylli.* (See Chapter 20.) Credit: Raymond I. Carruthers.

the effect of using animal manure over a long period of time (e.g., Lazarus et al. [1980], Odell et al. [1974], Smith [1942]; see also the summary by Brown [1989]), but an evaluation of LIP alternatives was not an explicit objective of these trials.

All researchers and institutions share the common problem of obtaining secure funding sources with a 5- to 10-year (or longer) planning horizon. With the exception of funding for long-term ecological research sites from the National Science Foundation, other funding sources, sometimes including experiment station commitment, tend to run on a year-to-year time frame. A national agricultural funding commitment to long-term research, such as USDA-funded long-term ecological research sites, is needed if these experiments are to continue. Most of these experiments are in their first or second cropping season, with the exception of those at the University of Nebraska (1990, year 15), Rodale Research Center (1990, year 10), and North Carolina State University (1990, year 5). Valuable data have been and continue to be collected at two or more sites simultaneously, that is, soil microbial studies at the University of Nebraska and Rodale Research Center (Fraser, 1984), $^{15}$N studies at Michigan State University and the Rodale Research Center (Harris et al., 1989), and organic matter studies at Ohio State University and the Rodale Research Center (M. Wander, Ohio State University, Columbus, personal communication), generally through the participation of graduate students. As more long-term experiments are initiated, the opportunities for collaboration among researchers at several sites that represent various climatic zones increase, and the chances for truly understanding the mechanisms important for pest control and nutrient cycling are enhanced.

## MEANINGFUL INTEGRATION OF FARMER AND RESEARCHER INFORMATION

Many of the sustainable agriculture programs at various locations around the country are also linked to networks of farmers who advise researchers on the direction of on-station research and who function as research collaborators by conducting experiments and demonstration trials on their farms. Researchers involved with the cropping systems experiment at Cornell University currently conduct research on reduced-tillage techniques, cover crop overseeding, reduced herbicide use through banded applications, cultivation for weed control, and manure management with six New York State farmers (Mt. Pleasant, 1990). At the Rodale Research Center, members of the agronomy department serve as technical advisers to the Midwest on-farm research network (McNamara and Janke, 1988; McNamara et al., 1987) and also work with local farmers in southeastern Pennsylvania who are exploring overseeding options, crop rotation diversification, soybean relay cropping (Peters, 1989), and no-till planting into cover crops.

Some farmers are able to articulate the successes of their farming practices to researcher and farmer audiences without the benefit of a structured network to disseminate the results (Kirschenmann, 1988; Thompson and Thompson, 1989), but even well-known spokespeople such as the Thompsons benefit from their affiliation with networks of other farmer-researchers, such as the Rodale Institute Network and the Practical Farmers of Iowa. Networks facilitate information dissemination by organizing and publicizing field days and farm tours and by organizing panels of farmers to speak at events cosponsored by the Cooperative Extension Service. Networks of farmer-researchers also facilitate information flow back to researchers as to what works (e.g., hairy vetch is great for nitrogen production and soil improvement), what does not work (e.g., no-till planting into incompletely killed rye grain), the fine-tuning of systems that is needed (e.g., do not fertilize wheat when you are relay cropping soybeans), and the next set of questions that need to be addressed (e.g., why do my soil phosphorus and potassium levels seem to be going up in fields with cover crops, even without manure or mineral fertilizer additions?). Farmer-directed on-farm research empowers farmers because it encourages them to answer their own questions (Janke and McNamara, 1988; Janke et al., 1990) as they go through the transition period. It allows them to cut back on chemical and fertilizer inputs gradually, testing the results at each step. Many of the farmers in the Rodale Institute Network compare their usual rate of nitrogen fertilizer application with a reduced rate. The Practical Farmers of Iowa have been comparing low versus high nitrogen application rates and mechanical cultivation versus herbicide use for corn and soybean production in ridge-tillage systems (Practical Farmers of Iowa Newsletter, 1989).

Limitations of on-farm research include the difficulty of doing extensive sampling (this is easier to accomplish at research station sites) and the amount of time required by the farmer to conduct complex, long-term trials. On-farm trials of 2 or 3 years or similar 1-year trials repeated for 2 or more consecutive years are more feasible. Workable designs include two or three treatments replicated four to six times. There remains a role, however, for testing of farmer and researcher questions in complex, long-term systems trials at research stations. On-farm research complements, but does not replace, work done at research stations.

## SUMMARY

The support of long-term cropping systems research from 1988 to the present by the LISA program of USDA is to be applauded. However, long-term experiments require secure funding sources to attract the necessary intellectual talent and commitment of researchers and other field station personnel. Farmers serve a valuable role as advisers to those conducting studies at research stations and as collaborators who find some answers but who also raise many questions and generate valuable hypotheses that need to be tested in long-term, replicated trials.

One example includes use of the Rodale conversion experiment to test the commonly held belief that yields go down during the first 3 to 5 years after the withdrawal of chemical inputs from the cropping system. The cause of the lower yields was thought to be related to changes in biological processes in the soil. The Rodale trial has shown that yields were, in fact, lower in the LIP systems compared with those in the CONV system for the first 4 years, but only for corn. The yield suppression was found to be largely due to the low levels of nitrogen available to the crop, a situation that changed in year 5, once legumes were in the rotation for at least two cycles. Soil biology appeared to play a role in the amount and timing of nitrogen availability in both types of LIP rotations, and greater microbial activity and soil fauna populations were found in the LIP plots compared with those in the CONV plots. Studies are under way to explore the role of the labile fraction of soil organic matter in nutrient bioavailability in these systems.

The second experiment described in this chapter was also designed to answer questions generated by farmers. In New York State, farmers believe that cool soils limit corn silage production and that no-till farming will not work for them. The Cornell trial tests the reduced-tillage technique of ridge tillage in a long-term cropping systems context. Ridge tillage is a technique that leaves the crop residue on the field for the duration of the winter and spring (protecting the field from soil erosion) and that requires no primary tillage before crops are planted in the spring (energy savings); the well-drained ridges warm up faster than do the cool wet soils. These farmers also feel locked in to growing corn several years in a row on their best land while they use their other land for pastures and woodlots. Crop diversification does not seem to be a viable option, but the Cornell trial is demonstrating the benefits of red clover and ryegrass overseeding to improve tilth and help suppress weeds in what would otherwise have been a strictly monoculture situation.

The third trial tests a technique that was tried on farms long before research stations had the nerve to attempt it—that is, no-till cropping systems without herbicides. Farmers in a number of midwestern states have been experimenting with rye and hairy vetch, and some were even beginning to plant soybeans by the no-till method into rye with some success. The LIP reduced tillage experiment at Rodale is attempting to combine information from farmers about reduced tillage, including chisel/disk-based systems, ridge-tillage, and especially no-till (mow-sow) planting into cover crops, with what researchers have learned from the conversion experiment about how to grow crops without purchased fertilizers or pesticides (diversified crop rotations that include legumes). This long-term cropping systems trial is, possibly, the only site in the country where not only are tillage regimes in similar cropping systems being compared but where two of the three cropping systems in the experiment are LIP (no-pesticide) systems.

Once the experiment has gone through one or two rotation cycles (4 to 8 years) and the tillage systems are well established, this trial will be a valuable testing ground for hypotheses regarding soil organic matter, soil biology, and plant-animal-soil-microbe interactions. In addition, this will be a site where agronomic questions can be answered about system feasibility, and energy and economic comparisons can be calculated from a realistic, multiyear data set.

## ACKNOWLEDGMENTS

The authors express their gratitude to Richard Harwood, William Liebhardt, and Martin Culik, who began the farming systems trial at the Rodale Research Center. The authors continue to benefit from their foresight in planning and implementing this trial.

The authors also acknowledge the various funding sources that have made this research possible, including the LISA program of USDA, the Charles Stewart Mott Foundation, the Pennsylvania Energy Development Authority, the Cornell Agricultural Experiment Station and Rodale Press.

## REFERENCES

Andrews, R. W., S. E. Peters, R. R. Janke, and W. W. Sahs. 1990. Converting to sustainable farming systems. Pp. 281–313 in Sustainable Agriculture in Temperate Zones, C. A. Francis, C. B. Flora, and L. D. King, eds. New York: John Wiley & Sons.

Brown, J. R. 1989. A Listing of Long-Term Field Research Sites. Department of Agronomy Miscellaneous Publication No. 89-04. Columbia, Mo.: Missouri Agricultural Experiment Station.

Crookston, K., and W. Nelson. 1989. The University of Minnesota's Koch Farm. University of Minnesota, Saint Paul, Minn. Mimeograph.

Culik, M. N. 1983. The conversion experiment: Reducing farming costs. Journal of Soil and Water Conservation 38:333–335.

Doran, J. W., D. G. Fraser, M. N. Culik, and W. C. Liebhardt. 1987. Influence of alternative and conventional agricultural management on soil microbial processes and nitrogen availability. American Journal of Alternative Agriculture 2:99–106.

Duffy, M., R. Ginder, and S. Nicholson. 1989. An Economic Analysis of the Rodale Conversion Project: Overview. Staff Paper No. 212. Ames, Iowa: Iowa State University Economics Department.

Edwards, C., and N. Creamer. 1989. The Sustainable Agriculture Program at The Ohio State University. Columbus, Ohio. Mimeograph.

Fraser, D. G. 1984. Effects of Conventional and Organic Management Practices on Soil Microbial Populations and Activities. M.S. thesis. University of Nebraska, Lincoln.

Gardner, J. C. 1989. Evaluation of Integrated Low Input Crop-Livestock Production Systems. North Central LISA Program 1988–89 Progress Report. Fargo, N.D.: Carrington Center, North Dakota State University.

Hanson, J. C., D. M. Johnson, S. E. Peters, and R. R. Janke. 1990. The Profitability of Sustainable Agriculture in the Mid-Atlantic Region—A Case Study Between 1981 and 1989. Working Paper No. 90-12. College Park, Md.: Department of Agricultural and Resource Economics, University of Maryland.

Harris, G. H., O. B. Hesterman, R. R. Janke, and S. E. Peters. 1989. Fate and cycling dynamics of nitrogen in sustainable and conventional agricultural systems. ASA Abstracts, October 15–20, 1989.

Harwood, R. R. 1984. Organic farming research at the Rodale Research Center. Pp. 1–17 in Organic Farming: Current Technology and Its Role in a Sustainable Agriculture. Proceedings of ASA Symposium, November 29–December 3, 1981. D. F. Bezdicek, J. F. Power, D. R. Keeney, M. J. Wright, D. M. Kral, and S. L. Hawkins, eds. ASA Special Publication No. 46. Madison, Wis.: American Society of Agronomy.

Harwood, R. R. 1985. The integration efficiencies of cropping systems. Pp. 64–75 in 1984 Conference Proceedings, Sustainable Agriculture & Integrated Farming Systems, T. C. Edens, C. Fridgen, and S. L. Battenfield, eds. East Lansing, Mich.: Michigan State University Press.

Helmers, G. A., M. R. Langemeier, and J. Atwood. 1986. An economic analysis of alternative cropping systems for east-central Nebraska. American Journal of Alternative Agriculture 1:153–158.

Janke, R., and K. McNamara. 1988. Using replicated on-farm research trials to answer farmer's questions about low-input cropping systems. In Contributions of FSR/E Towards Sustainable Agricultural Systems, Proceedings of Farming Systems Research/Extension Symposium, October 1988. FSR Paper Series Paper No. 17. Fayetteville, Ark.: University of Arkansas and Winrock International Institute of Agricultural Development.

Janke, R., D. Thompson, K. McNamara, and C. Cramer. 1990. A Farmer's Guide to On-Farm Research. Emmaus, Pa.: Rodale Institute.

King, L. D. 1990. Reduced Chemical Input Cropping System Experiment. Progress Report. North Carolina State University, Raleigh. March 3. Mimeograph.

Kirschenmann, F. 1988. Switching to a Sustainable System. Windsor, N.D.: Northern Plains Sustainable Agriculture Society.

Lazarus, W. F., L. D. Hoffman, and E. J. Partenheimer. 1980. Economic Comparisons of Selected Cropping Systems on Pennsylvania Cash Crop and Dairy Farms with Highly Productive Land. Bulletin No. 828. University Park, Pa.: Agricultural Experiment Station, The Pennsylvania State University College of Agriculture.

Liebhardt, W. C., R. W. Andrews, M. N. Culik, R. R. Harwood, R. R. Janke, J. K. Radke, and S. R. Schwartz. 1989. Crop production during conversion from conventional to low-input methods. Agronomy Journal 81:150–159.

McNamara, K., and R. Janke. 1988. On-Farm Research: 1988 Summary of Results—Midwestern Region. Emmaus, Pa.: Rodale Institute.

McNamara, K., R. Janke, and R. Hofstetter. 1987. On-Farm Research: 1987 Summary of Results—Midwestern and Eastern Regions. Emmaus, Pa.: Rodale Institute.

Mt. Pleasant, J. 1990. Alternative Cropping Systems for Low-Input Agriculture in the Northeast. LISA Program Progress Report. Ithaca, N.Y.: Cornell University. Mimeograph.

National Research Council. 1989. A mixed crop and livestock farm in Pennsylvania: The Kutztown Farm. Pp. 286–307 in Alternative Agriculture. Washington, D.C.

Odell, R. F., W. M. Walker, L. V. Boone, and M. G. Oldham. 1974. The Morrow Plots: A Century of Learning. Bulletin No. 775. Urbana, Ill.: University of Illinois Agricultural Experiment Station.

Palada, M. C., S. Ganser, R. Hofstetter, B. Volak, and M. Culik. 1982. Association of interseeded legume cover crops and annual row crops in year-round cropping systems. Pp. 193–213 in Environmentally Sound Agriculture, Selected papers from the 4th International Convention of the IFOAM, August 18–20, 1982, W. Lockeretz, ed. New York: Praeger Special Studies.

Peters, S. 1989. On-Farm Research Results of Soybean Interseeding, Lancaster County, Pa. Technical Report No. 89/33. Kutztown, Pa.: Rodale Research Center.

Peters, S., R. Andrews, and R. Janke. 1988. Rodale's Farming Systems Experiment 1981–1987. Technical Report. Kutztown, Pa.: Rodale Research Center.

Peters, S., M. Böhlke, and R. Janke. In preparation. Low-Input Reduced Tillage Experiment: 1988–1989 Results. Technical Report. Kutztown, Pa.: Rodale Research Center.

Practical Farmers of Iowa Newsletter. 1989. Achievements of 1989, opportunities for 1990, vol. 4. Boone, Iowa: Practical Farmers of Iowa.

Radke, J. K., R. W. Andrews, R. R. Janke, and S. E. Peters. 1988. Low input cropping systems and efficiency of water and nitrogen use. Pp. 193–217 in Cropping Strategies for Efficient Use of Water and Nitrogen. ASA-CSSA-SSSA Special Publication No. 51. Madison, Wis.: American Society of Agronomy.

Rodale Research Center. In preparation. Proceedings of a Working Conference on Long Term Cropping Systems Research, March 14–16, 1990. Kutztown, Pa.: Rodale Research Center.

Sahs, W., and G. Lesoing. 1985. Crop rotations and manure versus agricultural chemicals in dryland grain production. Journal of Soil and Water Conservation 40:511–516.

Scott, T. W., J. Mt. Pleasant, R. F. Burt, and D. J. Otis. 1987. Contributions of ground cover, dry matter, and nitrogen from intercropped cover crops in a corn polyculture systems. Agronomy Journal 79:792–798.

Smith, G. E. 1942. Sanborn Field: Fifty Years of Field Experiment with Crop Rotation, Manure, and Fertilizers. Bulletin No. 458. Columbia, Mo.: Missouri Agricultural Experiment Station.

Thompson, D., and S. Thompson. 1989. The 1989 Report. Boone, Iowa. Mimeograph.

U.S. Department of Agriculture. 1980. Report and Recommendations on Organic Farming. Washington, D.C.: U.S. Department of Agriculture.

Vargas, A. M., R. R. Janke, W. N. Musser, D. K Israel, M. D. Shaw, J. M. Hamlett, S. E. Peters, and F. Higdon. 1989. Comparison of Energy Use in Conventional and Low-Input Reduced Tillage Cropping Systems. Staff Paper No. 169. University Park, Pa.: Agricultural Economics and Rural Sociology Department, The Pennsylvania State University.

Waltman, W. J., and T. W. Scott. 1989. Soils of the Rodale Research Farm. Kutztown, Pa.: Rodale Research Center. Mimeograph.

Werner, M. R. 1988. Impact of Conversion to Organic Agricultural Practices on Soil Invertebrate Ecosystems. Ph.D. dissertation. College of Environmental Science and Forestry, State University of New York, Syracuse.

# 19

# Perspectives for Sustainable Agriculture from Plant Nutrient Management Experiences in Pennsylvania

*Les E. Lanyon*

Research and education in plant nutrient management have a long history in Pennsylvania. More than 90 percent of the cash receipts for field crops and animals on Pennsylvania farms result from the sale of animal products (Pennsylvania Agricultural Statistics Service, 1989). This emphasis on livestock in the state's agriculture has encouraged the production of a range of crops, usually on the farms where the animals are located. Forage production has been a major enterprise on the many ruminant-based farms. However, present trends are for a decreased emphasis on production of the complete ration on the farm and increased reliance on off-farm sources of feeds, especially feed concentrates. McSweeny and Jenkins (1989) reported that, on average, approximately 50 percent of the rations for dairy cows in Pennsylvania was purchased feed. At the extreme end of the feed supply spectrum, a trend has developed, primarily with nonruminant animals, for 100 percent of the feed to be provided by off-farm sources. Whether the feed is produced on the farm or purchased, manure is generated that must be distributed away from the livestock facilities.

The production of forages, especially forage legumes, and the requirements for manure handling have created the foundations for nutrient management in Pennsylvania. Although the idea of nutrient management on farms in Pennsylvania has extensive historic roots, the goals and challenges of nutrient management have not remained the same with time.

Historically, low inputs of plant nutrient-containing materials, such as purchased feeds or fertilizers, necessitated the use of internal farm nutrient sources and the organization of many farm activities to deal with this oligotrophy. With the changes in off-farm sources of nutrients following World

War II, especially nitrogen fertilizers and the availability of off-farm feeds, the emphasis on efficient use of internal nutrient sources diminished in the face of net farm nutrient loading. In this process of nutrient loading, deficiencies of plant nutrients in the soil and on the farm were eliminated; and the nutrient factors that limit crop production shifted from the amount of nutrients available to utilization of the host of on- and off-farm sources. In recent times, the reported use of fertilizer plant nutrients in Pennsylvania has actually decreased to only 80 percent of the 1955 levels (Berry and Hargett, 1987). This change in fertilizer use reflects in part the transformation of Pennsylvania farms from nutrient-poor to nutrient-rich status.

The linkage of animal numbers on particular farms, and by association, the amount of manure produced, to the potential nutrients that could be applied to crops was significantly diminished following World War II by the ready availability of off-farm feeds and the technology used to house large numbers of animals in high-density facilities. The shift in emphasis for manure management went from the use of an essential nutrient source to the least direct cost distribution of manure. The costs of distribution are not exclusively economic, but they could include a host of on-farm factors related to the practice.

The least direct cost distribution of manure continues to be an effective foundation for nutrient management strategies on many farms in Pennsylvania. In some cases, however, the external effects of manure management practices on the most eutrophic farms could be substantial when nutrients are discharged into the environment. Therefore, current approaches to nutrient management must focus on procedures for supplying nutrients in a timely fashion and at rates that are appropriate for the crop and soil conditions, but perhaps not by the least direct cost method. The most eutrophic farms must transfer manure away from the farm to comply with the nutrient utilization potential of the farm where manure is generated.

As farm conditions have changed, the role of the immediate community around the farm has changed from being a primary market and the source of limited inputs to being a stopover point in the process of long-distance market and input supply transportation. The surrounding community is currently concerned with the impacts of farm operations on the environment. An as yet undefined community must be developed in the future to cooperate with the farms to distribute manure within an acceptable area. This possible future cooperation will involve the geographic area where manure will be applied and the information on which to base the appropriate transfer of nutrients, if not direct financial support to do so. This new community will likely be made up of different people and a different geographic region from the one that is the source of the farm inputs and the one in which the products of the farm are distributed.

In one part of the nutrient management research and extension program in Pennsylvania, the emphasis has been to understand nutrient management strategy implications for crop and livestock farms, to relate the strategies to environmental and economic performance indicators, and to formulate and support a whole-farm nutrient management process with the appropriate tools. Field studies of complete farm operations are under way to monitor the actual dynamics of nutrients in nutrient management pathways on Pennsylvania farms and to develop techniques for on-farm nutrient management. The goals are to recognize and understand the site specificity (specific characteristics of the farm site) and information richness (wealth of available information about the farms). These goals are similar to those of sustainable agriculture.

In Pennsylvania, the experience of dealing with whole farms, especially those that focus on livestock production, has been a source of insight into the science of agricultural research and education. It is the inspiration for a review of normal science approaches to research and for the need to work cooperatively with farmers in the context of a complex, interconnected, biological managed system.

## PLANT NUTRIENT MANAGEMENT STRATEGY IMPLICATIONS

Westphal et al. (1989) investigated the consequences of several strategies for dealing with manure nitrogen (N) utilization, residual legume N, and manure application approaches on a simulated dairy farm. Each of these was evaluated by assuming either that the available N supply was limited to that which could be used by the crop or that the available N could be oversupplied. In the first case, the N was managed so that groundwater contamination by leaching excess N beyond the plant root zone was avoided. In the second case, the risk of N contamination of groundwater was much greater because of the potential for applications in excess of the potential amounts that could be used by plants. Crop sequences of 2 years of corn (C) followed by 3 years of alfalfa (A) (C-C-A-A-A) or 3 years of corn preceding 3 years of alfalfa (C-C-C-A-A-A) were considered.

The combination of the best agronomic recommendation (a high efficiency of manure N utilization and the full amount of residual N available from the preceding alfalfa crop) and the best environmental protection criterion (the amount of N supplied was restricted to that required by the crop) resulted in the lowest economic net return to the farm when compared with other less efficient or sensitive strategies (Table 19-1). The potential N utilization by the crops was the factor that restricted the amount of manure that could potentially be spread and thus restricted the number of cows on the farm. Since each 1,400-pound (lb) dairy cow produces approximately 21 tons of manure containing almost 210 lbs of total N (Midwest Plan Service, 1975) per year, from 42 to 110 lbs of N available from the

**TABLE 19-1** Relationship of Crop Sequence and Nutrient Management Strategies to Herd Size and Net Return on a Simulated 125-Acre Dairy Farm in Pennsylvania

| Crop Sequence | Manure N Efficiency | Available N Restriction | Herd Size (Number) | Limiting Factor | Net Return* |
|---|---|---|---|---|---|
| C-C-A-A-A | Low | Yes | 72 | Feed | 99 |
| | Low | No | 72 | Feed | 99 |
| | High | Yes | 41 | Crop N use | 23 |
| | High† | No | 72 | Feed | 100 ($36,838) |
| C-C-C-A-A-A | Low | Yes | 84 | Feed | 98 |
| | Low | No | 84 | Feed | 98 |
| | High | Yes | 69 | Crop N use | 79 |
| | High† | No | 84 | Feed | 100 ($44,347) |

NOTE: N, nitrogen; C, corn; A, alfalfa.

*Percentage of maximum return within a crop sequence (dollar amount of maximum return is given in parentheses).

†Original simulation used for comparison within a crop sequence.

SOURCE: P. J. Westphal, L. E. Lanyon, and E. J. Partenheimer. 1989. Plant nutrient management strategy implications for optimal herd size performance of a simulated dairy farm. Agricultural Systems 31:381–394.

total must be used in crop production for each cow that is added to the herd. The net return to the farm operation was reduced much less when the percentage of corn in the crop sequence was increased from 40 to 50 percent. About 25 percent more land was available for manure application, and higher amounts of manure could be applied to the corn since there was an additional year beyond the influence of the residual N from the alfalfa in the C-C-C-A-A-A crop sequence.

Application of manure to alfalfa in the last year of the crop sequence resulted in an increased number of cows on the farm for those strategies in which herd size was limited by the potential area where manure could be applied (Table 19-2). Inclusion of the off-farm purchase of grain for lactating dairy cows increased the net economic returns to the farm (Table 19-3). When the requirement that phosphorus (P) and potassium (K) application to the soil and utilization by the plants be balanced (constant maintenance by testing levels in soil) was relaxed, the numbers of cows and net returns again increased. Therefore, two practices that provided relatively little positive contribution to the crops, manure application to alfalfa and increases in the levels of P and K in soil (above the sufficiency level), resulted in the enhanced economic performance of the simulated dairy farm.

**TABLE 19-2**   Herd Size and Net Return to a Simulated 125-Acre Dairy Farm with and without Manure Applied to Alfalfa in the Last Year of Two Crop Sequences

| Crop Sequence | Manure N Efficiency | Available N Restriction | Manure on Alfalfa | Herd Size (Number) | Limiting Factor | Net Return* |
|---|---|---|---|---|---|---|
| C-C-A-A-A | High[†] | Yes | No | 41 | Crop N use | 100 ($8,429) |
| | High | Yes | Yes | 60 | Crop N use | 352 |
| C-C-C-A-A-A | High[†] | Yes | No | 69 | Crop N use | 100 ($35,213) |
| | High | Yes | Yes | 84 | Crop N use | 126 |

NOTE: N, nitrogen; C, corn; A, alfalfa.

*Percentage of net return with no manure application to alfalfa within a crop sequence (dollar amount of net return is given in parentheses).

[†]Original simulation used for comparison within a crop sequence.

SOURCE: P. J. Westphal, L. E. Lanyon, and E. J. Partenheimer. 1989. Plant nutrient management strategy implications for optimal herd size performance of a simulated dairy farm. Agricultural Systems 31:381–394.

**TABLE 19-3**   Herd Size and Net Return to a Simulated 125-Acre Dairy Farm as Influenced by Corn Grain Purchases and Restrictions on Soil Test Phosphorus and Potassium Levels for Two Crop Sequences

| Corn Grain Purchase | Manure N Efficiency | Restrictions Available N | Soil Tests | Number of Cows | Limiting Factor | Net Return* |
|---|---|---|---|---|---|---|
| No[†] | Low | Yes | Yes | 84 | Feed | 100 ($43,337) |
| Yes | Low | Yes | Yes | 94 | Crop P use | 114 |
| Yes | Low | Yes | No | 105 | Crop N use | 122 |
| Yes | Low | No | No | 135 | Manure application[‡] | 150 |
| No[†] | High | No | Yes | 84 | Feed | 100 ($44,347) |
| Yes | High | No | No | 135 | Manure application | 148 |

NOTE: N, nitrogen; P, phosphorus.

*Percentage net return within an available N restriction group (dollar amount of net return is given in parentheses).

[†]Original simulation used for comparison within an available N restriction group.

[‡]A practical manure application limit of 50 tons/acre was imposed.

SOURCE: P. J. Westphal, L. E. Lanyon, and E. J. Partenheimer. 1989. Plant nutrient management strategy implications for optimal herd size performance of a simulated dairy farm. Agricultural Systems 31:381–394.

The enhanced performance was most notable when outcomes were compared with base scenarios in which the nutrients were managed in an efficient and environmentally sensitive manner.

The energy economy of a dairy farm can be affected substantially by a variety of nutrient management strategies such as incorporation of manure into soil or application of manure to the soil surface and crop sequence selection (Vinten-Johansen et al., 1990). However, the significance of the practices can be influenced by other features of the crop production system such as tillage practices (no-till or conventional tillage, for example), the frequency and power requirements of the machinery operations, and machinery size. In a whole-farm linear program simulation, incorporation of manure into soil increased the fuel requirements slightly in all situations compared with those for surface application, but it significantly reduced the energy input to the farm embodied in fertilizer (Table 19-4). A reduction of tillage (no-till) for corn production had a greater energy conservation effect than did N conservation by manure incorporation when a C-C-A-A-A legume-based rotation was considered. When a forage legume in combination with corn rather than corn monoculture was used, the total energy requirements were reduced substantially through a reduction in the indirect energy embodied in fertilizer and pesticides. Nevertheless, the direct energy (fuel) consumption for the crop sequence with the forage crop increased as a result of the greater fuel requirements for forage crop production. The cultural practices for these crops require more trips across each field in a growing season than does corn production. A transition from monoculture corn to a mixed set of crops requiring different machines and field and management operations also has additional requirements for the farm that must be considered beyond the energy performance criteria of the nutrient management practices.

## MONITORING WHOLE FARMS

Unlike the relatively straightforward flow of materials on cash crop farms, the complex internal cycles of the flow of managed materials on crop and livestock farms are not readily adapted to accurate monitoring. Bacon et al. (1990) described an approach for measuring the flows of nutrients within a hierarchically nested set of management units on a Pennsylvania dairy farm. The essential elements of the approach were an effective record-keeping system based on data collected by the farmer, with sampling and data management provided by the research team. Internal flows of crops and manure were routed across a 25-ton-capacity mechanical scale, and samples were collected for nutrient concentration determinations. Material flow at the farm gate was determined from farm financial records, and input suppliers were interviewed to identify the geographic sources of many materials.

**TABLE 19-4** Selected Energy and Production System Inputs for a Simulated 125-Acre Dairy Farm with Different Crop Sequences, Tillage Methods, and Manure either Injected into the Soil or Surface Applied

| | Continuous Corn Crop Sequence | | | | C-C-A-A-A Crop Sequence | | | |
| | No-Till | | Conventional Till | | No-Till | | Conventional Till | |
| Input | Injected Manure | Surface Manure | Injected Manure | Surface Manure | Injected Manure | Surface Manure | Injected Manure | Surface Manure |
|---|---|---|---|---|---|---|---|---|
| **Energy (dfe)[*]** | | | | | | | | |
| Fuel | 787 | 771 | 1,251 | 1,235 | 1,224 | 1,203 | 1,402 | 1,394 |
| Fertilizer | 2,229 | 2,603 | 2,229 | 2,603 | 220 | 330 | 220 | 330 |
| Pesticides | 429 | 429 | 286 | 286 | 264 | 264 | 209 | 209 |
| Machinery | 870 | 870 | 1,387 | 1,387 | 1,348 | 1,348 | 1,348 | 1,348 |
| Purchased feeds | 1,497 | 1,497 | 1,497 | 1,497 | 1,172 | 1,040 | 1,172 | 1,172 |
| Miscellaneous | 748 | 721 | 737 | 765 | 969 | 1,123 | 1,178 | 1,178 |
| Total | 6,560 | 6,891 | 7,387 | 7,773 | 5,197 | 5,308 | 5,529 | 5,631 |
| Energy product$^{-1}$ (dfe cwt$^{-1}$)[†] | 0.82 | 0.86 | 0.92 | 0.97 | 0.65 | 0.66 | 0.69 | 0.70 |
| **Crop production** | | | | | | | | |
| Corn grain (acres) | 11.9 | 11.9 | 11.9 | 11.9 | 32.8 | 32.8 | 32.8 | 32.8 |
| Corn silage (acres) | 111.6 | 111.6 | 111.6 | 111.6 | 16.5 | 16.5 | 16.5 | 16.5 |
| Alfalfa (acres) | 0 | 0 | 0 | 0 | 74.1 | 74.1 | 74.1 | 74.1 |
| Fuel (gallons) | 787 | 771 | 1,251 | 1,235 | 1,224 | 1,203 | 1,402 | 1,394 |
| Nitrogen (lbs) | 14,046 | 17,023 | 14,046 | 17,023 | 1,204 | 2,037 | 1,169 | 1,958 |
| Herbicides (lbs) | 639 | 639 | 463 | 463 | 381 | 381 | 311 | 311 |

NOTE: C, corn; A, alfalfa.

[*]Diesel fuel equivalent (42.9 Mcal/gallon).

[†]Diesel fuel equivalent per hundred weight of milk.

SOURCE: C. J. Vinten-Johansen, L. E. Lanyon, and K. Q. Stephenson. 1990. Reducing external inputs to a simulated small dairy farm. Agriculture, Ecosystems and Environment 31:225–242.

This approach to studying a farm was preferred to the description of a representative farm operation. First, because work was done on an actual farm, the real problems and conditions encountered by farmers were sure to become apparent. The monitoring program could then evolve to address these real problems. Because much of the information concerning the principles for utilizing a variety of on- and off-farm nutrient sources is well established, the on-farm work could provide insight into possible barriers to the implementation of this existing information. Second, the attempt to study a whole system implied that the integrity of the actual connections among the various parts of the system, the surroundings, and the manager would be maintained. These connections are just as critical to the definition of a whole system as are the components of the system. Finally, the tools developed to study simulated representative farms, while appropriate for such analyses, are not as well suited for application to specific farms. The site specificity and information richness of individual farms can overwhelm representative models. Part of the goal of studying an actual farm was to develop techniques that might be transferred to specific farms and to integrate the research results into appropriate tools for application at the individual farm level.

Even though the animal density on the farm was relatively high (approximately 1.1 animal units/acre; an example of an animal unit is a 1,000-lb dairy cow), the managed flow of nutrients to and from the fields was relatively well balanced (Table 19-5). Balance in the fields was achieved through crop management, resulting in high crop yields; through the application of manure on rented ground, even though it was planted to alfalfa; and through judicious purchases of fertilizer. The importance of animals to the whole farm nutrient flow was apparent in the primary nutrient loading of the farm from animal feed rather than from fertilizer purchased for crop production (Table 19-6). The set of crop and animal factors contributing to the management of on- and off-farm plant nutrient sources reflects the site-specific management by the farmer in both enterprise areas and the potential productivity of the soils. A simple description of the animal-to-land-area ratio was not adequate to characterize the nutrient balance status of the farm fields or the nutrient flow on the whole farm.

Site specificity is readily recognized as part of the internal structure and function of a system, but the surroundings of a farm create site-specific relationships as well. The study farm was located in an intensive agricultural area with an abundance of input suppliers that were supported by many input sources. The suppliers of inputs to the farm identified local, regional, and even international sources of materials that contributed to the nutrient loading of the farm. Many of the regional geographic sources of farm material inputs are identified on the map in Figure 19-1.

A linear program simulation of this same dairy farm compared costs

**TABLE 19-5**  Nutrient Balance of the Managed Flows Classified by Material or Source for all the Fields on a Pennsylvania Dairy Farm in 1985 and 1986

| Input or Output | 1985 | | | | 1986 | | | |
|---|---|---|---|---|---|---|---|---|
| | TN | AN | P | K | TN | AN | P | K |
| Input (lbs/yr) | | | | | | | | |
| Fertilizer | 900 | 900 | 977 | 514 | 2,335 | 2,335 | 785 | 0 |
| Manure | 19,847 | 6,099 | 4,086 | 16,015 | 16,429 | 4,269 | 3,076 | 16,948 |
| Residual | 377 | 377 | — | — | 686 | 686 | — | — |
| Fixation* | 7,918 | 7,918 | — | — | 6,447 | 6,447 | — | — |
| Total | 29,042 | 15,658 | 5,063 | 16,529 | 25,898 | 13,737 | 3,861 | 16,948 |
| Crop output (lbs/yr) | 20,414 | 20,414 | 3,195 | 17,856 | 20,460 | 20,460 | 3,312 | 16,052 |
| Nutrient balance (lbs/yr) | 8,628 | −4,756 | 1,868 | −1,327 | 5,438 | −6,723 | 549 | 896 |
| Nutrient balance (lbs/acre/yr) | 90 | -49 | 20 | −13 | 56 | -70 | 5 | 9 |

NOTE:  TN, total nitrogen; AN, available nitrogen; P, phosphorus; K, potassium. Available nutrient balance (inputs minus outputs) was calculated with the total nutrient outputs.

*Assuming 60 percent of the alfalfa herbage nitrogen content is from fixation.

SOURCE:  S. C. Bacon, L. E. Lanyon, and R. M. Schlauder, Jr. 1990. Plant nutrient flow in the managed pathways of an intensive dairy farm. Agronomy Journal 82:755–761.

**TABLE 19-6**  Nutrient Balance of the Managed Flows in 1985 and 1986 for a Pennsylvania Dairy Farm

| Input or Output | 1985 | | | 1986 | | |
|---|---|---|---|---|---|---|
| | N | P | K | N | P | K |
| Input (lbs/yr) | | | | | | |
| Fertilizer | 900 | 977 | 514 | 2,335 | 785 | 0 |
| Feed | 12,758 | 2,412 | 3,283 | 9,100 | 1,118 | 2,531 |
| Bedding | 926 | 126 | 895 | 75 | 11 | 163 |
| Total | 14,584 | 3,515 | 4,692 | 11,510 | 1,914 | 2,694 |
| Output (lbs/yr) | | | | | | |
| Livestock | 406 | 95 | 24 | 150 | 35 | 9 |
| Product | 5,656 | 942 | 1,321 | 5,532 | 922 | 1,290 |
| Manure | 646 | 287 | 919 | 0 | 0 | 0 |
| Total | 6,708 | 1,323 | 2,264 | 5,682 | 957 | 1,299 |
| Nutrient balance (lbs/yr) | 7,876 | 2,192 | 2,428 | 5,828 | 957 | 1,395 |

NOTE:  N, nitrogen; P, phosphorus; K, potassium. Nutrient balance is inputs minus outputs.

SOURCE:  S. C. Bacon, L. E. Lanyon, and R. M. Schlauder, Jr. 1990. Plant nutrient flow in the managed pathways of an intensive dairy farm. Agronomy Journal 82:755–761.

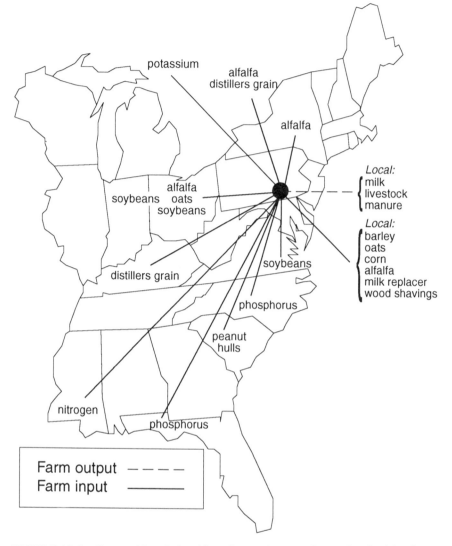

FIGURE 19-1 Geographic relationships of a southeastern Pennsylvania dairy farm and the sources of selected production inputs and the outlets for farm products.

associated with a material flow monitoring program that involved weighing of farm materials and several material sampling and analysis intensities (Lemberg, 1989). These operational parameters of the monitoring program were considered within the context of common property or open-access water resources perspectives (Runge, 1981). Under conditions of common property, all parts of society control the use and pollution of water resources, but under open access there are no constraints on individual behavior in relationship to the resource.

The net returns to the farm with the monitoring program in place were not significantly less than the returns in a comparable simulation without a monitoring program, except that there was no labor or analysis cost for the same information collected in the program. In both cases, the net returns were greater than those in the simulation in which manure was disposed of on the land and the complete plant nutrient requirements were provided by fertilizers. In contrast, the water resource relationship perspective of society had a considerable effect on the potential farm profitability when the amounts of nutrients applied were limited strictly to those required by the crops. These restrictions resulted in manure that was produced but unspread under the common property perspective compared with the open-access perspective. The complex and indivisible ratio of nutrients in the manure compared with the crop requirements precluded the use of manure and favored the application of fertilizers that contained the necessary nutrients in exact proportions.

In general, the method for monitoring farm material inputs and, consequently, on-farm nutrient utilization appears to be economically feasible. On the other hand, societal perspectives for acceptable management may deserve more attention. A consensus on the appropriate precision and range of performances that define responsible management may be very critical to the potential utilization of organic sources of plant nutrients on crop and livestock farms.

## PLANT NUTRIENT MANAGEMENT PROCESS

The difficulties in predicting the best management practices from general principles of crop nutrient supply and the need to deal with site specificity are being integrated into a nutrient management process in Pennsylvania (Figure 19-2). The stages of this process parallel those of tactical planning, day-to-day plan implementation, and control operations that have been used widely in business administration (Meredith and Gibbs, 1984).

This management process has several distinctive features. First, it approaches a farm at the farmer's tactical decision-making level. With this level of resolution for the human activity system (Checkland, 1981), it is a potentially viable heuristic for participation with the farmer. The usual facts about, for example, nutrient requirements of crops and the dynamics of N volatilization, are embedded in the process, but these facts do not constitute the entire process, nor do these facts determine the strategy within which the process operates. This approach to management facilitates the implementation of a particular nutrient management strategy, but it does not determine the strategy. The technical expert does not need to know

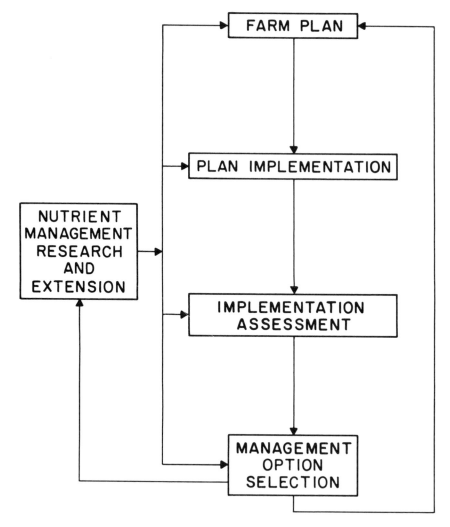

FIGURE 19-2   Proposed nutrient management process to integrate management advisers with the farmer in on-farm management activities. Source: L. E. Lanyon and D. B. Beegle. 1989. The role of on-farm nutrient balance assessments in an integrated approach to nutrient management. Journal of Soil and Water Conservation 44:164–168.

the nutrient management answer for a specific farm before becoming involved in the process.

Second, the management process is iterative. That is, the process does not need to achieve the right answer on the first attempt. The opportunity exists to monitor the system, to respond to performance indicators, and to innovate based on the specifics of the system functions. This openness to iteration is a mechanism by which existing practice can be integrated into the process. Because farmers create the farm based on their own particular rationales, the likely path of action on a particular farm is most closely related to the action of the preceding year rather than to some completely new selection from a wide spectrum of possibilities. The opportunity to incorporate the history of the farm explicitly may be an effective bridge between old and new management systems.

Third, new tools must be developed to support the various stages of the process. These tools are a means to link the technical experts with the operation of the farm and to complement the management skills of the farmer. The ultimate goal of the process is to enhance the manager's understanding of the farm operation so that goals can be incorporated into the operation of the farm based on farm managers' knowledge of their operations rather than, simply, the requirements of good practices.

Computer-based tools have been developed to complement the management activities in several parts of the management process in the nutrient management research and education effort in Pennsylvania (Lanyon and Beegle, 1989). Planning tools are available that estimate manure production and that allocate the manure to fields based on the nutrient requirements of the crops (Beegle and Durst, 1989). A program has been developed to collect information about actual farm operations and to process that information (Lanyon and Meij, 1989). The information can be submitted directly to the farm manager or can be the basis of implementation assessment activities (Lanyon and Schlauder, 1988).

Further refinement of existing tools, the development of tools to complement the remaining parts of the process, and implementation of the process by personnel or institutions are still needed. Selected tools have been accepted and used in the field, but the integrated approach has not yet been accepted.

## NORMAL SCIENCE, LEARNING, AND PROFESSIONAL ACTIVITIES

The outcomes of normal science, like puzzle solving (Kuhn, 1970), can be embedded within the proposed management process, but normal science is unlikely to coopt the process. The puzzle-solving metaphor implies that the ruling paradigm is known and that a missing piece can be discovered. Nutrient management, on the other hand, incorporates site specificity and information richness, that is, the particular elements of place and management. It does not address the missing pieces of a management system. For normal science to deal with site specificity and information richness, thin chains of logic (hypotheses) must be developed with as many factorial treatment combinations as are needed, and experimental measurements must be made to test (disprove) the hypotheses. Farmers faced with nutrient management decisions are more likely to encounter a complex web of interconnections rather than a thin chain with a weak link to be stressed until it breaks. Farmers do not have the flexibility to wait until all possible hypotheses have been tested before they make each management decision.

Modern science suggests that the surroundings or experimental conditions also influence the outcomes of the experiments themselves. The measurements cannot be separated from the experiment or the surroundings. Consequently, even complete factorial experiments that must hold all other factors constant may be inadequate to address the much different condition of site specificity on an actual farm. Other factors may be relevant, to an unknown degree, to the efficacy of the experimental results in the setting outside the experiment. Nevertheless, these experiments should not be abandoned, as they contribute to the formulation of universal principles, but their limitations should be recognized when the focus of the effort is the particular and concrete elements of site specificity and information richness.

Modern science also recognizes the impact of surroundings on the characteristics of internal system structure and function, as emphasized in irreversible thermodynamics. This revision of the traditional reversible thermodynamic perspective has transformed the focus from the singularity of ideal reversible phenomena to the multiplicity of real irreversible phenomena (Nicolis and Prigogine, 1989). Irreversible phenomena can follow multiple paths in the transformation of a given state, such as a change in pressure and volume of a unit of gas, compared with the single reversible path. Irreversible transformations affect and are affected by the surroundings, while reversible transformations are unaffected by the surroundings. The irreversibility of a process leads to the requirement for an infinite amount of information to predict the behavior of a dynamic system with multiple potential paths. This situation is often described as the sensitivity of a system to initial conditions. It is known, however, that an infinite amount of information for predicting the behavior of particular real systems is not attainable. Therefore, approaches to dealing with the possibilities of these systems that are outside the domain of traditional science must be considered.

Contemporary cognitive theories recognize the elementary left brain/right brain orientations of fact-based/art-based humans (Hampden-Turner, 1981) or the less simplistic paradigm that views the mind as a process shaped by individual physiology and a unique set of personal experiences (Minsky, 1986). Farmers who participate in the management process possess a certain perspective and certain abilities, that is, distinctive cognitive styles. Each style affects the selection of management strategies, the efficacy of implementation activities, and the acceptability of system performance. Thus, the additional site specificity and information richness of the manager must be included in an approach to complex (i.e., real-world) systems such as farms.

The proposal of a formal management process may not be compatible with the cognitive styles of all members of the farming population. However, it can be a common framework by which to approach both farms and farmers. With some effort, it could be adapted to a variety of management styles by sensitive buffering between the managers of the technical process and the farm managers. The incentive for farmer participation in such a process may be the feasibility of performance-based rather than specification-based farming. The establishment by farmers, at an acceptable level of confidence, of farming methods for their own farms may be a practical, more satisfying approach to management than farming according to a prescription from off-farm experts.

Many people may be more comfortable with compromise or satisfactory situations than scientists who, in their professional lives, strive for the black and white results of hypothesis testing. Off-farm experts generally deal best with judgments about static and unchanging physical objects. Generally, experts perform poorly in situations in which dynamic systems that are linked to human behavior are confronted (Shanteau, 1989). Farms and farmers, especially those focusing on site specificity and information richness, are more likely to be dynamic and complex.

Chambers (1983) further emphasizes that professionals tend to be inculcated with values and preferences during their education that influence the projects, clientele, and locations they select for future work. He advocates a willingness on the part of the professionals to deal with the last as well as the first of these preferences and to assume new roles as the outsiders. The professionals in these new roles do not emphasize the transfer of technology, but attempt to empower farmers to learn, adapt, and do better (Chambers, 1989).

This brief overview of modern science, contemporary cognitive theories, and an alternative approach for practicing professionals suggests that the activities of professional agriculturalists be crafted within these evolving contexts. The harnessing of ever more powerful computers to achieve mechanistic solutions of natural system relationships based on normal science will not adequately incorporate these new perspectives. Ulanowicz (1986) contends that in a phenomenological perspective of complex ecological systems, such as the operation of a specific farm, one is less concerned with true and false than with degrees of adequacy of process description. This perspective may be the basis for a contemporary paradigm by which to understand particular farms with their site specificity and information richness.

Focusing instead on an integrative management process could facilitate a mode of learning about farms for farmers, professionals, and perhaps even society. Such a process will be enhanced with ongoing application. Its conceptualization and implementation may serve to integrate universal principles that are developed in the course of normal science with the "messiness" that characterizes real systems. The approaches of normal science and management of information about the world and the people in it are quite different, but they can be complementary and synergistic.

## CONCLUSION

An approach to farm systems that is sensitive to the contributions of normal science and the opportunities of an integrative management process will be able to take advantage of both science and management alternatives. The site specificity and information richness that are essential in contemporary plant nutrient management also have been heralded as the hallmarks of sustainable agriculture. Therefore, the utility of normal science in addressing issues of sustainability may be as limited as it is for nutrient management. There are no missing pieces in the existing landscape of the world in which sustainable agriculture must make its place. Sustainability is a complex strategy, not a fact that can be experimentally measured or perhaps even defined. Sustainability as a strategy is a guide for real-world operations. It depends on a value decision of what is right or wrong. Facts about nature and the world do not answer value questions (Gould, 1982).

Normal science may be inadequate as the source of sustainable agriculture. Likewise, by itself, a site-specific, information-rich management process cannot ensure that a sustainable agriculture system will be implemented. The evolution of real systems depends too much on the surroundings to be specified simply by understandings of or intentions for the system. For sustainability to become a guiding principle of agriculture, it must become a widespread and consistent strategy reflected in informed management actions and reinforced by society. Anthropologist R. A. Rappaport (1979) emphasizes that detailed knowledge alone cannot replace the human expression of respect as the critical element in our relationship with the environment.

## REFERENCES

Bacon, S. C., L. E. Lanyon, and R. M. Schlauder, Jr. 1990. Plant nutrient flow in the managed pathways of an intensive dairy farm. Agronomy Journal 82:755–761.

Beegle, D. B., and P. T. Durst. 1989. Farm Nutrient Management Worksheet. Version 2.01, ECS AAG-0103 v2.01. University Park, Pa.: Department of Agronomy, The Pennsylvania State University.

Berry, J. T., and N. L. Hargett. 1987. 1986 Fertilizer Summary Data. Bulletin No. Y-197. Muscle Shoals, Ala.: National Fertilizer Development Center, Tennessee Valley Authority.

Chambers, R. 1983. Rural Development. New York: Longman Inc.

Chambers, R. 1989. Reversals, institutions and change. Pp. 181–195 in Farmer First, Intermediate Technology Publications, R. Chambers, A. Pacey, and L. A. Thrupp, eds. New York: The Bootstrap Press.

Checkland, P. 1981. Systems Thinking, Systems Practice. Chichester, England: John Wiley & Sons.

Gould, S. J. 1982. Non-moral nature. Natural History 91(2):19–26.

Hampden-Turner, C. 1981. Maps of the Mind. New York: Macmillan.

Kuhn, T. S. 1970. The Structure of Scientific Revolutions, 2nd ed. Chicago: The University of Chicago Press.

Lanyon, L. E., and D. B. Beegle. 1989. The role of on-farm nutrient balance assessments in an integrated approach to nutrient management. Journal of Soil and Water Conservation 44:164–168.

Lanyon, L. E., and H. K. Meij. 1989. FINFO: Field and Farm Technical Information Management Program—Users Guide. Agronomy Series No. 106. University Park, Pa.: Department of Agronomy, The Pennsylvania State University.

Lanyon, L. E., and R. M. Schlauder, Jr. 1988. Nutrient Management Assessment Worksheets. Version 2.0L. Agronomy Series No. 103. University Park, Pa.: Department of Agronomy, The Pennsylvania State University.

Lemberg, B. 1989. An Economic Evaluation of a Method to Obtain Farm-Specific Nutrient Information on a Lancaster County Dairy Farm Under Two Nutrient Management Strategies. M.S. thesis, Department of Agricultural Economics and Rural Sociology, The Pennsylvania State University, University Park, Pa.

McSweeny, W. T., and L. C. Jenkins. 1989. 1988 Pennsylvania Dairy Farm Business Summary. Extension Circular 374. University Park, Pa.: College of Agriculture, The Pennsylvania State University.

Meredith, J. R., and T. E. Gibbs. 1984. The Management of Operations, 2nd ed. New York: John Wiley & Sons.

Midwest Plan Service. 1975. Livestock Water Facilities Handbook MWPS-18. Ames, Iowa: Midwest Plan Service, Iowa State University.

Minsky, M. 1986. The Society of Mind. New York: Simon & Schuster.

Nicolis, G., and I. Prigogine. 1989. Exploring Complexity. New York: W. H. Freeman.

Pennsylvania Agricultural Statistics Service. 1989. Statistical Summary 1988–89. Harrisburg, Pa.: Pennsylvania Agricultural Statistics Service.

Rappaport, R. A. 1979. Ecology, Meaning, and Religion. Berkeley, Calif.: North Atlantic Books.

Runge, C. F. 1981. Common property externalities: Isolation, assurance, and resource depletion in a traditional grazing context. American Journal of Agricultural Economics 63:595–606.

Shanteau, J. 1989. Psychological characteristics of agricultural experts: Application to expert systems. Pp. 163–179 in Proceedings of Climate and Agriculture Systems Approaches to Decision Making, A. Weiss, ed. Washington, D.C.: American Meteorological Society.

Ulanowicz, R. E. 1989. Growth and Development. New York: Springer-Verlag.

Vinten-Johansen, C. J., L. E. Lanyon, and K. Q. Stephenson. 1990. Reducing external inputs to a simulated small diary farm. Agriculture, Ecosystems and Environment 31:225–242.

Westphal, P. J., L. E. Lanyon, and E. J. Partenheimer. 1989. Plant nutrient management strategy implications for optimal herd size performance of a simulated dairy farm. Agricultural Systems 31:381–394.

# 20

# Use of Fungal Pathogens for Biological Control of Insect Pests

*Raymond I. Carruthers, Alan J. Sawyer,*
*and Kirsten Hural*

Although the vast majority of insects are either beneficial or harmless to humans, control of a few pest species has been a challenge since the beginning of time. Actually, fewer than 1 percent of known insect species are considered pests (Davidson and Lyon, 1979). These insects destroy crops, damage dwellings, eat food from the table, and even attack people. In response to these pests and their insults, some of the most lethal toxins known have been developed and spread throughout the environment. Although they are effective in the short term, chemical pesticides are expensive and typically provide only temporary relief, as the explosive reproductive and evolutionary capacities of the insects allow them to develop mechanisms of resistance to these and other control strategies (Metcalf, 1980).

Secondary effects (human health hazards, damage to nontarget organisms, environmental pollution, etc.) produced by the application of pesticides to residential and agricultural lands for control of insects, plant diseases, and weeds suggest that some of these control strategies have become self-defeating (Perkins, 1982). Clearly, however, chemical pesticides are valuable tools that must be used wisely to combat insect pests. The challenge, then, is to use them only when necessary and in consort with other more ecological methods of pest control. In response to this need, the discipline of integrated pest management (IPM) (Allen, 1980) has evolved a philosophy of controlling pest species based on intimate knowledge of pest population dynamics and associations with the surrounding ecosystem. Multiple tactics, including chemical, cultural, and biological methods, are used collectively with respect for the environment, to maintain pest numbers below economically significant levels (Allen, 1980; Hoy and Herzog, 1985).

Biological control, which is the use of natural enemies to help regulate pest populations, has been the mainstay of many IPM programs. In most cases, biological control involves the use of parasites, predators, or pathogens that attack, injure, or kill the target pest. To that end, the U.S. Department of Agriculture (USDA) has been involved in the development and implementation of biological control programs for insect and weed pests for over 100 years (Coppel and Mertins, 1977; King and Coulson, 1988). Most of these efforts have focused on the collection and use of exotic parasites and predators for control of insect pests. Several classical examples of biological control successes have been documented by Huffaker (1974) and Coppel and Mertins (1977). Although some early attempts were made to use insect pathogens as biological control agents, much less emphasis has been placed on microbial control than on other areas of biological control. Recently, more emphasis has been placed on the use of insect pathogens (bacteria, viruses, and fungi) as biological control agents. These pathogens are known to produce widespread epizootics in nature, often decimating their insect host populations, but scientists have been relatively unsuccessful in producing insect disease outbreaks at will (Carruthers and Soper, 1987). Most research efforts have focused on the isolation, development, and production of highly pathogenic microbes with characteristics favoring long-term storage and application, as if they were chemical insecticides. Although this approach has been successful with some bacteria, the use of microbes as replacements for chemical insecticides has usually resulted in unpredictable and inadequate responses under field conditions (Fuxa, 1987). One of the primary reasons for these failures is the lack of understanding of the natural dynamics of host-pathogen life systems. Diseases develop in complex ways based on biological and physical associations between the host, the pathogen, and the environment. Understanding of these associations in general and more specifically on a system-by-system basis should provide a significant amount of assistance in the ability to manipulate insect pathogens for IPM purposes.

Fungal pathogens are important natural biological control agents of many insects and other arthropods and frequently cause epizootics that significantly reduce host populations (Burges, 1981; Carruthers and Soper, 1987; MacLeod, 1963; McCoy et al., 1988). Because of the frequency of natural epizootics and the conspicuous symptoms associated with fungus-induced mortality (McCoy et al., 1988; Steinhaus, 1963), the significance of fungi in regulating insect populations was noted early in recorded history by the ancient Chinese (Roberts and Humber, 1981). In fact, *Beauveria bassiana,* an insect-pathogenic fungus, was the first microbe recognized as a pathogen and was used by Agostino Bassi (1835) to demonstrate his germ theory of disease.

Approximately 750 species of entomopathogenic fungi are known from

85 genera (exclusive of the 115 genera in the order *Laboulbeniales*) found throughout the classes of fungi (Gillespie and Moorhouse, 1989; McCoy et al., 1988; Roberts and Humber, 1981).   The majority of the entomopathogenic species are classified in the classes *Hyphomycetes, Zygomycetes* (order *Entomophthorales*), and *Ascomycetes* (in particular, the genera *Cordyceps* and *Torubiella*).   These pathogens cause mycoses in many different taxa of arthropods and in almost every order of the *Insecta* (Bell, 1974; Gillespie and Moorhouse, 1989).   They are known to infect all life stages of insects and are commonly found in aquatic, terrestrial, and subterranean habitats (Ferron, 1978).   Although fungal pathogens have much in common with viruses, bacteria, and other insect-pathogenic microbes, they are unique in many ways (Ferron, 1978).   Perhaps the most significant difference lies in the mode of infection; whereas most entomopathogens infect their hosts through the gut following consumption, fungi typically penetrate the insect cuticle and thus are the only major pathogens known to infect insects with sucking mouthparts, orders *Hemiptera* and *Homoptera* (Roberts and Humber, 1981).

Attempts to manipulate fungi as biological control agents of insects began in the late nineteenth century with only poor to moderate success (Krassilistchik, 1888; Metchnikoff, 1879; Snow, 1895; Steinhaus, 1956). Little basic or applied research was conducted on entomopathogenic fungi from the late nineteenth century until the late 1960s, when interest in the use of fungi as biological control agents increased because of problems with chemical control (Roberts, 1979).   Although several successful programs have been developed over the past 20 years (Burges, 1981; Ferron, 1978; Gillespie and Moorhouse, 1989), very few fungi have been used extensively for biological control of insect pests.   The reasons for this are numerous, but a general lack of understanding of fungal population biology and the environmental factors that limit disease development have certainly restricted their use (Allen et al., 1978; Carruthers and Soper, 1987; McCoy et al., 1988).

Naturally occurring epizootics caused by fungal pathogens, particularly those caused by fungi in the order *Entomophthorales,* are noted frequently in both natural and managed ecosystems (Carruthers and Soper, 1987; Carruthers et al., 1985b; McCoy et al., 1988; Mohamed et al., 1977; Nordin et al., 1983; Pickford and Riegert, 1964; Soper et al., 1976; Wilding, 1975, 1981).   Because of the catastrophic impacts these pathogens have on their host populations, they hold significant potential for biological control of some pest species (Carruthers and Soper, 1987; Ferron, 1978; McCoy, 1981; Wilding, 1981).   Definite limitations on their manipulation as biological control agents exist, and some researchers feel that fungus-induced epizootics are too dependent on high host densities and environmental conditions, particularly moisture (Bucher, 1964).   The majority of researchers who study insect pathogens, however, believe that fungi will play a vital role in IPM systems in the near future (Allen et al., 1978; Carruthers and Soper, 1987; Fuxa, 1987; McCoy et al., 1988).

This chapter discusses naturally occurring fungal diseases, basic host and pathogen biology, disease development and spread, environmental factors that allow or restrict disease progression, and the application of fungal pathogens in specific biological control situations. This is not meant to be an exhaustive review but, rather, an overview of the subject. Readers interested in more comprehensive reviews on fungal pathology in general or on specific taxa of fungal pathogens are referred to Steinhaus (1963), Burges and Hussey (1971), Burges (1981) and McCoy et al., (1988). Those interested in additional information on epizootiology, biological control, and use of fungi in IPM should refer to Tanada (1963), Ferron (1978), Wilding (1981), Carruthers and Soper (1987), Fuxa (1987), and McCoy et al. (1988).

## FUNGAL PATHOGEN LIFE CYCLES AND BIOLOGY

Despite the taxonomic diversity of entomopathogenic fungi, there are many similarities among the major groups in their basic life histories and ecology. In general, the pathogen life cycle begins with spore germination and penetration of the host's cuticle, followed by a rapid proliferation of fungal cells which ultimately results in the death of the host. Host death may be followed by the production of infective spores which can immediately repeat the cycle, or by the production of resting spores or other resistant structures which require a period of dormancy.

Infective spores of entomopathogenic fungi may be asexual propagules such as conidia of the order *Entomophthorales* and the class *Deuteromycetes,* or they may be the result of sexual recombination such as the ascospores of the genus *Hypocrella* (the class *Ascomycetes*) (Evans, 1982) or the biflagellate zygote of the genus *Coelomomyces* (*Chytridiomycetes*) (Whisler et al., 1975b). After contact with a potential host, infective spores adhere to the insect cuticle if host recognition is positive (Al-Aidroos and Roberts, 1978). Adhesive processes have not yet been intensively studied in entomogenous fungi; however, both physical and chemical interactions are probably important (Fargues, 1984). Epicuticular compounds such as fatty acids, amino acids, and glucosamines are thought to play a significant role in determining the specificity and pathogenicity of entomopathogenic fungi (Boucias and Pendland, 1984; Kerwin, 1983; Saito and Aoki, 1983; Smith and Grula, 1982; Woods and Grula, 1984).

Spore germination is highly dependent on moisture and probably requires free water (Kramer, 1980; Newman and Carner, 1975; Roberts and Campbell, 1977; Shimazu, 1977), but this requirement may be met by moisture conditions of the microclimate in the absence of measurable precipitation (Ben-Zev and Kenneth, 1980; Hall and Dunn, 1957; Kramer, 1980; Mullens et al., 1987; Tanada, 1963). Penetration of the cuticle is accomplished by the germ tube itself or by the formation of an appressorium which attaches to the cuticle and gives rise to a narrow penetration peg (Boucias and Pendland, 1982; Roberts and Humber, 1981; Wraight et al., in press a; Zacharuk, 1973). Penetration is both a mechanical and an enzymatic process (Charnley, 1984; McCoy et al., 1988; St. Leger et al., 1987, 1988b).

Vegetative growth in the insect hemocoel is common to most entomopathogenic fungi (Roberts and Humber, 1981) and usually consists of discrete yeastlike structures or hyphal bodies. This form of growth, in contrast to the typical filamentous fungal mycelium, allows the entomopathogen to disperse rapidly and colonize the insect's circulatory system and increases the fungal surface area which is in contact with the nutrient medium. Several species of the order *Entomophthorales* produce vegetative protoplasts (cells without cell walls) within the hemocoel (Butt et al., 1981; Latge et al., 1988; Nolan, 1985; Tyrrel and MacLeod, 1972) which may help the pathogen to escape detection by the host's immune responses. The length of the incubation period varies among species; however, disease development during the vegetative stage is typically temperature dependent (Carruthers and Soper, 1987; Carruthers et al., 1985a; Hall, 1981). Fungi have been observed to elicit insect immune responses, but it is not known what role they may play in preventing or slowing the development of mycoses (Butt et al., 1988; Gupta, 1986; St. Leger et al., 1988a).

Pathogens may simply overcome their hosts by consumption of the available nutrients in the hemocoel, as do most lower fungi such as the genera *Coelomomyces* and *Lagenidium* and most genera of the order *Entomophthorales* (Roberts, 1981) or by digestion of host tissues and organs (Brobyn and Wilding, 1977). In cases in which fungi overcome their hosts after a relatively short period of vegetative growth, toxins produced by the pathogen are presumed to be the cause of death (Roberts, 1981). Compounds toxic to insects are produced by several entomopathogenic fungi both in vitro and in vivo (Roberts, 1981). However, their role in pathogen-induced mortality in nature is not well understood.

Shortly after host death, the fungal hyphae penetrate the host cuticle from within and terminate in the formation of sporophores (usually conidiophores) that yield asexual spores (conidia) which function as dispersive and infective units. In many species of fungi, the production of conidia is highly dependent on moisture (Millstein et al., 1983; Wilding, 1969). Conidia are the infective propagules of secondary infection and determine disease development and spread within a season. Environmental factors that control conidial production, survival, and germination are critical to the rate of epizootic development (Carruthers and Soper, 1987).

Entomopathogenic fungi survive adverse environmental conditions or the absence of their host by producing resting spores or other resistant structures, or they survive as dormant mycelia in dried insect mummies (Kenneth et al., 1972; Wilding, 1973). Most species of the order *Entomophthorales* produce a spherical, thick-walled resting spore (MacLeod, 1963) which may also be the sexual stage of these fungi (McCabe et al., 1984). The conditions required to break the dormancy of resting spores are not well understood (Wilding, 1981), although temperature, moisture, and photoperiod have been indicated as triggers in some species (Perry and Latge, 1982; Wallace et al., 1976). When dormancy requirements have been met and other environmental conditions are correct, resting spores germinate and produce germ conidia which then initiate mycoses in susceptible hosts. Some species of the order *Entomophthorales* such as *Pandora neoaphidis* apparently do not produce resting spores (Milner et al., 1983) and presumably overwinter as dormant mycelia inside mummified insects. Fungi in the classes *Ascomycetes* and *Deuteromycetes* may produce specialized masses of hardened vegetative tissue called sclerotia or stromata, as in the genera *Cordyceps* and *Torubiella,* or modified hyphae called chlamydospores, as in the genera *Beauveria* and *Metarhizium* (McCoy et al., 1988; Roberts and Humber, 1981).

Many variations of this basic life cycle occur in different species of entomopathogenic fungi. A unique feature within the order *Entomophthorales,* for example, is the active discharge of infective conidia and their ability to produce repetitional conidia in the absence of a suitable host (King and Humber, 1981). Primary conidia of *Zoophthora radicans,* a common entomophthoralean pathogen, can give rise to actively discharged secondary conidia, which in turn may produce tertiary conidia and so on until the protoplastic reserves of the spore have been consumed. Alternatively, they may produce capilloconidia, which are passively detached conidia borne atop elongate capillary conidiophores. Capilloconidia have a sticky substance at one end and are positioned at an angle which is thought to enable them to be picked up by insects walking over a surface (McCoy et al., 1988; Soper, 1985). The type of spore produced is dependent on host availability and environmental conditions during germination. Another variation on the basic theme is the heteroecious cycle or alternate hosts as in *Coelomomyces* infections of mosquitoes and copepods. A biflagellate zygote, a fusion product of gametes produced in the copepod alternate host, infects mosquito larvae by encysting on the cuticle (Zebold et al., 1979). Following infection, hyphae proliferate throughout the hemocoel and fat body, ultimately producing thick-walled multinucleate sporangia at their tips. Upon death of the larvae, sporangia are released into the water. Meiosis occurs prior to sporangial germination and results in the production of haploid meiospores which are capable of infecting copepod hosts. Each meiospore forms a gametophyte within the copepod and matures into a gametangium which releases gametes of a single mating type. The fusion of opposite mating types completes the cycle by producing mosquito-infective biflagellate zygotes (Federici, 1981; Whisler et al., 1975b).

## DISEASE EPIZOOTIOLOGY

Natural disease development and spread are governed by characteristics of both host and pathogen populations and by the environment in which their interactions occur. In managed ecosystems this scheme is further complicated as human intervention adds a fourth component. These four components (host, pathogen, environment, and humans) and their interactions have been represented by the disease tetrahedron (Carruthers and Soper, 1987; Zadoks and Schein, 1979). The factors encompassed by this model are necessarily interrelated, and although they may be discussed individually, it must remembered that their interactions are critical to overall system behavior.

### The Pathogen Population

Properties of the pathogen population that are important in epizootiology include virulence and pathogenicity, dispersal and survival in the host's environment, and inoculum density and spatial distribution (Tanada and Fuxa, 1987).

Virulence refers to the intensity of the disease caused by a pathogen, whereas pathogenicity refers to an organism's ability to cause disease. Most fungal pathogens are considered highly virulent relative to other pathogenic organisms because they typically have short incubation periods, produce copious amounts of secondary inoculum, and can cause a rapid increase in disease prevalence. A fungal species which is pathogenic on a wide range of hosts may be more likely to persist in an environment because of the availability of alternate hosts. With a greater number of susceptible hosts, there may be a greater reservoir of inoculum available to produce an epizootic (Tanada, 1963).

There are many reports of intraspecific variation in virulence among strains of a pathogen (Daoust and Roberts, 1982; Ignoffo and Garcia, 1985; Papierok and Wilding, 1981). Very often the isolates compared in these studies were collected from locations in different parts of the world. In order to understand the role of intraspecific variability with respect to virulence, it is necessary to measure the extent of variation in well-defined populations of hosts and pathogens. To our knowledge, no studies of this type have yet been published.

Pathogen survival in the host's environment is necessary for the long-term persistence of disease in the population; as previously mentioned, fungi accomplish this by producing various types of resting spores or structures. The ability to survive adverse conditions or periods of host absence may determine the frequency with which epizootics occur, because without some means of survival, infection is dependent upon the movement of inoculum into the host's habitat (Tanada, 1963). Some pathogens, such as *Beauveria bassiana* and *Metarhizium anisopliae,* are ubiquitous and cause mycoses frequently, suggesting that they have successfully persisted in the host's environment. A possible advantage of naturally occurring pathogens over microbial insecticides may be their ability to persist over long periods of time.

Dispersal is necessary for the rapid spread of disease. Abiotic factors such as rain or wind may carry spores, or the movement of infected and uninfected hosts may transport inoculum from one place to another (Hall and Dunn, 1957). Some fungal pathogens cause their hosts to climb to aerial locations just prior to death (MacLeod, 1963); such behavior may aid dispersal of infective conidia which may rain down on or blow to nearby hosts.

Pathogen population density and spatial distribution are key factors in the development of an epizootic, as they affect the likelihood of contact with viable hosts (Tanada and Fuxa, 1987). There has been substantial work on the relationship between pathogen dose and mortality in the laboratory, generally concluding that mortality increases with inoculum dose (Pinnock and Brand, 1981). Although this relationship has rarely been studied under realistic conditions, studies of *Entomophaga grylli* infection of the clear-winged grasshopper revealed that field infection reasonably followed a logistic response curve (see below and Figure 20-1).

## The Host Population

Important host population factors which must be considered are susceptibility, density, movement, and spatial distribution of individuals. There are very few reports of genetic variabilities in the susceptibilities of insects to fungal disease in natural populations. Paperiok and Wilding (1979) reported that one of two clones of the pea aphid (*Acyrthosiphon pisum*) was resistant to infection by *Conidiobolus obscurus* compared with the other highly susceptible clone. Milner (1982, 1985) was able to characterize field-derived clones of pea aphids as resistant or susceptible based on percent mortality when exposed to a particular isolate of *Pandora neoaphidis.* Undoubtedly, the population dynamics of variability in host resistance and pathogen virulence would have an impact on epizootiology and the long-term stability of disease incidence in the population; however, this is another area of research which lacks sufficient study.

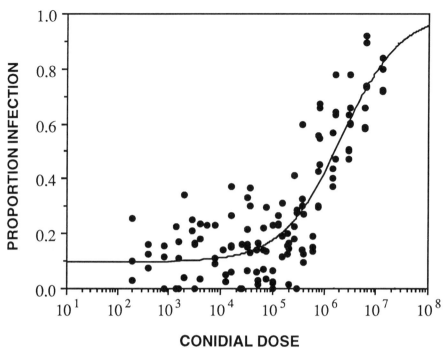

FIGURE 20-1    *Entomophaga grylli* doses (conidia per 0.5-square-meter cage) producing different levels of infection under field conditions.

Although climatic conditions may be limiting for disease development,
under nonlimiting conditions, host density may directly influence the rate of
disease buildup (Benz, 1987; Carruthers et al., 1985b; Watanabe, 1987). In
some situations the spatial arrangement of hosts may be more important
than the actual numbers of individuals. If hosts are highly aggregated, it
may be difficult for the pathogen to disperse between aggregates, unless
infected hosts or pathogen propagules are highly mobile.

### The Abiotic Environment

Abiotic environmental factors such as moisture, temperature, and solar
radiation affect many of the biotic factors mentioned above; moreover, they
may determine whether or not infection can occur. Atmospheric moisture is
often considered the most important abiotic factor in the epizootiology of
fungal diseases (Fuxa and Tanada, 1987; Nordin et al., 1983). As mentioned in the previous section, germination and sporulation of most fungi
are highly dependent on moisture. However, fungi are able to acquire

moisture from sources other than precipitation, such as dew or from the boundary layer of the host. The humidity of the microclimate, such as a dense plant canopy, may be much higher than that of the ambient air (Fuxa and Tanada, 1987; Kramer, 1980; Tanada, 1963).

Temperature-dependent processes that directly affect the progression of disease are the rate at which insects develop, the rate of fungal development within the insect, and the rate and quantity of spore production (Benz, 1987). Most fungal pathogens do well at all temperatures suitable for insect growth. The effects of temperature and humidity are intimately related: Some fungi may tolerate higher temperatures if there is more moisture in the air, as condensation may readily occur and water loss is minimized (Benz, 1987).

Conidia may be very sensitive to solar radiation (Carruthers et al., 1988a; Ignoffo et al., 1977), and spore longevity and germination may be improved if the microhabitat, such as a dense crop canopy, can protect the conidia from direct radiation (Figure 20-2).

Secondary infection, that is host-to-host infection, is crucial in the epizootiology of fungal diseases, in that repetitive cycles of infection result in a rapid increase in disease prevalence (Zadoks and Schein, 1979). Pathogens that have many secondary cycles during one or a few generations of the host are most likely to cause dramatic epizootics. Transmission efficiency has been shown theoretically to be an important parameter related to the rate at which secondary infections occur and thus of epizootic progress (Anderson and May, 1980). Transmission of fungal pathogens occurs primarily through the insect integument; this may be accomplished through direct host-to-host contact or by host contact with infective spores in the environment, such as conidia deposited on plant surfaces. Disease transmission is thus influenced by host and inoculum density and spatial distribution. Theoretical models have demonstrated that a pathogen can be maintained within a host population only if the host density exceeds a threshold value, and this value has been defined as being inversely proportional to transmission efficiency (Anderson and May, 1980). Brown and Nordin (1982) determined that the threshold density for *Zoophthora* spp. epizootics of alfalfa weevil populations is 1.7 weevils per stem. Subsequent field studies showed that larval densities below this threshold could not support a *Zoophthora* epizootic. Nordin et al. (1983) concluded that the initiation of disease was best correlated with degree-day accumulations, and that epizootic dynamics were controlled by atmospheric moisture levels, as long as the host population remained above the threshold density. Further methods of enhancing the development and spread of *Zoophthora* spp. are discussed in a subsequent section.

In recent years, systems analysis and modeling have been recognized as useful aids to understanding the complex dynamics of fungal epizootics

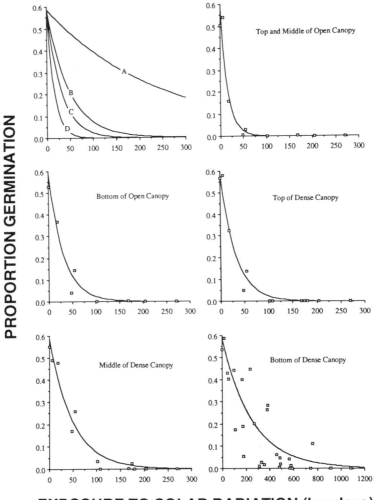

## EXPOSURE TO SOLAR RADIATION (langleys)

FIGURE 20-2   Proportion germination of *Entomophaga grylli* conidia (*p*) as predicted by cumulative solar radiation ($\int Ldt$) by the exponential decay model, $p = 0.558e^{b_i(\int Ldt)}$ (estimates of $b_i$ for each canopy area are provided below). Results from different canopy locations are shown both in a composite group for easy comparison and individually with the corresponding observed data (A, bottom of dense canopy, $b = -0.237$; B, middle of dense canopy, $b = -1.228$; C, top of dense canopy and bottom of open canopy, $b = -1.915$; D, middle and top of open canopy, $b = -3.804$). Source: R. I. Carruthers, Z. Feng, M. E. Ramos, and R. S. Soper. 1988. The effect of solar radiation on the survival of *Entomophaga grylli* conidia. Journal of Invertebrate Pathology 52:154.

(Anderson and May, 1980; Brown, 1987; Carruthers, 1985; Carruthers and Soper, 1987; Carruthers et al., 1988b; Onstad and Carruthers, 1990). Relatively simple models have been used to explore basic questions in epizootiology (Anderson, 1982; Anderson and May, 1980; Brown, 1984; Brown and Nordin, 1982; Hochberg, 1989; Regniere, 1984). More detailed simulation models of specific host-pathogen systems have also been developed for a few fungal pathogens (Carruthers et al., 1986, 1988b). These models and the techniques of systems science have proven to be very useful in guiding the collection, analysis, and synthesis of information about the dynamics of host and pathogen life systems. In addition to providing basic understanding, models can assist in pest management decision making and can be used to help design and evaluate strategies for employing fungal pathogens in biological control programs. Expert systems (Logan, 1988) and intelligent modeling systems (Larkin and Carruthers, 1990; Larkin et al., 1988) are becoming available to make the analytical power of systems techniques more accessible to scientists and pest management specialists who may have little experience with modeling.

Although epizootics of fungal disease are known to cause major declines in pest populations, very little information is available on the long-term regulatory ability of these pathogens. This is particularly true during periods of enzootic activity, when disease prevalence is low and signs of host infection are difficult to detect. Disease assessment is complicated further if pathogen-induced mortality occurs in habitats that are difficult to observe, as is the case in soil or aquatic ecosystems. Over the past two decades, however, several significant descriptive and experimental studies on disease dynamics have provided new insights into the population biology and possible use of fungi as biological control agents (Carruthers et al., 1985b; Kish and Allen, 1978; MacLeod et al., 1966; Nordin et al., 1983; Soper and MacLeod, 1981; Soper et al., 1976). As it is not possible to review all the significant literature, some relevant examples have been chosen to highlight the impact of fungal pathogens on their hosts, both in natural and managed ecosystems.

## COWORKERS AND COOPERATORS INVOLVED
## IN SPECIFIC RESEARCH PROJECTS

The Plant Protection Research Unit, Agricultural Research Service (ARS), USDA, has been involved in the development and use of fungal pathogens of insects in several different ecosystems, including forest, rangeland, and agricultural production systems. Cooperators are listed in Table 20-1, along with their involvement in specific projects, as indicated in the footnotes. Implementation projects have involved cooperators from action agencies, such as state departments of agriculture and markets, The U.S. Forest Ser-

**TABLE 20-1**   Research Team Members Involved in Development and Use of Fungal Pathogens

| Affiliation and Team Member | Affiliation and Team Member |
| --- | --- |
| USDA-ARS | Cornell University |
| R. I. Carruthers[*–#] | Z. Feng[‡,§] |
| H. Firstencel[§,#] | S. Galaini-Wraight[†,#] |
| R. A. Humber[*,§,ǁ] | K. Hural[*,§,#] |
| A. J. Sawyer[§,#] | T. S. Larkin[†,–ǁ] |
| R. S. Soper[*,†,§,ǁ,#] | D. G. Robson[†,§] |
| USDA-APHIS | Boyce Thompson Institute |
| R. N. Foster[§] | for Plant Research |
| J. L. Fowler[§] | M. E. Ramos[§,ǁ] |
|  | D. W. Roberts[†,‡,#] |
|  | S. P. Wraight[#] |
| Illinois Natural History Survey |  |
| J. V. Maddox[#] |  |
| S. Roberts[#] |  |

NOTE: USDA, U.S. Department of Agriculture; ARS, Agricultural Research Service; APHIS, Animal and Plant Health Inspection Service.

[*]Population genetics of the aphid pathogen *Pandora neoaphidis*.

[†]Microbial control of the Colorado potato beetle using *Beauveria bassiana*.

[‡]Microbial control of the European corn borer using *Beauveria bassiana*.

[§]Evaluation of *Entomophaga grylli*-like biological control agents for grasshopper management.

[ǁ]Introduction of *Entomophaga maimaiga* for gypsy moth biological control.

[#]Introduction of *Zoopthora radicans* for biological control of leafhoppers.

vice and the Animal and Plant Health Inspection Service (APHIS) of USDA, and cooperating ranchers and farmers in areas where actual field projects have been conducted.

## SPECIFIC EXAMPLES OF INSECT BIOLOGICAL CONTROL RESEARCH USING FUNGAL PATHOGENS

The use of entomopathogenic fungi for biological control in managed ecosystems has followed four basic strategies: (1) awareness of natural biological control impacts, (2) augmentation of natural enemies, (3) enhancement of impacts through active manipulations, and (4) introduction of exotic natural enemies (Ignoffo, 1985; McCoy et al., 1988). Examples of ongoing research that demonstrate the important aspects of and the levels of expertise that investigators have in each of these areas of biological control are provided. The first example comes from an on-going research project aimed at developing a basic understanding of natural host-pathogen dynamics in a rangeland ecosystem. This project has evolved into an active biological control project that is now aimed at introducing fungal pathogens (both native and exotic germ plasm) into areas where they do not currently exist.

## Awareness of Naturally Occurring Pathogens and Their Relevance to Biocontrol

### Entomophaga grylli *Mycosis of Grasshoppers*

Rangelands are low-energy ecological systems that provide resources to people primarily in the form of forage for grazing livestock. Although rangelands are low in productivity per unit area, their vastness (ca. 1.0 billion acres in the continental United States [Heath et al. 1973], which is the equivalent of five times the area of Texas) makes them extremely important in terms of food production and general resource management. Grasshoppers are the dominant group of herbivorous insect pests associated with rangelands worldwide. Their populations have plagued humans from the earliest days of recorded history, and they continue to do so today. In the United States alone, millions of federal dollars have been spent on grasshopper spray programs each year over the past decade, and populations are still expected to increase. In other areas of the world, grasshoppers and migratory locusts are commonly seen in outbreak numbers year after year.

In response to this pest problem, the ARS of USDA has initiated a variety of research programs to study grasshopper biology and management. One of these projects is focused on the development of grasshopper pathogens as long-term biological control agents. One such pathogen is *Entomophaga grylli,* a fungus that attacks several grasshopper species (Carruthers and Soper, 1987). This pathogen has caused widespread but sporadic epidemics in grasshopper populations and has been responsible for major population reductions of several economically important pest species. Although this pathogen is known to play a significant role in the regulation of natural grasshopper populations, very little was known about its ecological associations with different grasshopper hosts or with other components of the environment. Interest in manipulating this pathogen for use in IPM programs was the stimulus for initiating a long-term research project on the natural dynamics of *E. grylli* (Carruthers and Soper, 1987; Carruthers et al., 1988a, in press). For the past few years, descriptive and experimental studies have been conducted that are designed to determine key abiotic and biotic factors involved in regulating disease dynamics under natural field conditions.

*E. grylli* is an obligate fungal pathogen of grasshoppers and thus is nonpathogenic to humans, beneficial insects, and other nontarget organisms. Study of its parasitic and free-living stages is important in understanding its biology and interactions with grasshopper hosts. *E. grylli* has a complex life cycle. Resting spores overwinter in the soil and germinate each spring. The phenology of germination depends on local environmental conditions. Upon the breaking of dormancy, resting spores produce specialized spores (germ conidia) that are forcibly ejected from the soil into the lower plant canopy where they contact grasshoppers, produce germ tubes, and then penetrate the body wall of the insect through enzymatic and mechanical actions. Once it is inside the insect, the pathogen multiplies rapidly while digesting away the surrounding host tissues. Under normal field conditions, this disease retards egg laying and feeding damage prior to actual host death, which occurs approximately 7 to 10 days after initial exposure. Upon host death, the fungus produces either resting spores, which are released into the soil to initiate infection in subsequent years, or airborne conidia, which either germinate immediately (within 24 hours) or die from adverse environmental conditions.

Although each dead grasshopper may be the source of several million fungal conidia, these free-living spores are extremely ephemeral, as their survival is highly influenced by temperature, moisture, and solar radiation (Figure 20-2). To maximize their infection potential, *E. grylli* induces grasshopper mortality in midafternoon, followed by conidial production starting in the early evening. This synchronizes maximum spore production with dew formation and minimum solar radiation, conditions that increase the chance of germination and infection. The magnitude of host infection is directly related to the inoculum level (Figure 20-1) and microenvironmental conditions in the habitat where infection occurs (Feng et al., in press). If conidia are successful in contacting a susceptible host under appropriate environmental conditions, secondary infection (infection initiated from spores produced within the same season) will perpetuate the disease through another cycle. Since the incubation period of this disease is substantially shorter than the life cycle of the grasshoppers (Carruthers et al., in press), the pathogen may have several generations each season, with the exact number of cycles depending on environmental conditions. Natural epizootics of this disease commonly produce high levels of grasshopper mortality (Figure 20-3) and have been known to reduce outbreak populations of grasshoppers to nondamaging levels (Riegert, 1968). Based on research conducted on the natural dynamics of this pathogen in restricted ecosystems and detailed laboratory evaluations of important pathobiological characteristics, APHIS and ARS of USDA are now cooperating on a biological control implementation program by using *E. grylli* for the control of rangeland grasshoppers in North Dakota and

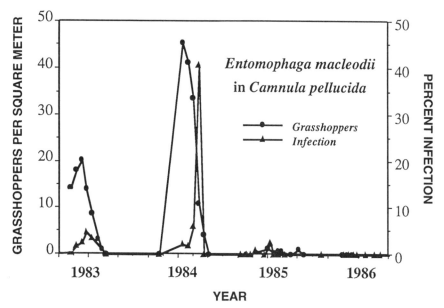

FIGURE 20-3   Grasshopper population density over a 4-year period and levels of *Entomophaga grylli* mycosis.   Grasshopper mortality caused by this pathogen was clearly responsible for the observed population declines.

Alaska.  If successful, this program is likely to expand into adjacent rangeland areas.

### Augmentation of Fungal Pathogens for Biological Control

Augmentation (increasing pathogen inoculum density) through the development of microbial insecticides has received substantial attention and is discussed here only briefly, as it is the focus of other articles (Burges, 1981; Burges and Hussey, 1971; McCoy, 1990).   It must be said, however, that augmentation has not only taken the form of microbials applied like pesticides with the goal of high acute host mortality but has also been used to initiate epizootics prematurely (Ferron, 1981; Ignoffo et al., 1976) or in situations in which epizootics would not develop naturally (McCoy, 1981; Riba, 1984).   In these cases, disease development is dependent not only on the efficacy of the fungal material originally applied to the target host but also on the pathogen's ability to become established in the environment and produce a secondary inoculum that is capable of polycyclic infection. These methods of pathogen augmentation require detailed information about host, pathogen, and disease dynamics that have been researched only recently (Allen et al., 1978; Carruthers and Soper, 1987; McCoy et al., 1988).

Beauveria bassiana *Mycosis of Colorado Potato Beetles*

The fungal pathogen *Beauveria bassiana* was investigated as a potential mycoinsecticide for control of the Colorado potato beetle, *Leptinotarsa decemilineata*. Increased resistance of this pest to various chemical insecticides and groundwater pollution associated with chemical use prompted investigations into the use of fungi as an alternative means of control. Soviet researchers (Lappa, 1978) found specific isolates of *B. bassiana* to be effective in controlling populations of the Colorado potato beetle in the Soviet Union. Based on their strategies, field trials were first conducted in potato-growing regions of Long Island, New York (Galaini, 1984), using spore preparations developed by Abbott Laboratories. Many difficulties were encountered in the initial stages of this project, including extremely high Colorado potato beetle populations, low efficacy of the fungal propagules, and improper application strategies. Over the course of 3 years of investigation, several improvements in the use of *B. bassiana* as a mycoinsecticide were developed. Application techniques were improved by altering equipment and spray strategies through field testing and computer modeling (Galaini-Wraight et al., in press a). Four applications of a commercial standard pesticide (Pydrin/piperonyl butoxide [PBO]) were compared with five applications of *B. bassiana* during the first Colorado potato beetle generation and two Pydrin/PBO sprays versus three *B. bassiana* applications in the second Colorado potato beetle generation. Results indicate that *B. bassiana* is capable of providing reasonable protection of potato foliage in the study plots (Table 20-2) (Galaini-Wraight et al., in press a). Mortality of immature Colorado potato beetles in *B. bassiana*-treated plots was only slightly lower than that in the insecticide-treated plots and provided total yields about 75 to 80 percent of those achieved in the chemical insecticide treatment (Table 20-3). Adjacent untreated plots were totally destroyed by the Colorado potato beetle and produced no measurable yield. Although *B. bassiana* was not capable of controlling Colorado potato beetles to the same degree as chemical insecticides were, it clearly provided substantial protection against Colorado potato beetle damage under some conditions. Further refinements in the use of this pathogen as a component of a Colorado potato beetle management program are clearly warranted. A subsequent study, using *B. bassiana* as a mycoinsecticide on a commercial scale, gave extremely variable results (Hajek et al., 1987).

Variable results have plagued the development and use of mycoinsecticides on a number of crops (McCoy, 1990). Several biological and technical limitations currently exist in using fungi as replacements for insecticides (McCoy, 1990). Although fungal propagules can be applied to crops as an inundative spray, strategies for their use and expectations in terms of the resulting pest mortality and crop protection should be drastically altered

**TABLE 20-2** Estimated Percent Mortalities* of *Leptinotarsa decemlineata* Larvae for *Beauveria bassiana* and Pydrin/PBO Treatments

Estimated Percent Mortalities

| | First Generation | | | | Second Generation | | | |
| | Early Immatures | | Late Immatures | | Early Immatures | | Late Immatures | |
| Plot No. | B. bassiana | Pydrin/PBO | B. bassiana | Pydrin/PBO | B. bassiana | Pydrin/PBO | B. bassiana | Pydrin/PBO |
|---|---|---|---|---|---|---|---|---|
| 1 | 77.3 | 85.6 | 82.2 | 84.7 | 86.1 | 82.6 | 21.2 | 23.0 |
| 2 | 74.7 | 88.4 | 82.8 | 86.6 | 79.3 | 79.0 | 54.8 | 43.9 |
| 3 | 76.6 | 87.8 | 81.1 | 85.0 | 82.2 | 81.4 | 30.4 | 51.7 |
| 4 | 81.7 | 88.2 | 81.6 | 90.0 | 80.9 | 81.4 | 30.3 | 25.3 |
| Mean† | 77.6a | 87.5b | 81.9a | 86.7b | 82.2a | 81.1a | 33.7b | 35.5b |
| 95 percent confidence interval | (72.7–82.2) | (85.4–89.5) | (80.7–83.1) | (82.4–90.4) | (77.3–86.7) | (78.7–83.4) | (13.7–57.4) | (15.4–58.8) |

*Percent mortality estimated graphically, as described by T. R. E. Southwood. 1978. Ecological Methods. New York: Chapman and Hall.
†Analysis of variance was performed on the mean percent mortalities (arcsine transformed) at alpha = 0.05. Means followed by the same letters within each life stage group of each generation are not significantly different.

**TABLE 20-3**   *Solanum tuberosum* var. Wauseon 1982 Yield
Data for *Beauveria bassiana* and Pydrin/PBO-Treated Field
Plots*

| | Yield (lbs/acre [$10^4$]) after the Following Treatments | | | |
| | *B. bassiana* | | Pydrin/PBO | |
| Plot No. | Salable | Total | Salable | Total |
|---|---|---|---|---|
| 1 | 1.67 | 1.84 | 1.25 | 1.43 |
| 2 | 1.24 | 1.42 | 2.67 | 2.88 |
| 3 | 1.83 | 2.02 | 2.80 | 3.03 |
| 4 | 2.03 | 2.25 | 2.12 | 2.27 |
| Mean | 1.69 | 1.88 | 2.21 | 2.41 |
| Standard error[†] | 0.168 | 0.175 | 0.352 | 0.362 |

*Planted on May 3 and harvested on August 23.
†Mean values for *B. bassiana* versus Pydrin/PBO were not significantly
different ($p > 0.05$, by analysis of variance).

SOURCE: S. Galaini-Wraight, R. I. Carruthers, D. W. Roberts, and M.
Semel. In press a. Comparative efficacy of foliar applications of *Beauveria
bassiana* conidia and a synthetic pyrethroid against *Leptinotarsa decemlineata*.
Journal of Economic Entomology.

from those for chemical insecticides. Fungal materials may be capable of
controlling pests under certain circumstances but not in others. Most importantly, their use should be developed and managed with other control
methods in mind, so that truly integrated methods of control can be
established rather than just substituting microbials for chemical insecticides.

### Enhancement of Naturally Occurring Fungal Pathogens

Depending on the specific details of the host and pathogen population
and the associated ecosystem, enhancement of disease development and
spread has been accomplished by a number of different methods (Hostetter
and Ignoffo, 1978; McCoy et al., 1988). Some specific methods include
habitat manipulation, altering cultural practices such as planting or harvesting (Brown and Nordin, 1986), and managing controlled inputs such as
irrigation or pesticides (Hamm and Hare, 1982; Kish and Allen, 1978;
McCoy et al., 1976; Sprenkel et al., 1979). The following discussion of
*Zoophthora* mycosis of the alfalfa weevil is an example of how early
harvesting has been used to enhance the effects of naturally occurring populations of this pathogen in an IPM program.

## Zoophthora *Mycosis of the Alfalfa Weevil*

The fungal pathogen *Zoophthora phytonomi* is known for its natural regulatory effects on populations of the cloverleaf weevil *Hypera punctata* in North America (Author, 1886a,b; U.S. Department of Agriculture, 1956) and a related weevil species, *Hypera variabilis*, in Israel (Ben-Zev and Kenneth, 1980). Following the introduction of the alfalfa weevil (*Hypera postica*) into North America, 17 years of extensive surveys revealed no infections in *H. postica* populations (Puttler et al., 1978) until 1973, when a pathogen similar to *Z. phytonomi* was first observed, causing significant epizootics in alfalfa weevil populations throughout southern Ontario, Canada (Harcourt et al., 1974). Harcourt et al. (1984) suggest that this pathogen was one of the primary reasons for the major alfalfa weevil decline seen in Ontario during the mid-1970s. Subsequently, this pathogen spread into other North American alfalfa production areas, where it also caused significant mortality of *H. postica* (Gardner, 1982; Nordin et al., 1983; Pienkowski and Mehring, 1983; Puttler et al., 1978). Although taxonomic questions still exist, current evidence suggests that the pathogen that infects alfalfa weevils is a different species of *Zoophthora* than is found to infect the cloverleaf weevil (Harcourt et al., 1981).

It is unclear whether the pathogen that causes mycosis in the alfalfa weevil was introduced into North America in conjunction with weevil introductions or exotic parasite releases or whether the pathogen switched from another North American host species onto *H. postica*. Although the origin of this pathogen is still unknown (Puttler et al., 1978), the fungus has become well established as a major natural biological control agent of the alfalfa weevil throughout much of its range. Epizootiological studies have shown that disease prevalence varies between sites and seasons, but levels from 30 to 70 percent are not uncommon at the time of peak larval occurrence (Puttler et al., 1980) and have approached 100 percent late in the host's developmental cycle, particularly when densities are high (Harcourt et al., 1974; Nordin et al., 1983). Disease levels of this magnitude produce larval mortalities of between 65 and 90 percent, with an additional 40 to 50 percent expressed in the less visible pupal stage (Harcourt et al., 1974). Brown and Nordin (1982) determined that a threshold host density (1.7 weevils/alfalfa stem) was required for *Zoophthora* spp. to induce epizootics. Although this threshold varied with environmental moisture, low host densities were thought to limit early-season spread of mycosis. Using laboratory, field, and simulation modeling experiments, Brown and Nordin (1986) developed early harvesting strategies that maximized the development and spread of *Zoophthora* spp. even when weevil populations were lower than critical levels for epizootic development. This was accomplished by harvesting the alfalfa when *Zoophthora* mycosis of weevil larvae was first observed. Early harvesting is thought to increase disease prevalence because it (1) concentrates larvae in windrows, (2) damages or stresses larvae, making them more susceptible, and (3) alters microenvironmental conditions (increases moisture) to enhance sporulation, germination, and thus, infection. Research fields managed by this technique showed higher disease prevalence, which demonstrates that enhancement of fungal diseases is not only possible but is practical and economical, as it is based on accepted production practices and equipment (Brown and Nordin, 1986). However, a better understanding of the spatial distribution and dynamics of this pathogen is necessary to implement this strategy effectively in a statewide IPM program (Brown and Nordin, 1986).

## Introduction of Exotic Fungal Pathogens

Classical biological control (introduction of exotic fungi, either new species or more pathogenic strains of a species that already exists) has received very little attention in the field of insect pathology (Roberts, 1978). Some reports of attempts to establish exotic fungi include the use of *Coelomomyces stegomyiae* for control of mosquitoes (Laird, 1967), *Entomophaga aulicae* for control of the browntail moth (Speare and Cooley, 1912), *Entomophthora erupta* for control of the green apple bug (Dustan, 1923), and *Zoophthora radicans* for control of the spotted alfalfa aphid (Hall and Dunn, 1957; Milner et al., 1982). Although releases were made in each of these cases, detailed population assessments documenting the effects of these fungi were rarely conducted. The introduction of *Zoophthora radicans* into the United States provides an example of current research on a introduction of a fungus and its possible impact on the target host population.

### *Introduction of* Zoophthora radicans *into the United States*

Alfalfa, *Medicago sativa,* is the world's most valuable forage crop, providing the best food value for all classes of livestock. Forages provide in excess of half the feed units for livestock, which in turn provides half of the nutrients consumed by Americans. In addition to its nutritional value, alfalfa is important because it has high nitrogen-fixing capabilities and because it is a soil-conserving perennial crop. Alfalfa is grown on over 27 million acres in the United States and is exceeded in production acreage only by corn, soybeans, and wheat. Alfalfa, however, exceeds each of these crops in protein output per unit area, producing an overall cash value estimated in excess of $6 billion per year (New York Crop Reporting Service, 1985).

Several insect pest species attack alfalfa throughout the United States, but currently, the potato leafhopper, *Empoasca fabae,* is considered to be the most serious economic threat (Gyrisco et al., 1978). Feeding by the potato leafhopper causes severe yellowing and stunting both in seedling and established alfalfa stands. Losses combined with control expenditures cost U.S. alfalfa producers millions of dollars annually. Entomologists in Illinois and Wisconsin estimate that in an average season, yields are decreased approximately 20 percent in terms of quantity, with additional losses of approximately 20 percent through reduced protein content. In addition, *E. fabae* is a major pest in numerous crops (potatoes, soybeans and other beans, clovers, and more than 100 other cultivated plants), causing significant losses (Gyrisco et al., 1978).

Successful biological control agents, both insect parasitoids and microbial pathogens, are currently helping to reduce populations of other major alfalfa pests (e.g., the alfalfa weevil, the pea aphid, and the alfalfa blotch leaf miner) to well below economic thresholds in many areas of the country (Brown and Nordin, 1982; Dunbar and Hower, 1976; Hower and Davis, 1984). No such biological control organisms are being used to manage potato leafhoppers. After 5 years of effort, the European Parasite Laboratory of ARS has terminated its program aimed at locating beneficial insects to control the potato leafhopper. Currently, the only control option for leafhopper management is the application of chemical insecticides, which is not only costly but also interferes with other highly successful biological control programs for alfalfa. More insecticides are now being applied to alfalfa in the central and eastern United States to control *E. fabae* than are being used to control all other insect pests combined. Microbial control of *E. fabae* may provide an alternate management tool in the alfalfa production system that would be more compatible with existing alfalfa pest management programs, provide less of a hazard to livestock, and in general, would be more sound environmentally.

Initial research on the effects of *Zoophthora radicans* on *Empoasca fabae* was conducted through cooperative efforts of various agencies including ARS, the Cooperative State Research Service, and the U.S. Agency for International Development. The Plant Protection Research Unit, ARS, USDA, joined in an effort with the Boyce Thompson Institute for Plant Research, the Illinois Natural History Survey, and Cornell University to explore the possibility of microbial control of the potato leafhopper by using *Z. radicans* isolates collected from a related leafhopper species in Brazil. Through efforts made in this program, it became clear that *Z. radicans* had significant potential for the biological control of *E. fabae* (Galaini-Wraight et al., in press b; McGuire et al., 1987a,b; Wraight et al., 1986).

Epizootic levels of *Z. radicans* are commonly seen in Brazilian leafhopper populations and are known to cause significant population reductions in both bean and cowpea crops (Galaini-Wraight et al., in press b).

Pathogen isolates from these epizootics were imported to the United States and tested against the potato leafhopper under laboratory conditions at the Boyce Thompson Institute for Plant Research. These studies show that pathogen levels as low as 3.3 spores/square millimeter cause significant levels of infection (greater than 70 percent) in late-instar nymphs (Figure 20-4). A 10-fold increase in dose (33.0 spores/square millimeter) repeatedly caused 100 percent mortality. Time from infection to host death and secondary spore production (pathogen multiplication) is very rapid, 2 to 4 days, depending on host age and environmental conditions.

*Z. radicans* is a very attractive pathogen for use as a biological control, as it may either be introduced in an inoculative release or be used more as a microbial pesticide (inundative release). This pathogen can be easily produced in large quantities and stored for relatively long periods of time by using the marcesence technology developed by the Plant Protection Research Unit of ARS (R. S. Soper and D. M. McCabe, USDA patent 4,530,834). Fungal mycelium can be grown in liquid medium, dried, milled, stored, and then applied in the field, where it rehydrates and produces infective conidia in patterns very similar to those of fungi grown directly from a newly killed host.

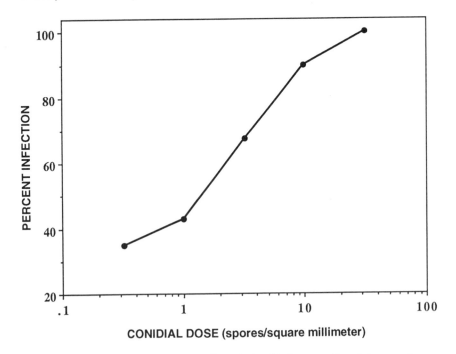

FIGURE 20-4   Laboratory dose-mortality relationship between *Zoophthora radicans* and the potato leafhopper *Empoasca fabae*.

This material has been applied to small field plots and evaluated for its sporulation potential and pathogenicity to caged populations of *E. fabae*. These studies not only showed primary infection levels as high as 70 percent in caged leafhoppers within 24 hours but also high infection levels in natural populations through inadvertent mycosis of leafhoppers outside the test cages. Inoculation densities of approximately 5.0 grams of dried hyphae per square meter were applied to a total of 12 square meters, inducing infection levels as high as 92 percent across a 0.75-acre test field over a 3-week sample period (Figure 20-5). The epizootic reduced leafhopper populations from 1.7 to 0.08 leafhoppers per stem—a 95 percent reduction. A second study conducted in an adjacent field gave similar results.

In Illinois, a similar release (using infected cadavers in a potato field) produced the same general results (McGuire et al., 1987b). Although not clearly documented in the year of release, *Z. radicans* became established in central Illinois and has caused substantial epizootics in leafhopper populations over the last few seasons. Infection levels as high as 90 percent have been recorded among leafhoppers in beans, where the epizootics have been most evident. Since its introduction into Illinois, *Z. radicans* has spread approximately 20 miles from the original release sites and has caused major leafhopper declines in bean, potato, and alfalfa crops. Additional introductions in Illinois have continued to cause similar epizootics.

In 1979, different isolates of this same pathogen were introduced into New South Wales, Australia, in a cooperative effort between the Plant Protection Research Unit, ARS, USDA, and Commonwealth Scientific and Industrial Research Organization (CSIRO) for biological control of the spotted alfalfa aphid (Milner and Soper, 1981). *Z. radicans* quickly became established in New South Wales, caused epizootics in the pest population, and spread over 200 miles from the original release site. In the third field season, *Z. radicans* was again found to induce epizootics (up to 74 percent prevalence) in sampled spotted alfalfa aphid populations (Milner and Lutton, 1986) and has spread over 186 miles from the original release sites (Milner and Lutton, 1983; Milner and Soper, 1981; Milner et al., 1980, 1982; R. J. Milner, CSIRO, personal communication, 1989). Clearly, a similar potential exists for *Z. radicans* to become a major biological control agent against *E. fabae* throughout its U.S. distribution.

Although a single isolate of *Z. radicans* has shown significant potential to control *E. fabae* under natural field conditions, several additional isolates have been collected from South America and Europe. A number of these isolates have shown equal or higher pathogenicity to *E. fabae* in laboratory screenings than the material originally tested and released in the field. These isolates may also possess other beneficial characteristics, such as a faster speed of kill or better overwintering ability, which are yet to be evaluated. Screening of this germ plasm in the laboratory is under way, and further evaluation of selected pathogen isolates will be conducted in the field following approval by APHIS, USDA, and the U.S. Environmental Agency. It

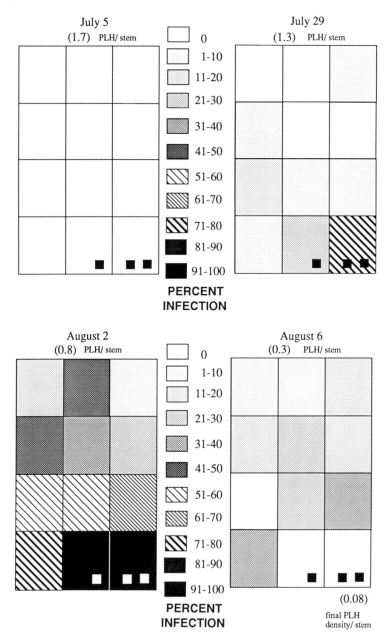

FIGURE 20-5 Example epizootic of *Erynia radicans* in a natural population of *Empoasca fabae* in a 0.75-acre alfalfa field. A total of 60 grams of dried mycelium was applied to the three blocks (12 square meters) marked in the lower right corner. This small amount of fungal material multiplied rapidly, causing almost a complete decline in the leafhopper population in this field. Leafhoppers in an adjacent 1.0-acre field were also decimated. PLH, potato leafhoppers.

is hoped that these studies will allow the introduction and establishment of *Z. radicans* as a biological control agent throughout the full geographic range of the potato leafhopper (Carruthers and Soper, 1987).

## CONCLUSION, RESEARCH NEEDS, AND FUTURE APPLICATIONS OF BIOTECHNOLOGY

Entomopathogenic fungi are important natural regulators of many arthropod populations, including several pest species. A variety of strategies have been used successfully to manipulate fungi in biological control programs. While there have been fewer documented biological control successes with fungal pathogens than with parasitoids and predators, much less effort and capital have been expended to understand and manage them. Most of the research conducted on fungal pathogens of insects has emphasized the use of these organisms as microbial insecticides. Application of pathogens by inundative release may indeed prove to be a useful tactic, but fungi should not be considered as direct replacements for chemical insecticides. Fungi are complex organisms that interact with their hosts and the environment in intricate ways. Intelligent use of fungi as biological control agents will require detailed knowledge of their pathobiology, epizootiology, and interactions with other components of the ecosystems in which they are to be used. This will become increasingly important as scientists begin to alter fungal pathogens genetically to improve their efficacy as biological control agents.

Admittedly, fungi are limited in their ability to control insect pests. For example, not all pest species are susceptible to fungal pathogens, and even if they are susceptible, the target hosts may live in an environment that is not conducive to fungal infection and transmission. As mentioned previously, fungal pathogens are highly dependent on moisture for spore germination and infection to proceed. They are also adversely affected by high temperatures. Pest control with any biological control agent will not be as certain or repeatable as the action of chemical insecticides. For this reason, expectations need to be altered, integrated control programs need to be developed by using multiple tactics, and efficiency in the use of fungi and other pathogens as biological control agents needs to be continued. Most importantly, a research agenda should be developed that is aimed at solving specific problems associated with the development and application of microbial agents for biological control.

Although the list of needed research on fungal pathogens and their use in biological control is long, five specific areas need additional attention.

1. It is important that fungal pathogen germ plasm be collected now and preserved for future evaluation and use.

2. Continued expansion of research on fungal epizootiology and ecology is needed so that disease dynamics, the impact of pathogens on host populations, and the factors that limit disease development and spread under field conditions can be understood.

3. The genetic aspects of host and pathogen populations that affect the establishment, spread, and maintenance of disease in insect populations need to be studied. It must be recognized that natural populations of fungal pathogens contain high levels of genetic variation that could contribute to their ability to regulate host populations. Furthermore, the potential for the evolution of resistance in the host population must be explored before it becomes a problem in the field.

4. Efforts to integrate the use of pathogens with other control tactics must be increased. Until control efforts use multiple tactics to manage a complex of pest species, IPM will remain in its infancy.

5. The use of innovative techniques in addressing problems associated with biological control agents needs to be expanded.

New biotechnologies provide the tools needed to answer questions that have never been able to be asked before. It is expected that innovative biological techniques can be used to manipulate desirable traits, and thus improve the effectiveness of some fungal pathogens. Recombinant DNA techniques are being used to study the mechanisms of pathogenicity and virulence at the molecular level. For example, fungal enzymes and associated genes involved in penetration of the insect cuticle have been identified (St. Leger et al., 1987, 1988b). Knowledge of these genes and gene products may eventually lead to genetic alteration of fungal pathogens. By using transformation systems developed for other types of fungi (Shimizu, 1987; Yoder et al., 1987), it may be possible to clone the genes responsible for these products and transfer them to fungi that may have poor penetration ability but that are well adapted to a particular environment. The genes involved in toxin production are also candidates for cloning.

Biotechnological methods are also being applied to improve methods for field monitoring of insect fungal diseases. Enzyme-linked immunosorbent assays are used to detect a variety of fungi in plant tissues (Mendgen, 1986) and have shown promise for detecting fungal cells of *Entomophaga maimaiga* in gypsy moth larval hemolymph (Hajek et al., 1988). Other techniques such as isoenzyme polymorphisms (Boucias et al., 1982; Micales et al., 1986) and restriction fragment length polymorphisms will become increasingly useful in detecting different fungal pathogen strains.

It is important, however, that researchers do not get lost in the techniques and lose sight of the important question: How can the ability to manage pest problems be improved while improving both the economic and environmental situation? Despite the promise of new technologies, successful biological control depends on a fundamental knowledge of host and pathogen biology, not only at the molecular level but at the cellular, organismal, population, and ecosystem levels as well. It is evident that specialists from many subdisciplines of entomology, genetics, mycology, systems science, and other fields of study will be required to identify, manipulate, and use fungi successfully for biological control purposes in the future.

## REFERENCES

Al-Aidroos, K., and D. W. Roberts. 1978. Mutants of *Metarhizium anisopliae* with increased virulence toward mosquito larvae. Canadian Journal of Genetic Cytology 20:211.

Allen, G. E., ed. 1980. Integrated pest management: A special issue. Bioscience 30:655–701.

Allen, G. E., C. M. Ignoffo, and R. P. Jacques, eds. 1978. Microbial Control of Insect Pests: Future Strategies in Pest Management Systems. Gainesville, Fla.: National Science Foundation, U.S. Department of Agriculture, and University of Florida.

Anderson, R. M. 1982. Theoretical basis for the use of pathogens as biological control agents of pest species. Parasitology 84:3–33.

Anderson, R. M., and R. M. May. 1980. The population dynamics of microparasites and their invertebrate hosts. Philosophical Transactions of the Royal Society of Britain 291:451.

Author, J. C. 1886a. Disease of clover-leaf weevil. *Entomophthora phytonomi* Author. Fourth Annual Report of the New York Agricultural Experiment Station, 1885 258:120. Ithaca, N.Y.: New York Agricultural Experiment Station.

Author, J. C. 1886b. A new larval *Entomophthora*. Botanical Gazette 11:14.

Bassi, A. 1835. Del mal del sengo, calcinaccioo moscardino, malattia che affligge i bachi de seta e sul modo di liberane le bigattaie anche le piu infestate. Part I. Teoria. Lodi, Italia: Orcesi.

Bell, J. V. 1974. Mycoses. In Insect Diseases, Vol. 1, G. Cantwell, ed. New York: Marcel Dekker.

Benz, G. 1987. Environment. In Epizootiology of Insect Diseases, J. R. Fuxa and Y. Tanada, eds. New York: John Wiley & Sons.

Ben-Zev, I., and R. G. Kenneth. 1980. *Zoophthora phytonomi* and *Conidiobolus osmodes*, two pathogens of *Hypera* species coincidental in time and place. Entomophaga 25:171.

Boucias, D. G., and J. C. Pendland. 1982. Ultrastructural studies on the fungus, *Nomuraea rileyi*, infecting the velvetbean caterpillar, *Anticarsia gemmatalis*. Journal of Invertebrate Pathology 39:338.

Boucias, D. G., and J. C. Pendland. 1984. Host recognition and specificity of entomopathogenic fungi. In Infection Processes of Fungi, A Bellagio Conference, D. W. Roberts and J. R. Aist, eds. New York: The Rockefeller Foundation.

Boucias, D. G., C. W. McCoy, and D. J. Joslyn. 1982. Isozyme differentiation among 17 geographical isolates of *Hirsutella thompsonii*. Journal of Invertebrate Pathology 39:329.

Brobyn, P. J., and N. Wilding. 1977. Invasive and developmental processes of

*Entomophthora* species infecting aphids. Transactions of the British Mycology Society 69:349.

Brown, G. C. 1984. Stability in an epizootiological model with age-dependent immunity and seasonal host reproduction. Bulletin of Mathematical Biology 46:139–153.

Brown, G. C. 1987. Modelling. In Epizootiology of Insect Diseases, J. R. Fuxa and Y. Tanada, eds. New York: John Wiley & Sons.

Brown, G. C., and G. L. Nordin. 1982. An epizootic model of an insect-fungal pathogen system. Bulletin of Mathematical Biology 44:731.

Brown, G. C., and G. L. Nordin. 1986. Evaluation of an early harvest approach for induction of *Erynia* epizootics in alfalfa weevil populations. Journal of the Kansas Entomological Society 59:446.

Bucher, G. E. 1964. The regulation and control of insects by fungi. Annals of the Societé Entomologique Quebec 9:30.

Burges, H. D., ed. 1981. Microbial Control of Pests and Plant Diseases, 1970–1980. London: Academic Press.

Burges, H. D., and N. W. Hussey, eds. 1971. Microbial Control of Insects and Mites. New York: Academic Press.

Butt, T. M., A. Beckett, and N. Wilding. 1981. Protoplasts in the in vivo life cycle of *Erynia neoaphidis*. Journal of General Microbiology 127:417.

Butt, T. M., S. P. Wraight, S. Galaini-Wraight, R. A. Humber, D. W. Roberts, and R. S. Soper. 1988. Humoral encapsulation of the fungus *Erynia radicans* by the potato leafhopper. Journal of Invertebrate Pathology 52:49.

Carruthers, R. I. 1985. The use of simulation modeling in insect-fungal disease research. Proceedings of the Summer Computer Simulation Conference. San Diego, Calif.: Society of Computer Simulation.

Carruthers, R. I., and R. S. Soper. 1987. Fungal diseases. In Epizootiology of Insect Diseases, J. R. Fuxa and Y. Tanada, eds. New York: John Wiley & Sons.

Carruthers, R. I., Z. Feng, D. S. Robson, and D. W. Roberts. 1985a. In vivo temperature-dependent development of *Beauveria bassiana* mycosis of the European corn borer. Journal of Invertebrate Pathology 46:305.

Carruthers, R. I., D. L. Haynes, and D. M. MacLeod. 1985b. *Entomophthora muscae* mycosis of the onion fly, *Delia antiqua*. Journal of Invertebrate Pathology 45:81.

Carruthers, R. I., G. W. Whitfield, R. L. Tummala, and D. L. Haynes. 1986. A systems approach to research and simulation of insect pest dynamics in the onion agroecosystem. Ecological Modelling 33:101–121.

Carruthers, R. I., Z. Feng, M. E. Ramos, and R. S. Soper. 1988a. The effect of solar radiation on the survival of *Entomophaga grylli* conidia. Journal of Invertebrate Pathology 52:154.

Carruthers, R. I., T. L. Larkin, and R. S. Soper. 1988b. Simulation of insect disease dynamics: An application of SERB to a rangeland ecosystem. Simulation 51:101–109.

Carruthers, R. I., T. S. Larkin, H. Firstencel, and Z. Feng. In press. Influence of thermal ecology on the mycosis of rangeland grasshoppers. Ecology.

Charnley, A. K. 1984. Physiological aspects of destructive pathogenesis in insects by fungi: A speculative review. In Invertebrate-Microbial Interactions, J. M. Anderson, A. D. M. Raynor, and D. W. H. Walton, eds. Cambridge: Cambridge University Press.

Coppel, H. C., and J. W. Mertins. 1977. Biological Insect Pest Suppression. New York: Springer-Verlag.

Daoust, R. A., and D. W. Roberts. 1982. Virulence of natural and insect-passaged strains of *Metarhizium anisopliae* to mosquito larvae. Journal of Invertebrate Pathology 40:107.

Davidson, R. H., and W. F. Lyon. 1979. Insect Pests of Farm, Garden, and Orchard, 7th ed. New York: John Wiley & Sons.

Dunbar, R. B., and A. A. Hower. 1976. Relative toxicity of insecticides to the alfalfa weevil parasite, *Microctonus aethipos* and the influence of parasitism on host susceptibility. Environmental Entomology 5:311–315.

Dustan, A. G. 1923. Studies on a new species of *Empusa* parasitic on the green apple bug (*Lygus communis* var. *novascotiensis* Knight) in the Annapolis Valley. Proceedings of the Acadian Entomological Society 9:14.

Evans, H. C. 1982. Entomogenous fungi in tropical forest ecosystems: An appraisal. Ecology and Entomology 7:47.

Fargues, J. 1984. Adhesion of the fungal spore to the insect cuticle in relation to pathogenicity. In Infection Processes of Fungi, A Bellagio Conference, D. W. Roberts and J. R. Aist, eds. New York: The Rockefeller Foundation.

Federici, B. A. 1981. Mosquito control by the fungi *Culicinomyces, Lagenidium* and *Coelomomyces*. In Microbial Control of Pests and Plant Diseases, H. D. Burges, ed. London: Academic Press.

Feng, Z. In press. Laboratory and field dose-mortality responses of *Entomophaga grylli* mycosis of the clearwinged grasshopper, *Camnula pellucida*. Environmental Entomology.

Ferron, P. 1978. Biological control of insect pests by entomogenous fungi. Annual Review of Entomology 23:409.

Ferron, P. 1981. Pest control by the fungi *Beauveria* and *Metarhizium*. In Microbial Control of Pests and Plant Diseases, 1970–1980, H. D. Burges, ed. London: Academic Press.

Fuxa, J. R. 1987. Ecological considerations for the use of entomopathogens in IPM. Annual Review of Entomology 32:225.

Fuxa, J. R., and Y. Tanada, eds. 1987. Epizootiology of Insect Diseases. New York: John Wiley & Sons.

Galaini, S. 1984. The Efficacy of Foliar Application of *Beauveria bassiana* Conidia against *Leptinotarsa decemlineata*. M.S. thesis, Cornell University, Ithaca, N.Y.

Galaini-Wraight, S., R. I. Carruthers, D. W. Roberts, and M. Semel. In press a. Comparative efficacy of foliar applications of *Beauveria bassiana* conidia and a synthetic pyrethroid against *Leptinotarsa decemlineata*. Journal of Economic Entomology.

Galaini-Wraight, S., S. P. Wraight, R. I. Carruthers, B. P. Magalhaes, and D. W. Roberts. In press b. Description of a *Zoophthora radicans* epizootic in a population of *Empoasca krameri* on beans in Brazil. Journal of Invertebrate Pathology.

Gardner, W. A. 1982. Occurrence of *Erynia* sp. in *Hypera postica* in central Georgia. Journal of Invertebrate Pathology 40:146.

Gillespie, A. T., and E. R. Moorhouse. 1989. The use of fungi to control pest of agricultural and horticultural importance. In Biotechnology of Fungi for Improvement of Plant Growth, J. M. Whipps and R. D. Lumsdon, eds. London: Cambridge University Press.

Gupta, A. P. 1986. Hemocytic and Humoral Immunity in Arthropods. New York: John Wiley & Sons.

Gyrisco, G. G., D. Landman, A. C. York, B. J. Irwin, and E. J. Armbrust. 1978. The literature of arthropods associated with alfalfa. Part IV. A bibliography of the potato leafhopper, *Empoasca fabae*. University of Illinois Agricultural Experiment Station Special Publication No. 51. Urbana, Ill.: University of Illinois.

Hajek, A. E., R. S. Soper, D. W. Roberts, T. E. Anderson, K. D. Biever, D. N. Ferro, R. A. LeBrun, and R. H. Storch. 1987. Foliar applications of *Beauveria bassiana* for control of the Colorado potato beetle, *Leptinotarsa decemlineata*: An overview of pilot test results from the northern United States. Canadian Entomology 119:959–974.

Hajek, A., T. M. Butt, and L. I. Strelow. 1988. Development of an enzyme-linked immunosorbent assay for detection of *Entomophaga maimaga* in Gypsy Moth, *Lymantria dispar*, larvae. Proceedings, New Directions in Biological Control. New York: Alan R. Liss.

Hall, I. M., and P. H. Dunn. 1957. Fungi on spotted alfalfa aphid. California Agriculture 11:5.

Hall, R. A. 1981. The fungus *Verticillium lecanii* as a microbial insecticide against aphids and scales. In Microbial Control of Pests and Plant Diseases, 1970–1980, H. D. Burges, ed. London: Academic Press.

Hamm, J. J., and W. W. Hare. 1982. Applications of entomopathogens in irrigation water for control of fall armyworms and corn earworms. Journal of Economic Entomology 75:1074.

Harcourt, D. G., J. C. Guppy, D. M. MacLeod, and D. Tyrrell. 1974. The fungus *Entomophthora phytonomi* pathogenic to the alfalfa weevil, *Hypera postica*. Canadian Entomology 106:1295.

Harcourt, D. G., J. C. Gupp, D. M. MacLeod, and D. Tyrrell. 1981. Two *Entomophthora* species associated with disease epizootics of the alfalfa weevil, *Hypera postica* in Ontario. Great Lakes Entomology 14:55.

Harcourt, D. G., J. C. Guppy, and M. R. Binns. 1984. Analysis of numerical change in subeconomic populations of the alfalfa weevil, *Hypera postica*, in eastern Ontario. Environmental Entomology 13:1627.

Heath, M. E., D. S. Metcalfe, and R. F. Barnes. 1973. Forages: The science of grassland agriculture, 3rd ed., Ames, Iowa: Iowa State University Press.

Hochberg, M. E. 1989. The potential role of pathogens in biological control. Nature 337:262–265.

Hostetter, D. L., and C. M. Ignoffo. 1978. Induced epizootics: Fungi. In Microbial Control of Insect Pests: Future Strategies in Pest Management Systems, G. E. Allen, C. M. Ignoffo, and R. P. Jacques, eds. Gainesville, Fla.: National Science Foundation, U.S. Department of Agriculture, and University of Florida.

Hower, A. A., and G. A. Davis. 1984. Selectivity of insecticides that kill the potato leafhopper and alfalfa weevil and protect their parasite, *Microctonus aephipoides*. Journal of Economic Entomology 77:1601–1607.

Hoy, M. A., and D. G. Herzog. 1985. Biological Control in Agricultural IPM Systems. Orlando, Fla.: Academic Press.

Huffaker, C. B., ed. 1971. Biological Control. New York: Plenum/Rosetta.

Ignoffo, C. M. 1985. Manipulating enzootic-epizootic diseases of arthropods. In Biological Control in Agricultural IPM Systems, M. A. Hoy and D. C. Herzog, eds. Orlando, Fla.: Academic Press.

Ignoffo, C. M., and C. Garcia. 1985. Host spectrum and relative virulence of an Ecuadoran and a Mississippian biotype of Nomuraea rileyi. Journal of Invertebrate Pathology 45:346.

Ignoffo, C. M., N. Marston, D. L. Hostetter, B. Puttler, and J. V. Bell. 1976. Natural and induced epizootics of Nomuraea rileyi in soybean caterpillars. Journal of Invertebrate Pathology 27:191.

Ignoffo, C. M., D. L. Hostetter, P. P. Sikorowski, G. Sutter, and W. M. Brooks. 1977. Inactivation of representative species of entomopathogenic viruses, a bacterium, fungus, and protozoan by an ultraviolet light source. Environmental Entomology 6:411.

Kenneth, R., G. Wallis, U. Gerson, and H. N. Plant. 1972. Observation and experiments on Triplosporium floridanum (Entomophthorales) attacking spider mites in Israel. Journal of Invertebrate Pathology 19:366.

Kerwin, J. L. 1983. Fatty acid regulation of the germination of Erynia variabilis conidia on adults and puparia of the lesser housefly, Fannia canicularis. Canadian Journal of Microbiology 30:158.

King, D. S., and R. A. Humber. 1981. Identification of Entomophthorales. In Microbial Control of Pests and Plant Diseases, 1970–1980, H. D. Burges ed., London: Academic Press.

King, E. G., and J. R. Coulson. 1988. ARS National Biological Control Program, Proceedings of a Workshop on Research Priorities. Beltsville. Md.: Agricultural Research Service, U.S. Department of Agriculture.

Kish, L. P., and G. E. Allen. 1978. The biology and ecology of Nomuraea rileyi and a program for predicting its incidence on Anticarsia gemmatalis. Florida Agricultural Experiment Station Bulletin No. 795. Gainesville, Fla.: Florida Agricultural Experiment Station.

Kramer, J. P. 1980. The house-fly mycosis caused by Entomophthora muscae: Influence of relative humidity on infectivity and conidial germination. New York Entomological Society Journal 88:236.

Krassilistchik, I. M. 1888. La production industrielle des parasites vegetaux pour la destruction des insedtes nuisibles. Bulletin Scientifique Française Belgique 19:461.

Laird, M. 1967. A coral island experiment: A new approach to mosquito control. World Health Organization Chronicles 21:18.

Lappa, N. V. 1978. Practical applications of entomopathogenic muscardine fungi. Pp. 51–57 in Proceedings of the First Joint US/USSR Conference on the Production, Selection and Standardization of Entomopathogenic Fungi, C. M. Ignoffo, ed. Washington, D.C.: U.S. Department of Agriculture.

Larkin, T. S., R. I. Carruthers, and R. S. Soper. 1988. Simulation and object-oriented programming: The development of SERB. Simulation 51:93–100.

Larkin, T. S., and R. I. Carruthers. 1990. Development of a hierarchical simulation environment for research biologists. In Object Oriented Simulation, A. Guasch, ed. San Diego, Calif.: Simulation Council Inc.

Latge, J. P., J. Eilenberg, A. Beauvais, and M. Prevost. 1988. Morphology of *Entomophthora muscae* protoplasts grown in vitro. Protoplast 146:166.

Logan, J. A. 1988. Toward an expert system for development of pest simulation models. Environmental Entomology 17:359–376.

MacLeod, D. M. 1963. Entomophthorales infections. In Insect Pathology: An Advanced Treatise, Vol. 2, E. A. Steinhaus ed. New York: Academic Press.

MacLeod, D. M., J. W. Cameron, and R. S. Soper. 1966. The influence of environmental conditions on epizootics caused by entomogenous fungi. Revue Roumaine de Biologie 11:125.

McCabe, D. M., R. A. Humber, and R. S. Soper. 1984. Observation and interpretation of nuclear reductions during maturation and germination of entomophthoralean resting spores. Mycologia 76:1104.

McCoy, C. W. 1981. Pest control by the fungus *Hirsutella thompsonii*. In Microbial Control of Pests and Plant Diseases, 1970–1980, H. D. Burges ed. London: Academic Press.

McCoy, C. W. 1990. Entomogenous fungi as microbial pesticides. In New Directions in Biological Control: Alternatives for Suppressing Agricultural Pests and Diseases, UCLA Symposium on Molecular and Cellular Biology New Series, Vol. 112, R. R. Baker and R. E. Dunn, eds. Los Angeles: University of California.

McCoy, C. W., R. F. Brooks, J. C. Allen, and A. G. Selhime. 1976. Management of arthropod pests and plant diseases in citrus agroecosystem. Proceedings of the Tall Timbers Conference on Ecology and Animal Control Habitat Management 6:10.

McCoy, C. W., R. A. Samson, and D. G. Boucias. 1988. Entomogenous fungi. In Handbook of Natural Pesticides, Vol. 5, Microbial Insecticides, Part A, Entomogenous Protozoa and Fungi, C. M. Ignoffo and N. B. Mandava, eds. Boca Raton, Fla: CRC Press.

McGuire, M. R., J. V. Maddox, and E. J. Armbrust. 1987a. Host range studies of an *Erynia radicans* strain isolated from *Empoasca fabae*. Journal of Invertebrate Pathology 50:75–77.

McGuire, M. R., M. J. Morris, E. J. Armbrust, and J. V. Maddox. 1987b. An epizootic caused by *Erynia radicans* in an Illinois *Empoasca fabae* population. Journal of Invertebrate Pathology 50:78–80.

Mendgen, K. 1986. Quantitative serological estimations of fungal colonization. In Microbiology of the Phyllosphere, N. J. Fokkema and J. van den Heuvel, eds. Cambridge: Cambridge University Press.

Metcalf, L. 1980. Changing roles of insecticides in crop protection. Annual Review of Entomology 25:219–256.

Metchnikoff, E. 1879. Diseases of the larva of grain weevil. Insects harmful to agriculture. Issue 3. The grain weevil. Odessa, Russia: Commission Attached to the Odessa Zemstro Office for the Investigation of the Problem of Insects Harmful to Agriculture. (In Russian.)

Micales, J. A., M. R. Bonde, and G. L. Peterson. 1986. The use of isozyme analysis in fungal taxonomy and genetics. Mycotaxon 27:405.

Millstein, J. A., G. C. Brown, and G. L. Nordin. 1983. Microclimatic moisture and conidial production in *Erynia* sp.: In vivo moisture balance and conidiation phenology. Environmental Entomology 12:1339.

Milner, R. J. 1982. On the occurrence of pea aphids, *Acyrthosiphon pisum*, resistant to isolates of the fungal pathogen *Erynia neoaphidis*. Entomology and Experimental Applications 32:23.

Milner, R. J. 1985. Distribution in time and space of resistance to the pathogenic fungus *Erynia neoaphidis*. Entomology and Experimental Applications 37:235.

Milner, R. J., and G. G. Lutton. 1983. Effect of temperature on *Zoophthora radicans* (Brefeld) Batko: An introduced microbial control agent of the spotted alfalfa aphid *Therioaphis trifolii* (Monell) f. *maculata*. Journal of the Australian Entomological Society 22:167.

Milner, R. J., and G. G. Lutton. 1986. Microbial control of insects: Biological control of spotted alfalfa aphid. CSIRO Biennial Report 1983–85. Canberra, Australia: Commonwealth Scientific and Industrial Research Organization.

Milner, R. J., and R. S. Soper. 1981. Bioassay of *Entomophthora* against the spotted alfalfa aphid *Therioaphis trifolii* (Monell) f. *maculata*. Journal of Invertebrate Pathology 37:168.

Milner, R. J., R. E. Teakle, G. G. Lutton, and F. M. Dare. 1980. Pathogens of the blue-green aphid *Acyrthosiphon kondoi* Shinji and other aphids in Australia. Australian Journal of Botany 28:601.

Milner, R. J., R. S. Soper, and G. G. Lutton. 1982. Field release of an Israeli strain of the fungus *Zoophthora radicans* (Brefeld) Batko for biological control of *Therioaphis trifolii* (Monell) f. *maculata*. Journal of the Australian Entomological Society 21:113.

Milner, R. J., R. J. Mahon, and W. V. Brown. 1983. A taxonomic study of the *Erynia neoaphidis* group of insect pathogenic fungi, together with a description of the new species *Erynia kondoiensis*. Australian Journal of Botany 31:173.

Mohamed, A. K. A., P. P. Sikorowski, and J. V. Bell. 1977. Susceptibility of *Heliothis zea* larvae to *Nomuraea rileyi* at various temperatures. Journal of Invertebrate Pathology 30:414.

Mullens, B. A., J. L. Rodriguez, and J. A. Meyer. 1987. An epizootiological study of *Entomophthora muscae* in muscoid fly populations on southern California poultry facilities, with emphasis on *Musca domestica*. Hilgardia 55:1.

Newman, G. G., and G. R. Carner. 1975. Environmental factors affecting conidial sporulation and germination of *Entomophthora gammae*. Environmental Entomology 4:615.

New York Crop Reporting Service. 1985. Corn, wheat, oats and hay 1978–1984. Albany, N.Y.: Department of Agriculture and Markets.

Nolan, R. A. 1985. Protoplasts from Entomophthorales. In Fungal protoplasts, Applications in Biochemistry and Genetics, J. F. Peberdy and L. Ferenczy, eds. New York: Marcel Dekker.

Nordin, G. L., G. C. Brown, and J. A. Millstein. 1983. Epizootic phenology of *Erynia* disease of the alfalfa weevil, *Hypera postica*, in central Kentucky. Environmental Entomology 12:1350.

Onstad, D., and R. I. Carruthers. 1990. Epizootiological models of insect diseases. Annual Review of Entomology 35:399–419.

Papierok, B., and N. Wilding. 1979. Mise en evidence d'une difference de sensibilité entre 2 clones du Puceron du Pois, *Acyrthosiphon pisum*, exposes a 2 souches du champignon Phycomycete: *Entomophthora obscura*. Comptes Rendus Académie Scientifique Paris, Series D 288:93.

Papierok, B., and N. Wilding. 1981. Étude du comportement de plusieurs souches de *Conidiobolus obscurus* vis-à-vis des pucerons *Acyrthosiphon pisum* et sitobion avenae. Entomophaga 26:241.

Perkins, J. H. 1982. Insects, Experts, and the Insecticide Crisis. New York: Plenum Press.

Perry, D. F., and J. P. Latge. 1982. Dormancy and germination of *Conidiobulus obscurus* azygospores. Transactions of the British Mycological Society 78:221.

Pickford, R., and P. W. Riegert. 1964. The fungus disease caused by *Entomophthora grylli* Fres., and its effect on grasshopper populations in Saskatchewan in 1963. Canadian Entomology 96:1158.

Pienkowski, R. L., and P. R. Mehring. 1983. Influence of avermectrin B and carbofuran on feeding by alfalfa weevil larvae. Journal of Economic Entomology 76:1167.

Pinnock, D. E., and R. J. Brand. 1981. A quantitative approach to the ecology of the use of pathogens for insect control. In Microbial Control of Pests and Plant Diseases, 1970–1980, H. D. Burges, ed. London: Academic Press.

Puttler, B., D. L. Hostetter, and S. H. Long. 1978. *Entomophthora phytonomi,* a fungal pathogen of the alfalfa weevil in the Mid-Great Plains. Environmental Entomology 7:670.

Puttler, B., D. L. Hostetter, S. H. Long, and A. A. Borski. 1980. Seasonal incidence of the fungus *Entomophthora phytonomi* infecting *Hypera postica* larvae in central Missouri. Journal of Invertebrate Pathology 35:99.

Regniere, J. 1984. Vertical transmission of diseases and population dynamics of insects with discrete generations: A model. Journal of Theoretical Biology 107:287–301.

Riba, G. 1984. Application en essais parcellaires de plein champ d'un mutant artificial du champignan entomopathogene *Beauveria bassiana* contre la pyrale du mais, *Ostrinia nubilalis*. Entomophaga 29:41.

Riegert, P. W. 1968. A history of grasshopper abundance surveys and forecasts of outbreaks in Saskatchewan. Memoirs of the Entomological Society of Canada, No. 50. The Entomological Society of Canada, Ottawa, Ontario.

Roberts, D. W. 1978. Introduction and colonization: Fungi. In Microbial Control of Insect Pests: Future Strategies in Pest Management Systems, G. E. Allen, C. M. Ignoffo, and R. P. Jacques, eds. Gainesville, Fla.: National Science Foundation, U.S. Department of Agriculture, and University of Florida.

Roberts, D. W. 1979. Past history and current status of the development of entomopathogenic fungi in the United States. In Proceedings of Project V: Microbial Control of Insect Pests, C. M. Ignaffo, ed. Washington, D.C.: American Society of Microbiology.

Roberts, D. W. 1981. Toxins of entomopathogenic fungi. In Microbial Control of Pests and Plant Diseases, 1979–1980, H. D. Burges, ed. London: Academic Press.

Roberts, D. W., and A. S. Campbell. 1977. Stability of entomopathogenic fungi. Miscellaneous Publication of the Entomology Society of America 10:19.

Roberts, D. W., and R. A. Humber. 1981. Entomogenous fungi. In Biology of Conidial Fungi, G. T. Cole and B. Kendrick, eds. New York: Academic Press.

Saito, T., and J. Aoki. 1983. Toxic components on the larvae surfaces of two lepidopterous insects towards *Beauveria bassiana* and *Paecilomyces fumosorosea*. Annals of Entomology and Zoology 18:225.

Shimazu, M. 1977. Factors affecting conidial germination of *Entomophthora del phacis* Hori. Applied Entomology and Zoology 12:260.

Shimazu, M. 1987. Protoplast fusion of insect pathogenic fungi. In Biotechnology in Invertebrate Pathology and Cell Culture, K. Maramorosch, ed. New York: Academic Press.

Smith, R. J., and E. A. Grula. 1982. Toxic components on the larvae surface of the corn earworm *(Heliothis zea)* and their effects on germination and growth of *Beauveria bassiana.* Journal of Invertebrate Pathology 36:15.

Snow, F. H. 1895. P. 46 in Fourth Annual Report of the Director of the Agricultural Experiment Station of the University of Kansas, 1894. Lawrence, Kans.: University of Kansas.

Soper, R. S. 1985. Pathogens of leafhoppers and planthoppers. In The Leafhoppers and Planthoppers, L. R. Nault and J. G. Rodriguez, eds. New York: John Wiley & Sons.

Soper, R. S., and D. M. MacLeod. 1981. Descriptive epizootiology of an aphid mycosis. USDA Technical Bulletin No. 1632. Washington, D.C.: U.S. Department of Agriculture.

Soper, R. S., L. F. R. Smith, and A. J. Delyzer. 1976. Epizootiology of *Massopora levispora* in an isolated population of *Okanagana rimosa.* Annals of the Entomological Society of America 69:275.

Southwood, T. R. E. 1978. Ecological Methods. New York: Chapman and Hall.

Speare, A. T., and R. H. Cooley. 1912. The artificial use of the brown-tail fungous in Massachusetts with practical suggestions for private experiment, and a brief note on a fungous disease of the gypsy caterpillar. Boston: Wright and Potter Printing Co.

Sprenkel, R. K., W. M. Brooks, J. W. Van Duyn, and L. Deitz. 1979. The effects of three cultural variables on the incidence of *Nomuraea rileyi,* phytophagous Lepidoptera, and their predators on soybeans. Environmental Entomology 8:334.

Steinhaus, E. A. 1956. Microbial control—The emergence of an idea: A brief history of insect pathology through the nineteenth century. Hilgardia 26:107.

Steinhaus, E. A., ed. 1963. Insect Pathology: An Advanced Treatise, Vol. 2. New York: Academic Press.

St. Leger, R. J., R. M. Cooper, and A. K. Charnley. 1987. Production of cuticle-degrading enzymes by the entomopathogen *Metarhizium anisopliae* during infection of cuticles from *Calliphora vomitoria* and *Manduca sexta.* Journal of General Microbiology 133:1371.

St. Leger, R. J., R. M. Cooper, and A. K. Charnley. 1988a. The effect of melanization of *Manduca sexta* cuticle on growth and infection by *Metarhizium anisopliae.* Journal of Invertebrate Pathology 52:459.

St. Leger, R. J., P. K. Durrands, A. K. Charnley, and R. M. Cooper. 1988b. Role of extracellular chymoelastase in the virulence of *Metarhizium anisopliae* for *Manduca sexta.* Journal of Inverterbate Pathology 52:285.

Tanada, Y. 1963. Epizootiology of infectious diseases. In Insect Pathology: An Advanced Treatise, Vol. 2, E. A. Steinhaus, ed. New York: Academic Press.

Tanada, Y., and J. R. Fuxa. 1987. The pathogen population. In Epizootiology of Insect Diseases, J. R. Fuxa and Y. Tanada, eds. New York: John Wiley & Sons.

Tyrrel, D., and D. M. MacLeod. 1972. Spontaneous formation of protoplasts by a species of *Entomophthora*. Journal of Invertebrate Pathology 19:354.

U.S. Department of Agriculture. 1956. The clover leaf-weevil and its control. Farmer's Bulletin 1484. Washington, D.C.: U.S. Department of Agriculture.

Wallace, D. R., D. M. MacLeod, C. R. Sullivan, D. Tyrrell, and A. J. De Lyzer. 1976. Induction of resting spore germination in *Entomophthora aphidis* by long-day light conditions. Canadian Journal of Botany 54:1410.

Watanabe, H. 1987. The host population. In Epizootiology of Insect Diseases, J. R. Fuxa and Y. Tanada, eds. New York: John Wiley & Sons.

Whisler, H. C., S. L. Zebold, and J. A. Shemanchuk. 1975a. Alternate host for mosquito parasite *Coelomomyces*. Nature 251:715.

Whisler, H. C., S. L. Zebold, and J. A. Shemanchuk. 1975b. Life history of *Coelomomyces psorophorae*. Proceedings of the National Academy of Sciences USA 782:693.

Wilding, N. 1969. Effect of humidity on the sporulation of *Entomophthora aphidis* and *Entomophthora thaxteriana*. Transactions of the British Mycology Society 53:126.

Wilding, N. 1973. The survival of *Entomophthora* sp. in mummified aphids at different temperatures and humidities. Journal of Invertebrate Pathology 21:309.

Wilding, N. 1975. *Entomophthora* species infecting pea aphids. Transactions of the Royal Entomological Society, London 127:171.

Wilding, N. 1981. Pest control by Entomophthorales. In Microbial Control of Pests and Plant Diseases, 1970–1980, H. D. Burges, ed. London: Academic Press.

Woods, S. P., and E. A. Grula. 1984. Utilizable surface nutrients on *Heliothis zea* available for growth of *Beauveria bassiana*. Journal of Invertebrate Pathology 43:259.

Wraight, S. P., S. Galaini-Wraight, R. I. Carruthers, and D. W. Roberts. 1986. Field transmission of *Erynia radicans* to *Empoasca fabae* leafhoppers in alfalfa following application of a dry, mycelial preparation. In Fundamental and Applied Aspects of Invertebrate Pathology, R. Samson, ed. Wageningen, The Netherlands: Foundation of the Fourth International Colloquium of Invertebrate Pathology.

Wraight, S. P., T. M. Butt, S. Galaini-Wraight, L. L. Allee, and D. W. Roberts. In press. Germination and infection processes of the entomophthoralean fungus *Erynia radicans* on the potato leafhopper, *Empoasca fabae*. Journal of Invertebrate Pathology.

Yoder, O. C., K. Weltring, B. G. Turgeon, R. C. Garber, and H. D. Van Etten. 1987. Prospects for development of molecular technology for fungal insect pathogens. In Biotechnology in Invertebrate Pathology and Cell Culture, K. Maramorosch, ed. New York: Academic Press.

Zacharuk, R. Y. 1973. Electron-microscope studies of the histo-pathology of fungal infections by *Metarhizium anisopliae*. Miscellaneous Publication of the Entomological Society of America 9:112.

Zadoks, J. C., and R. D. Schein. 1979. Epidemiology and Plant Disease Management. New York: Oxford University Press.

Zebold, S. L., H. C. Whisler, J. A. Shemanchuk, and L. B. Travland. 1979. Host specificity and penetration in the mosquito pathogen *Coelomomyces psorophorae*. Canadian Journal of Botany 57:2766.

# 21

# Reactors' Comments

## Time Frame for Sustainable Agriculture and Pollution Prevention Research

*Clayton W. Ogg*

The research on the Northeast region presented in this volume emphasized the complexities of managing biologically interrelated systems in a dynamic farm setting. Earlier chapters in this volume describing other regions of the United States found similar challenges; but organic approaches, eutrophic livestock operations, and insect management using fungi addressed in the three chapters on research in the Northeast region present special challenges. Clearly, a longer-term research focus is required in these most challenging areas of sustainable agriculture research.

Other sustainable agriculture projects offer more immediate dividends. An earlier chapter in this volume by Paul Johnson described an aggressive, short-term research and education program in Iowa focusing on selected sustainable agriculture activities. Iowa's program provides substantial resources to calibrate one of the more promising sustainable agriculture technologies for statewide use, focusing on soil testing for nitrogen.

Without in any way challenging the need for longer-term research, it may be useful to explore nearer-term roles for sustainable agriculture research and education programs and to compare them with other policies currently under consideration that address pollution from agriculture. This review describes potential sustainable agriculture contributions to pollution prevention that are perhaps capable of avoiding some risks and remediation costs even before these costs are fully quantified. A brief discussion of short-term versus long-term research strategies is also provided.

## SUSTAINABLE AGRICULTURE AS A PREVENTION STRATEGY

Sustainable agriculture practices are targeted to farmers who can benefit from reductions in the use of potential pollutants. Sustainable agriculture offers an opportunity for pollution prevention because it prevents pollution from happening in the first place rather than pinpointing the source of an existing environmental problem and targeting action on the basis of where the problem has occurred. The two forms of targeting (targeting of receptive farmers versus targeting of existing environmental problems) work well together, but prevention initiatives can move forward even when costs or the time required to locate pollution sources precisely are prohibitive.

The pollution prevention strategy becomes most attractive when (1) one source may be implicated in multiple environmental concerns, (2) costs are low or actually favor adoption of the practices associated with prevention, (3) the problem is pervasive, and (4) considerable uncertainty exists as to the environmental risks posed by the existing practices. In agriculture, these four conditions appear to exist widely.

With regard to pesticides, reduction of their use may clearly alleviate multiple concerns, fulfilling condition 1 above. These concerns include potential health risks to farmers and workers, potential ecological effects, and potential human health effects from the consumption of residues on food or in the drinking water. A prevention approach may address several problems by reducing the need for a particular chemical.

Condition 2, economic feasibility, is fundamental to the sustainable agriculture concepts presented throughout this volume and needs no elaboration.

Condition 3 for favoring a prevention strategy is the presence of a pervasive problem. In assessing pollution problems in response to Section 319 of the Clean Water Act (P.L. 100-4), the states find that over half of the water bodies that have been assessed so far (which include about 41 percent of the total water bodies in the United States) are impaired by nutrients (U.S. Environmental Protection Agency, 1989); the Resources Conservation Act study conducted by the U.S. Department of Agriculture (1989) estimated that 70 percent of phosphorus now in streams originates from agricultural activities. Phosphorus presents the main nutrient problem in surface water. Yet, in some intensively farmed regions, nitrogen pollution of groundwater is also pervasive, with 5 to 20 percent of wells tested exceeding health advisory levels for nitrates (Madison and Brunett, 1985). For pesticides, a U.S. Geological Survey study (Goolesby et al., 1989) cited in other chapters in this volume and another recent study by Richards and

Baker (1989) suggest that pesticides occur above health advisory levels in streams, larger rivers, and according to Richards and Baker (1989), drinking water. All of these problems are concentrated in certain regions, where prevention becomes a particularly appropriate strategy: Farms that potentially benefit from prevention research and education in such regions likely contribute to one or several of the more common pollution problems.

Finally, uncertainties regarding the health effects and the ensuing remediation costs (condition 4 above) favor a prevention strategy. Many of the occurrences of substances in water described above are at levels at which considerable uncertainty exists as to their effects on human health, leading to ongoing study and debate as to what level of the substances should be permitted in drinking water. Lower-bound cost estimates for carrying out requirements under the Safe Drinking Water Act (P.L. 99-339) are nearly $1 billion per year. Costs will escalate to several billion dollars a year if safety standards for nitrates are lowered or if pesticides in surface water systems prove to be a widespread problem (Wade Miller Associates, 1989a,b). A prevention strategy has particular appeal when difficult choices can be avoided through the development and widespread adoption of low-chemical-input farming methods. The low-input sustainable agriculture research and education program is proving to be effective in developing farming systems that increase net farm income while advancing a wide range of environmental goals.

## TIME FRAMES FOR SUSTAINABLE
## RESEARCH AND EDUCATION

The pervasiveness of chemical contamination problems and the uncertainties regarding their health ramifications and potential costs of remediation (Wade Miller Associates, 1989b) lend an urgency to prevention efforts. However, past and ongoing long-term research programs in such areas as soil testing and integrated pest management provide some of the most promising sustainable farming methods available today. The long-term research projects identified in the chapter by Rhonda R. Janke and colleagues and other investigators are needed, but so are the more aggressive shorter-term research and education programs, such as those currently under way in Iowa.

Pennsylvania farmers reduced their use of nitrogen fertilizers state-wide by 52 percent (Berry and Hargett, 1984, 1988) between 1982 and 1988. This was a very welcome development in a state that is located on the Susquehanna River and that is above the ecologically rich and economically valuable waters of the Chesapeake Bay. Much of this reduction may have resulted from a variety of technologies introduced in Pennsylvania to more accurately account for the nitrogen available in the soil from the previous crop year and from applied manure (Fox and Piekielek, 1983). Research by Fox and colleagues has resulted in a soil test that was introduced in 1989 that will lead to an even greater efficiency of nitrogen fertilizer use (Fox and Piekielek, 1978a,b, 1984; Fox et al, 1989; Iversen et al., 1985; Michrina et al., 1981, 1982). Some fertilizer reductions have apparently been accomplished by farmers who learned that the available nitrogen from animal waste on their farms was more than adequate to meet crop needs.

The costs of developing and bringing these and many other sustainable technologies to farmers are very low relative to cost estimates for remediation (Wade Miller Associates, 1989a). Within the context of existing research and education programs, calibration of existing sustainable technologies, such as appropriate soil tests and a number of closely related nitrogen management technologies, merits particular consideration. Also important are expanded education programs that deliver these and other sustainable technologies to farmers.

## REFERENCES

Berry, J. T., and N. L. Hargett. 1984, 1988. Fertilizer Summary Data. Mussel Shoals, Tenn.: National Fertilizer Development Center, Tennessee Valley Authority.

Fox, R. H., and W. P. Piekielek. 1978a. Field testing of several nitrogen availability indexes. Soil Science Society of America Journal 42:747–750.

Fox, R. H., and W. P. Piekielek. 1978b. A rapid method for estimating the nitrogen supplying capability of a soil. Soil Science Society of America Journal 42:751–753.

Fox, R. H., and W. P. Piekielek. 1984. Relationships among anaerobically mineralized nitrogen, chemical indexes, and nitrogen availability to corn. Soil Science Society of America Journal 48:1087–1090.

Fox, R. H., and W. P. Piekielek. 1983. Response of Corn to Nitrogen Fertilizer and the Prediction of Soil Nitrogen Availability with Chemical Tests in Pennsylvania. Pennsylvania Agricultural Experiment Station Bulletin No. 843. University Park, Pa.: Pennsylvania State University.

Fox, R. H., G. W. Roth, K. V. Iversen, and W. P. Piekielek. 1989. Soil and tissue nitrate tests compared for predicting soil nitrogen availability to corn. Agronomy Journal 81:971–974.

Goolesby, D. A., and E. M. Thurman. 1990. Herbicides and Pesticides in Rivers and Streams of the Upper Midwestern United States. Proceedings of the 46th Annual Meeting of the Upper Mississippi River Conservation Committee. Washington, D.C.: U.S. Geological Survey.

Iversen, D. V., R. H. Fox, and W. P. Piekielek. 1985. The relationships of nitrate concentrations in young corn (*Zea mays* L.) stalks to soil nitrogen availability and grain yields. Agronomy Journal 77:927–932.

Madison, R. J., and J. O. Brunett. 1985. Overview of the occurrence of nitrate in ground water of the United States. Pp. 993–1105 in USGS National Water Summary, 1984. USGS Water Supply Paper No. 2275. Washington D.C.: U.S. Government Printing Office.

Michrina, B. P., R. H. Fox, and W. P. Piekielek. 1981. A comparison of laboratory, greenhouse and field indicators of nitrogen availability. Communications in Soil Science and Plant Analysis 12:519–535.

Michrina, B. P., R. H. Fox, and W. P. Piekielek. 1982. Chemical characterization of two extracts used in the determination of available soil nitrogen. Plant and Soil 64:331–341.

Richards, R. P, and D. B. Baker. 1989. Potential for reducing human exposures to herbicides by selective treatment of storm runoff water at municipal water supplies. In Proceedings of a National Conference on Pesticides in Terrestrial and Aquatic Environments, Diana Weigmann, ed. Blacksburg, Va.: Virginia Polytechnic Institute and State University.

U.S. Department of Agriculture. 1989. The Second RCA Appraisal: Soil, Water and Related Resources on Nonfederal Land in the United States—Analysis of Conditions and Trends. Washington, D.C.: U.S. Government Printing Office.

U.S. Environmental Protection Agency. 1989. National Water Quality Inventory—1988 Report to Congress. Washington, D.C.: U.S. Government Printing Office.

Wade Miller Associates. 1989a. Regulatory impact analysis of proposed national primary drinking water regulation for inorganic chemicals. Prepared for Office of Drinking Water, U.S. Environmental Protection Agency, Washington, D.C.

Wade Miller Associates. 1989b. Regulatory impact analysis of proposed national primary drinking water regulation for synthetic organic chemicals. Prepared for Office of Drinking Water, U.S. Environmental Protection Agency, Washington, D.C.

# Sustainable Agriculture Research and Education in the Northeast

*James F. Parr*

By most definitions, sustainable agriculture is viewed as a concept that comprises two major components: i.e. economic sustainability and environmental sustainability (some even emphasize the importance of social and political sustainability). For example, a farming system may be economically sustainable, but if it contributes to environmental degradation, it is not truly a sustainable system. By the same token, a farming system may be environmentally sustainable, but if it is not profitable, then, by definition, it is not a sustainable system.

Sustainability can also be thought of as a long-term goal that seeks to overcome the problems and constraints that afflict both U.S. agriculture and agriculture worldwide. How and whether this goal is achieved depends on the development of alternative management practices that are resource-conserving, energy-saving, economically viable, environmentally sound, and protecting of human and animal health.

Many have developed rather strong opinions on just what a sustainable farming system should be. However, the concept of sustainability involves a time dimension that will most certainly bring about changes that will test the sustainability of farming systems in the years ahead. What is judged to be a sustainable farming system today may not be sustainable in the future because of increased energy costs, global warming, increased soil salinization, and issues of food safety and quality.

As the world population increases and with the continuing decline in the per capita production of food in many developing countries, the natural resource base in the United States and throughout the world will come under greater pressure than ever before. Those farming systems that currently sustain the world's population may very likely be inadequate to do so in the future. There must begin to be a more futuristic attitude about what research and education programs for sustainable agriculture are needed now, so that entirely new and sustainable farming systems can be developed for the future.

## LONG-TERM, LOW-INPUT CROPPING
## SYSTEMS RESEARCH

The long-term cropping systems research study, also referred to as a *conversion* or *transition experiment*, was implemented at the Rodale Research Center in 1981 when it became apparent that farmers experienced problems when shifting from chemical-intensive farming to low-input (or low-chemical) systems. According to the 1980 *USDA Report and Recommendations on Organic Farming* (U.S. Department of Agriculture, 1980) the first 3 years of such a transition were often critical and the most difficult to cope with. Weeds were often cited as the main problem, but other chemical and biological factors were also suggested as possible causes.

Good research begins by asking the right questions, and the transition experiment described by Rhonda R. Janke and colleagues was designed to do exactly that. There was also a need to study farming systems holistically so that the interactions of the components could be evaluated. The following farming systems are being studied:

1. low-input/sustainable, with animals;
2. low-input/sustainable, cash grain; and
3. conventional cash grain.

Long-term experiments such as these are essential because significant changes in the chemical, physical, and biological components may not be detectable over the short term. The systems approach that this study uses allows researchers to know not just what happens, but why it happens.

Herein is the very basis for controlling and manipulating the system to the best advantage. A thorough economic analysis of this study is anxiously awaited.

## PERSPECTIVES FOR SUSTAINABLE AGRICULTURE FROM NUTRIENT MANAGEMENT EXPERIENCES IN PENNSYLVANIA

This chapter reported on the management practices of some dairy farms in the Northeast region where there has been a steady increase in nutrient levels on the farm because of a one-way flow of off-farm purchased inputs. According to the author, Les E. Lanyon, many of these farms now import all feed sources for the dairy cows, with little or no on-farm production of feed grains or forages. This has resulted in an excess of manure that is then applied back to the land as a disposal medium. Thus, over a period of time, nutrient-poor farms have become nutrient-rich farms or, indeed, eutrophic farms.

The chapter presents some strategies for dealing with this problem which, if it is not a real pollution problem, it is certainly a potential pollution hazard, especially to groundwater from excess nitrates. The chapter describes crop rotation scenarios that can help to utilize the accumulated nutrients while demonstrating economic benefits.

Agriculture is a system of inputs (some of which are purchased) and outputs (some of which are sold and removed from the farm). Generally, there is a net removal of nutrients during the production cycle. The dairy farms described in this chapter have a net gain of nutrients and a gross imbalance that must be managed properly to avoid a serious pollution problem. When excessive amounts of organic nitrogenous materials such as manure are applied to soil, serious water pollution problems can result, just as they can from improper use of chemical nitrogen fertilizers. These farms may be economically viable, but they definitely are not sustainable farming systems.

The question is, to what extent are these farms already polluting groundwater? Nitrate concentrations should be monitored in both soils and well water, and if they are excessive, remedial action should be taken to alleviate this situation.

## USE OF FUNGAL PATHOGENS FOR BIOLOGICAL CONTROL OF INSECT PESTS

Raymond I. Carruthers and colleagues point out that certain fungal pathogens have the potential to control insect pests and can greatly enhance integrated pest management programs and reduce pesticide use. Such fungal pathogens must be environmentally acceptable, cost-effective, reliable, and dependable and must not attack other beneficial natural predators.

It may be necessary to manipulate fungal pathogens to achieve the most effective control measures. Strategies for biological manipulation and control include germplasm maintenance, disease dynamics, use of population genetics of pathogens and hosts, and integration with other control strategies such as integrated pest management.

The new biotechnologies should be useful tools in future efforts to enhance the reliability and effectiveness of this unique biological control strategy.

### REFERENCE

U.S. Department of Agriculture. 1980. USDA Report and Recommendations on Organic Farming. Washington, D.C.: U.S. Government Printing Office. 164 pp.

# Sustainable Agriculture Research and Education in the Field

*Neil H. Pelsue, Jr.*

I am impressed by the information presented in this volume. It is especially impressive to know that this work is representative of a much larger body of work going on throughout the United States.

I applaud the discussion by Michael Duffy presented in this volume. His comments were especially appropriate because system sustainability needs to receive much more attention than it has up to now. That is my bias in my assessment of the potential contributions of projects to sustainable agriculture goals.

In the following sections, I address the three projects in the Northeast described by Rhonda R. Janke and colleagues, Les E. Lanyon, and Raymond I. Carruthers and colleagues. I will not discuss project methodology, as that should have been well covered in the project evaluation and selection process. Rather, I will focus on four other aspects that I believe are important criteria for low-input sustainable agriculture (LISA):

1. whole-farm interactions,
2. economic performance,
3. environmental impact, and
4. information delivery.

To my way of thinking, the fourth essential aspect—delivery of project information to farmers and other users in readily usable form—was not adequately discussed in any of the three presentations.

## LONG-TERM, LOW-INPUT CROPPING SYSTEMS RESEARCH

*Whole-Farm Interactions:* This project includes several cash grain systems and incorporates a livestock component to provide comparisons with conventional grain-livestock systems. The project makes good use of farmer inputs in project development and system assessment.

*Economic Performance:* This project makes a good attempt at estimating the conversion costs for farmers as they move from conventional to alternative systems. I would encourage that component analyses (corn) be integrated into a systems analysis to estimate interaction effects. This is a good use of FINPACK (Hawkins et al., 1987) (or other computer software) for analyzing the financial implications of alternative farm management decisions.

*Environmental Impact:* Important objectives of the project are to reduce chemical use, observe nitrogen recycling, and determine the water-handling capacities of soils.

## PERSPECTIVES FOR SUSTAINABLE AGRICULTURE FROM NUTRIENT MANAGEMENT EXPERIENCES IN PENNSYLVANIA

*Whole-Farm Interactions:* This project studies farm management activity flows and the interactions of production components.

*Economic Performance:* While the project compares on-farm manure use in financial terms, it lacks the necessary comprehensive economic assessment.

*Environmental Impact:* The project recognizes the need to identify the environmental effects of on-farm manure use both on and off the farm.

This chapter provided a good overview of the process that was used to select components to improve the system, but it did not provide specifics about the nutrient management project.

## USE OF FUNGAL PATHOGENS FOR BIOLOGICAL CONTROL OF INSECT PESTS

*Whole-Farm Interactions:* This chapter provided some of the necessary, basic information that will need to be integrated into whole-farm systems.

*Economic Performance:* The project needs to demonstrate the economic effectiveness of biological control technology before large commitments are made to further research.

*Environmental Impact:* The chapter stressed the importance of evaluating how alternative pest control methods are used and their interactive effects.

## ADDITIONAL ECONOMIC FEASIBILITY STUDIES

Additional work is needed to assess adequately the ecomomic feasibility of sustainable agriculture studies and projects currently being carried out. This economic analysis can be divided into five categories.

### On-Farm Analysis

There will continue to be a need for traditional partial and enterprise budgeting analyses to determine the economic impact of a particular practice (Osburn and Schneeberger, 1978). Essential component analysis is also needed to assess the overall impact of input changes to the farm business operation.

### Infrastructural Analysis

Infrastructural analysis refers to assessing the nature and extent of the economic impact on those sectors that provide farm production inputs: the manufacturers and suppliers. At the other end of the production process, the economic effects on marketing activities, processing, and commodity handling also need to be assessed. This analysis also includes government policies, because of the powerful impact of government actions on economic viability.

### Consumer Analysis

Other factors that must be taken into account are the reactions of consumers to changes in product prices and to the quantity, quality, and variety of the available food and fiber products. The nature and extent of consumer demand is as important a determinant of economic viability as is production efficiency. However, consumer demand often gets overlooked or taken for granted in analyses of alternative agriculture practices.

### Societal Analysis

Another important, but conceptually difficult, aspect is the determination of socioeconomic costs as communities and rural areas are affected by changes in the structure of the agricultural industry.

## Environmental Analysis

It is necessary to determine the economic impact of agricultural practices as they apply to environmental considerations of both on-farm and off-farm aspects of the production and marketing of food and fiber products. What are now referred to as externalities need to be incorporated into the economic modeling systems (Barlowe, 1986).

## INFORMATION DELIVERY

One of the most important objectives of LISA and related programs is to get study results to users in a timely and usable fashion. Farmers, researchers, and members of industry must be able to take advantage of proven and available information delivery systems, developing new or modified systems only as the existing systems are shown to be inadequate.

## REFERENCES

Barlowe, R. 1986. Land Resource Economics, 4th ed. Englewood Cliffs, N.J.: Prentice-Hall.

Hawkins, R. O., D. W. Nordquist, R. H. Craven, J. A. Yates, and K. S. Klair. 1987. An Overview of FINPACK. St. Paul, Minn.: Center for Farm Financial Management, Minnesota Extension Service, University of Minnesota.

Osburn, D. D., and K. C. Schneeberger. 1978. Modern Agriculture Management. Reston, Va.: Reston Publishing Co., Prentice-Hall.

# PART SIX
## Summary

# 22

# Assessing the Progress of Sustainable Agriculture Research

*Jonathan H. Harsch*

The more than 20 scholarly reports from all parts of the United States presented in this volume firmly established one fact: Despite the continuing chorus of criticism aimed at low-input sustainable agriculture (LISA), there already is an impressive body of thoroughly documented scientific evidence relating to all aspects of sustainable or alternative agriculture. This much is established, and this much marks a significant milestone in the latest campaign to improve the performance of U.S. agriculture.

What remains in doubt is interpretation of the scientific evidence. As Michael Duffy explained in his presentation on economic considerations, manure and crop rotations have been proved to be effective. The question is how they fit the goals of farmers and society. In other words, to what extent will the individual farmer be willing to sacrifice higher short-term profits for the rewards of environmental stewardship and other intangibles? To what extent will an environmentally aroused public and the U.S. Congress provide cash incentives, at least for a transitional period, to help farmers cut back on their use of purchased commercial fertilizers and pesticides?

The chapters in this volume have provided many important research findings. Some particular points from particular chapters are presented below.

North Dakota's John Gardner and colleagues, in their paper "Overview of Current Sustainable Agriculture Research," and U.S. Department of Agriculture (USDA) Assistant Secretary for Science and Education Charles Hess brought out a key point: there is a great reluctance within the scientific community to challenge accepted wisdom and a great reluctance for scientists to challenge their department heads.

This reluctance will be a continuing constraint, one that needs to be counterbalanced through greater efforts to pursue new research avenues.

A second key point is that no matter how overwhelming the evidence, interpretations will continue to differ. Dale Darling discussed both points clearly. During the workshop, Darling summed up his position by commenting that when he was a boy, his family's dairy farm in Vermont was 100 percent organic "because we didn't know any better."

No matter what the evidence may be, interpretation of this evidence will differ. Some differing opinions on the value of fertilizers of farm origin and current research findings were presented by Harold Reetz. This volume includes strong voices calling for going slowly enough in areas such as new groundwater protection legislation to avoid the overkill that is inevitable if public policy decisions are based on public fears rather than on sound scientific evidence. The need is to avoid a situation in which economic costs to the farm sector and its customers might far outweigh the economic and environmental gains sought by society as a whole.

Charles Hess provided a very strong response in that key area, a response that demonstrated that the Bush administration has come a very long way from the days of the Reagan administration. Hess indicated that there is a very narrow window, that the farm sector is under pressure to respond quickly and responsibly to public concerns in currently explosive areas such as food safety and water quality. He warned that there must be a positive response to the issues being raised. He also indicated that it is time to be proactive rather than defensive. To do otherwise would be to invite legislation and regulation that may remove the decision-making power and constrain the flexibility to adopt management practices that best fit each farming situation.

Raymond E. Frisbie, from Texas A&M University, and Iowa State Legislator Paul Johnson, in his remarks to workshop attendees, reinforced the point that agriculture must change its ways voluntarily or agriculture will inevitably find itself burdened with a mandated regulatory straitjacket tailored by other groups with little understanding of farm-sector realities.

The farm-sector response will not be easy. It will require broad agreement on often hard-to-accept voluntary constraints. In addition, because public funds are limited at all levels—federal, state, and local—the farm sector faces the challenge of working out a priority list of goals that will be acceptable to the public, to the U.S. Congress, and to the whole range of often competing government agencies. This challenge is complicated by the fact that along with achieving consensus support for more environmentally benign farming practices, these general practices necessarily will include an immense range of options in order to respond to site-specific problems.

How far has agriculture gone along this challenging response road? The research findings presented in this volume prove that there has been progress, both in the area of scientific research and in the practical changes that have been made.

Both the scientific and regulatory communities have made progress in identifying goals, specifically in recognizing the importance of focusing sustainable agriculture systems on three equally important goals:

1. improving farm-level profitability,
2. improving the U.S. farm sector's international competitiveness, and
3. at the same time, reducing environmental damage caused by farming practices.

Agreement on the equal importance of profitability, competitiveness, and environmental concerns is a major step forward in itself and is part of a three-stage progression. September 1989 marked stage one. Publication of the National Research Council's (1989) report *Alternative Agriculture* consolidated what had been learned. This important report stated the issues clearly and established that sustainable agriculture "isn't just for breakfast anymore"—that sustainable agriculture is a real meat-and-potatoes system capable of moving U.S. production agriculture, as a whole, closer to the three goals of profitability, competitiveness, and environmental stewardship.

The workshop on which this volume is based was part of stage two, that is, the creation of a vast data base of sound, replicated research findings that increasingly will make it possible for any commercial farmer in any part of the country, with any crop or livestock mixture, to improve his or her own operation's profitability, competitiveness, and environmental stewardship.

Because this seminar has proved that stage two is well established, stage three—when the sustainable practices detailed in *Alternative Agriculture* will be in widespread use throughout production agriculture—cannot be far off.

These points can be illustrated by a statement made at the workshop by Charles Hess:

> Overall, today's agriculture is being challenged to operate in an environmentally responsible fashion, while at the same time continuing to produce abundant supplies of food and fiber both economically and profitably. . . . On one hand, agriculture needs to be highly efficient and internationally competitive in order to be economically viable. On the other hand, it needs a system of production which is environmentally sensitive, sustainable, and whose products are viewed as safe.

Basing his conclusion on the solid research findings already available,

Hess stated firmly that he believes that both the economic and environmental goals are achievable. At the workshop, he explained:

> We feel it is the department's responsibility to provide farmers with a range of options that best fit their economic and environmental situation. The choices range from the optimal use of fertilizers, pesticides, and other off-farm purchases in conjunction with best management practices to operations which actively seek to minimize their off-farm purchases and emphasize crop rotation, integration of livestock and crop production, and mechanical or biological weed control.

To support this significant policy change—and to guard against the regulatory overkill that he and other speakers warned against—Hess pointed to the need for further government investment in research, because "we must work to get more hard data so we can make informed decisions based in science rather than in emotion." It is an important step forward to have this explicit USDA support for sustainable agriculture research.

As we learned from key speakers at the workshop, such as Michael Duffy and Robert Papendick, researchers themselves must take the next step and, in fact, are doing that in a big way. Throughout the country, both farm- and laboratory-level researchers and their bureaucratic funding sources now recognize that what is needed is thorough interdisciplinary research focusing on far more than just agronomics.

Speakers explained that the big research payoffs are coming because individual LISA research projects today integrate the full range of interrelated factors including agronomics, economics, social policy, and government farm program biases. Researchers today, as described in this volume by R. James Cook and Gail Richardson, are not just confining themselves to academic research. Instead, they are doing what must be done to satisfy the site-specific requirements of a truly sustainable system: they are working with commercial farmers in the fields and are giving these farmers an active role in the research process.

Researchers are also deeply involved in trying to unravel the complex and often conflicting effects of both current commodity programs and the variety of proposed 1990 farm bill changes.

Perhaps most important for the evolution of LISA research was definite evidence that the spirit of scholarly objectivity is replacing crusading zeal. Thomas Dobbs and colleagues straightforwardly explained their findings that after a 5-year comparison of a conventional farm with an alternative LISA farm, average net returns were higher for the conventional system when the premium prices for organically grown crops were replaced with conventional prices. Dobbs noted that the premiums for organically grown products actually received by the case study farm brought its profit above that of the comparison conventional farm.

Another two-pronged sign of LISA researchers' maturity was revealed. First, no one tried to sell his or her own specific definition of LISA. That must be considered progress, because as stated repeatedly, LISA needs to operate in very different ways, depending not only on the part of the country and the type of crop or livestock mixture but also depending on the specific characteristics of the individual farm and the individual farm manager, on his or her own management skills, and on the local availability of farm labor.

Second, it also must be considered a sign of progress that consensus seemed to emerge that it is time to replace the term *LISA* with the term *sustainable agriculture*. No one wants to abandon this attractive acronym; however, a number of speakers pointed to problems with the low-input connotation, noting that the low-input label automatically attracts the easy criticism that LISA is antichemical or even antitechnology per se.

LISA proponents appear ready to drop the acronym in order to shift attention away from the secondary effect, which in many, but not all, cases will be reduced use of purchased inputs. Instead, their aim is to emphasize what matters most: sustainable agriculture's primary goal of enhancing profitability, competitiveness, and environmental stewardship.

Charles Benbrook summed up this shift, emphasizing that "the goal of LISA systems really need not be viewed as reducing the use of pesticides and fertilizers. Lessened reliance on agrichemicals, though, is often one of the positive outcomes of successful adoption of LISA systems." As he and others have explained, fully sustainable, site-specific best management practices may call for increased rather than decreased use of chemical inputs under the right conditions.

From James Cook's key research on the 65 percent boost in wheat yields through the use of a 3-year rotation to break the cycle of harmful pathogens in the root zone, to Gail Richardson's use of neglected, low-technology gypsum blocks for monitoring soil moisture, the workshop and this volume on which it is based have broken important ground.

As has been stressed throughout, there is much more work left to be done. That is the challenge: to continue building on the sound scientific foundation revealed here.

To sum up both what research has already accomplished and the major challenges for future research, as well as for legislation over the years ahead, a key passage from Charles Benbrook is appropriate:

> Often, farmers are confronted with choices and sacrifices because of seemingly unavoidable trade-offs. An investment in a conservation system may improve soil and water quality, yet do so at some sacrifice in near-term economic performance. Diversification may increase the efficiency of resource use and bring within reach certain biological benefits, yet may require additional machinery and a more stable and

versatile supply of labor. Indeed, a major challenge for agricultural researchers, and those designing and administering farm policy is to seek ways to alleviate seemingly unwelcomed trade-offs by developing new knowledge and technology and, when warranted, new policies.

## REFERENCE

National Research Council. 1989. Alternative Agriculture. Washington, D.C.: National Academy Press.

# Appendix A

# Poster Sessions

## Low-Input Sustainable Agriculture
## Farm Decision Support System

*John E. Ikerd*

U.S. farmers are faced with growing environmental concerns and rising costs associated with highly specialized farming operations. They are searching for farming systems that are ecologically sustainable as well as productive and profitable. Many are motivated by perceived risks that the inputs on which they depend today may not be available, may not be effective, or may cost much more in the future. Such farmers are searching for ways to reduce their dependence on external purchased inputs while maintaining their productivity and profits through more intensive management of their internal resources.

The current search for sustainability and profitability in U.S. agriculture is centered on helping farmers develop more ecologically sound and economically viable farming systems with existing technology while searching for even more sustainable and profitable alternatives for the future.

A short-term objective is to improve the input efficiency of current farming systems. However, long-term sustainability may require more diversified systems of farming that include commodities that can be produced with more ecologically benign systems. Diversified farming systems traditionally use crop rotations to control pests, conserve soil, and maintain productivity. Integrated cropping and livestock systems have been used to reduce feed costs, recycle waste, and stabilize the incomes of U.S. farmers.

The current hope for future success lies in finding ways of combining new technologies such as microcomputers and biotechnology with the tried and proven principles of management by objectives and diversification—old principles with new technologies.

Such farming systems will be more complex and thus will require more intensive hands-on resource management than do higher-input, specialized systems. However, synergistic gains from effective integration of enterprises and activities in diversified farming systems represent the best hope for achieving long-term sustainability with a minimum of government regulation.

A microcomputer-based farm decision support system[1] is being developed under a project funded jointly by the Extension Service and Cooperative State Research Service, U.S. Department of Agriculture, to integrate the concept of sustainability into farm planning and to implement farm management strategies for sustainability.

The Low-Input Sustainable Agriculture Farm Decision Support System (LISA-FDSS) project was approved in 1988 for funding through November 1990.

## CHARACTERISTICS OF LISA-FDSS[2]

The LISA-FDSS system has six basic functions that are supported by two microcomputer-based program components, two farming systems data bases, and several specialized data bases to support the budgeting process. LISA-FDSS is designed to be compatible with a national financial planning project, FINPACK, and a national linear programming project that emphasizes labor and machinery management.

The six basic functions of the LISA-FDSS system are as follows: (1) resource management strategy (RMS) budgeting, (2) whole-farm planning, (3) environmental checking, (4) financial checking, (5) risk checking, and (6) resource checking. The two data bases are (1) default RMS budgets and (2) customized RMS budgets. Additional data bases include soil types and characteristics, fertilizer and pesticide characteristics, correlation coefficients, and energy conversion units.

### RMS Budgeting

The RMS associated with a cropping system consists of a crop sequence or rotation, an irrigation system (if any), a tillage system, a fertility system, and a pest management system. An RMS budget reflects the resource requirements, input requirements, input costs, expected production, expected returns, potential conservation impacts, and potential environmental impacts of the individual crops as components of a cropping system. An RMS budget contains all non-site-specific information needed to calculate expected soil losses, water-quality risks, resource use, gross margin over purchased inputs, and revenue risks.

The default RMS budget data base will contain budgets for cropping and livestock systems deemed appropriate for the geographic region of application. These data bases will be constructed by extension specialists using a basic FINPACK financial budget format augmented by additional resource and environmental (R&E) components. Development of R&E budget components will be facilitated by a budgeting program developed as a part of the LISA-FDSS project. The R&E budgeting program is one of the two basic microcomputer program components of the LISA-FDSS system.

Default data bases should include budgets for a wide range of cropping systems deemed appropriate for the geographic region where the LISA-FDSS program is to be used. A cropping system might include from 1 to 12 different crops. A monocrop system would have the same budget for each year. A given crop following different crops in different rotations might have a different budget for each rotational position. Different crops, of course, would have different budgets.

Each cropping system will be budgeted for up to four alternative input systems. An input system will reflect a specific fertility and pest management system. Most systems would be budgeted with unrestricted-input, reduced-input, and low-input RMS alternatives.

Unrestricted-input RMS budgets will reflect the use of typical fertilizer and pesticide inputs for a particular cropping system on fields with no significant fertilizer or pesticide leaching or runoff risk potential. Reduced-input RMS budgets will reflect some lower level of inputs suggested for fields with significant nutrient or pesticide risk potentials. Split applications and banding of fertilizers and pesticides might be a logical reduced-input system, for example. A low-input system should reflect minimum levels of external inputs that specialists deem feasible for commercial production on fields with high nutrient loading or pesticide risks.

Each cropping system will also be budgeted for alternative tillage levels. Tillage options will range from unrestricted tillage to minimum tillage. Unrestricted tillage would be the suggested system for fields without erosion problems, with minimum tillage suggested for highly erodible fields. Each tillage system should be matched with an appropriate complement of inputs. Consequently, some systems may have no low-input, minimum-tillage RMS, if such a combination of tillage and inputs is not considered feasible for a given cropping system.

In general, the alternative input systems will be designed to reduce water-quality and other environmental risks by moving to lower-input alternatives. In general, the alternative tillage systems will be designed to reduce soil erosion risks by moving to lower tillage levels. Irrigation systems, if any, will be specified as a part of each input system.

Farmers who use an unrestricted system for one crop would likely use an unrestricted system for another crop in the same rotation, although an unrestricted system might imply different tillage and input regiments for different crops in the same rotation. Likewise, a farmer interested in a low-input commercial alternative for one crop likely would be interested in a similar system for other crops in the same rotation. Thus, the levels of inputs and tillage will be identified for whole cropping systems rather than individual crops.

## Whole-Farm Planning

The Whole-Farm Planner (WFP) is a microcomputer-based decision support system that allows farmers to evaluate the potential impact of using various cropping systems or RMSs on their specific farms. The WFP is a field-based system. It allows farmers to plan their farms field by field and year by year and to assess the RMS implications for each field and each year for the whole farming system, including livestock as well as crops.

A typical FDSS user would begin with the whole-farm planner component of the system. An agent working with a farmer should have determined the basic rotations used by the farmer and have those RMSs available in the default data base at the time of the first planning session. Otherwise, the farmer and agent would have to add those budgets to the default data base before the planning process could begin. Most farmers will want to begin with an assessment of their current system before they begin to evaluate alternatives.

All site-specific information and the associated yield and environmental impact estimates are calculated within the whole-farm planner program. Thus, the whole-farm planning process begins with a field-by field inventory of the land or soil resources of the farm. Much of the information related to soil erosion and environmental vulnerability can be derived from the Soil Conservation Service (SCS) data base of soil types. Soil texture, pesticide leachability, pesticide surface loss potential, and average slope and slope length are identified in the SCS data base of U.S. soils. However, the farmer will be asked to verify yield potentials, soil characteristics, and environmental impact estimates in the planning process.

Environmental and conservation impacts will be evaluated field by field over a 12-year planning period. Thus, estimates of soil loss, water-quality risks from pesticides and fertilizers, and input toxicity will be evaluated for cropping systems rather than individual crops.

Financial and resource implications of alternative systems will be evaluated for the whole farm for each year in the planning period. The acreage of each crop, pasture, set-aside or conservation reserves, expected revenues, input costs, gross margins, revenue risks, corn equivalents produced and needed, hay equivalents produced and needed, and nonrenewable energy use will be summarized for each year.

The ecological vulnerability of each field will be identified by color-highlighted codes for high, medium, and low levels of vulnerability to soil loss, pesticide leaching, and residue runoff. Each cropping system and RMS will likewise be color-coded with respect to its potential for soil loss and water-quality risks. These two sets of codes, one for the field and the other for the RMS, will be combined to yield a similar color-coded set of implications for using a given RMS on a given field.

Each combination of field and RMS will have a color-coded indicator of soil loss, water-quality risk from pesticide and nitrogen use, and input toxicity. A set of red H's for a given RMS on a given field, for example, could indicate severe ecological problems. Such problems would be associated with the use of a particular RMS on a particular field. The same RMS might be acceptable on another field, but a different RMS might be indicated for the particular field being examined.

There will be relatively few alternatives for correcting the ecological vulnerability of a given field. Exceptions would be to contour till, terrace, strip crop, or ridge till a field to reduce soil loss potential. In most cases, farmers will have to change RMSs to correct ecological problems.

Each RMS will be identified with a code indicating the tillage and input levels associated with the particular strategy. A farmer with an erosion problem might consider an RMS with less tillage. If, instead, the farmer is faced with a water-quality problem, he or she might select a lower-input RMS. If the farmer has an erosion and water-quality problem, he or she could select a longer crop rotation that included meadow or some other soil-conserving crop.

A similar approach will be used in the financial, risk, and resource sections of the program. An unacceptable income level for a given year would be color coded with a red H or some similar sign. The farmer could first consider shifting rotations to get more high-income crops in a given year, if the problem occurred only for 1 or 2 years. However, if the problem occurs for several years, he or she may consider some more intensive RMSs that will generate more income in more years.

Inconsistencies between labor needs and availability would be flagged. Seasonal labor problems may be addressed by shifting rotations, changing to lower-labor RMSs, or hiring labor during peak need periods, if it is feasible. Feed needs and production would be handled in a similar manner. An unacceptable level of risk might suggest that diversity be added by selecting alternative cropping systems, adding livestock to the system, or, possibly, considering off-farm employment for income stability.

Changes in RMSs to solve financial, risk, or resource problems may generate ecological problems. However, no attempt will be made to calculate an optimum system for a given farm. Farmers will simply attempt to solve their ecological and economic problems by matching alternative resource management strategies (cropping systems with alternative tillage and input levels and livestock enterprises) with their internal resources (land, labor, and machinery).

Information describing each RMS, including any specialized machinery requirements, will be available from the whole-farm planner program. For example, a farmer may want to know what type of fertility program, tillage system, pest control system, and labor requirements are assumed for a low-input soybean alternative in a corn-soybean rotation in field number three in year 4 of the plan currently on the screen. He or she would indicate with some set of key strokes the basic data he or she wants to review for this particular alternative.

The whole-farm planner program assumes that a farmer has multiple objectives that include both ecologic and economic factors. The ecologic factors are soil loss, water quality, input toxicity, and nonrenewable energy use. Standards for the ecological factors will be predetermined. The economic factors are net returns or income; income risks; and utilization of land, labor, and machinery. Farmers will be asked to develop their own income objectives from overall farm financial information.

Some farmers may be willing to settle for a whole-farm plan with a large number of red, or warning, indicators on the ecologic factors to achieve green, or safe, indicators in the financial and resource areas. Others may be willing to tolerate lower economic results to achieve safe indicators (green color codes) in the ecologic areas. Others will continue to explore alternatives until they have all ecologic and economic indicators in acceptable ranges or they will not farm. These choices are to made by the individual farmer.

## Custom Budgets

Each farmer would need to work with his or her agent or specialist in customizing the default RMS budgets to reflect inputs and resources for tillage and cropping systems that the farmer actually expects to use on his or her farm. The WFP program would allow the farmer to greatly narrow the range of budgets that might be considered to be logical for his or her operation. However, the customized alternatives need not be limited to those for a single best farm plan identified by the farm planner.

Changes from default values to customized values for environmental and economic impacts may significantly change the estimated outcomes of a given farm plan. Thus, once the customization process is completed, the farmer would be expected to return to the WFP program. He or she would simply repeat the earlier iterative planning process with the customized sets of budgets until a satisfactory customized plan is achieved.

## OBJECTIVES OF LISA-FDSS

The RMS budgeting process will allow agricultural specialists to reflect the full range of existing and future research results and information in a form that is readily usable by farmers. For example, ecologic and economic impacts of cover crops, intercropping, and relay cropping in various rotations can be reflected in alternative RMS budgets. Uses of legumes and livestock manure for fertilizers as well as alternative systems of fertilizer application can be included among the RMS alternatives to be considered.

Impacts of alternative tillage systems and residue management programs on potential soil loss will be an integral part of the budgeting process. Alternative weed, insect, and other pest control systems, including specific pesticide uses and their potential risks to humans and water quality, will be reflected directly in the environmental components of each RMS budget.

The whole-farm planning process will allow farmers to synthesize profitable and sustainable farming systems by integrating relevant RMSs with their particular set of land, labor, machinery, and management resources. They can select RMSs that are well-suited for their soils, climate, and location-specific pest problems. They can integrate systems of livestock and crop RMSs that tighten or complete nutrient cycles, facilitate energy flows, and enhance the ecologic and economic viability of their farming systems.

Farmers who use the whole-farm planner can evaluate potential impacts of using various levels of various chemical fertilizers and pesticides on specific fields. They can match tillage systems and soil-conserving practices with specific slope and soil characteristics of fields to reduce erosion. They can assess risks through evaluation of diversification effects of alternative farming systems and develop systems that are resistant, resilient, and regenerative.

The LISA-FDSS will not result in a recipe for success. LISA-FDSS is just a tool to facilitate farm planning and management. Farmers who choose an alternative to their current system will be advised to gather as much additional information as is available before they adopt a new farming enterprise or practice. Farmers will be strongly encouraged to talk with other farmers who have experience with the practice under consideration. They will be encouraged to visit other farms where the practice is used before they change their own operation. They will be advised to work into any new system slowly, so they can learn as they go.

The LISA-FDSS will not ensure a more profitable or sustainable farming system. However, it will allow farmers to evaluate the potential impact of alternative LISA technologies and strategies within the context of their particular farming situation without doing the necessary research and testing on their own. Use of the LISA-FDSS will not ensure success, but the LISA-FDSS can be a valuable and important aid in taking the first step toward the goals of economic and ecologic sustainability.

## NOTES

1. This farm decision support system was developed by the LISA-FDSS Task Force: John Ikerd, Columbia, Missouri; Richard Levins, St. Paul, Minnesota; Larry Bond, Logan, Utah; Mike Duffy, Ames, Iowa; Don Tilmon, Newark, Delaware; Tim Hewitt, Marianna, Florida; and Patrick Madden, Glendale, California; special funding was provided by the Extension Service of USDA.

2. LISA-FDSS has been renamed Sustaining and Managing Agricultural Resources for the Future—Farm Resource Management System (SMART-FRMS). Further development and support of the system is being carried out by the Center for Farm Financial Management, University of Minnesota, St. Paul, Minnesota.

# Voisin Controlled Grazing Management:
# A Better Way to Farm

*William M. Murphy*

Permanent pastures in the northeastern United States typically have low productivity, producing only about 2 tons of moderate-to-poor-quality forage per acre during a 3- to 4-month grazing season. A proven method exists that enables these kinds of pastures to produce 4 tons or more of excellent-quality (23 percent crude protein, 0.72 Mcal/pound of net energy lactation) dry forage per acre during a 6- to 7-month grazing season. The method is controlled grazing management, as described by Andre Voisin (1959) (see also Murphy, 1987). This method, which is also known as short-duration grazing, intensive rotational grazing, and rational grazing, has been used for many years in New Zealand and for 8 years in Vermont. New Zealand's highly productive and profitable agriculture depends almost entirely on permanent pastures that are grazed under controlled management. New Zealanders raise 70 million sheep, 8 million cattle, 1 million deer, and 1 million goats, without grain supplements, on only 37 million acres of pastureland, which is the size of Iowa. This proves that the method works.

Many American dairy farmers, in contrast, use a system of zero pasturing or year-round confinement feeding that involves large amounts of purchased feed and supplements, huge capital investments in facilities and equipment, large cash flows, and low profitability. Partly because of this, many U.S. dairy farmers are experiencing a financial crisis that may eliminate many family farms, because feeding of livestock in confinement can cost six times as much as it does on well-managed pastures.

American farmers who do not use year-round confinement feeding put their livestock to pasture, where the animals are grazed continuously or rotated through a few large pasture divisions with little control and less planning. Invariably, by late June or early July the pastures are depleted and worn out. These dairy farmers generally do not feel that their pastures have much feed value and use the same ration all year, regardless of what the pastures produce. Therefore, pastures have been a wasted resource in the United States.

One way to increase profitability of a farm is to reduce feed costs. The permanent pastures that exist on most farms produce biomass far below their potential because of poor grazing management. Pastures managed under controlled grazing conditions can be some of the most valuable areas on a farm, producing high yields of excellent-quality forage. When incorporated into livestock feeding programs, this homegrown forage can reduce feed costs and increase the profitability of many northeastern farms. First-year costs of materials, maintenance, and labor for the grazing management method range from $1,500 to $2,000 for a 40-cow herd. Its use has returned $3.75 in benefits for each $1 invested by dairy farmers (Jones and Burns, 1988; Pillsbury and Burns, 1989).

Voisin grazing management is a simple system of controlling grazing by dividing pastures into small areas (paddocks) that are grazed on a rotational basis. This method minimizes the waste of forage and protects the plants from overgrazing.

## GUIDELINES FOR VOISIN CONTROLLED GRAZING MANAGEMENT IN VERMONT

The essentials of the Voisin grazing method and what its use has meant to three farmers in terms of increased profitability and improved quality of life are illustrated in a 33-minute video produced as part of a low-input sustainable agriculture (LISA) project (Murphy et al., 1989).

### Recovery Periods

The recovery periods between grazings must vary with the plant growth rate. This usually means that recovery periods must increase as the plant growth rate decreases as the season progresses. In Vermont, for example, this means that a 10- to 18-day recovery period is needed during May and June. This gradually increases to 36- to 42-day recovery periods by the end of September. Another way to look at it would be as follows: 10 to 12 days of recovery time in late April to early May, 15 to 18 days by June 1, 24 days by July 1, and 36 days by September 1. These are guidelines only; longer or shorter recovery times may be needed, depending on the local growing conditions.

Recovery periods reflect the pre- and postgrazing pasture mass (total forage) relationships shown in Figure A-1. Pasture mass influences most the net harvested forage production at the extremes of low postgrazing and high pregrazing masses. At a low pasture mass, the lack of leaf surface area limits solar interception and photosynthesis. At a high pasture mass, shading of lower leaf surfaces blocks solar interception, while respiration of shaded plant parts consumes the carbohydrates that are produced, until death and decomposition of the shaded parts occur, with consequent loss to net production.

Based on these relationships, the forage should not be taller than 6 to 8 inches when cows are turned into a paddock (the forage should not be taller than 4 inches for sheep, because sheep-grazed swards are more dense)

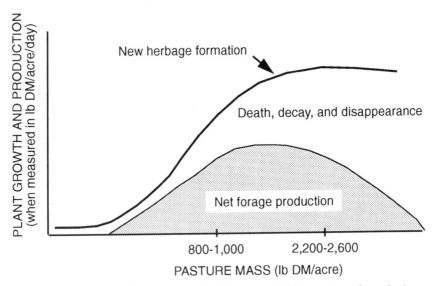

FIGURE A-1   Effect of pasture mass (as dry matter [DM]) on rates of new herbage formation, net forage production, and forage losses. Forage losses through death and decay result mainly from shading of lower plant parts, and these losses increase as pasture mass increases. Source: C. J. Korte, A. C. P. Chu, and T. R. O. Field. 1987. Pasture production. Pp. 7–20 in Feeding Livestock on Pasture, A. M. Nichol, ed. Occasional Publication No. 10. Hamilton, New Zealand: New Zealand Society of Animal Production.

and should be grazed down to 1 to 1.5 inches from the soil surface before the animals are removed. If the animals do not eat enough to keep up with the rapidly growing forage in the spring, some of the paddocks should be removed from the rotation and should be harvested for hay or silage. Usually, about one-half of the pasture area must be saved for machine harvesting, because too much forage is produced in May and June. This means, for example, that if there are 20 paddocks, 10 of the most level ones should be saved for machine harvesting. After harvesting, the paddocks should be rested until the plants regrow adequately before they are included in the next rotation. The larger number of paddocks then available for grazing in late July and early August automatically increases the recovery periods of all paddocks.

If, at any time, the paddocks have not fully recovered by their turn in the rotation, all of the animals should be removed from the pasture and should temporarily be fed elsewhere (e.g., they could be grazed on the hayland aftermath or fed green chop, hay, or silage harvested from the excess earlier in the season) until recovery periods are adequate before the animals are turned back into the pasture system. By strictly observing this need for adequate recovery times, permanent pastures in areas such as Vermont may be able to be grazed from mid-April to mid-November. In contrast, pastures that are not under Voisin controlled management can be grazed for a much shorter time, typically from mid-May to mid-August.

## Periods of Occupation

The total time that animals occupy a paddock in any one rotation must be less than 6 days, to prevent grazing of regrowth in the same rotation. Paddocks must be small enough so that all or most forage in each paddock is grazed down to about 1.5 inches from the soil surface within this time limit. If two separate groups of animals are grazed (e.g., milking cows or heifers, dry cows, and lambs and ewes), each group should not be in a paddock for longer than 3 days, because forage palatability and availability decrease too much after 3 days for each group.

In practice, the shorter are the periods of occupation, the better it is for optimum plant and livestock production. If animals are grazed as one group, they should not be in a paddock for longer than 2 days for the best livestock production. If two groups graze a paddock, each one should be in the paddock for only 1 day. Milking or fattening animals should not be in a paddock for longer than 1 day, so that they can be kept on a consistently high level of nutrition. Milking cows produce the most if they are given a fresh paddock after every milking. Growing lambs and beef cattle should be moved to a fresh paddock once a day for the best results.

## Paddock Size

Paddock sizes must be adjusted according to the desired intensity of management. For example, if milking cows are to be moved from a paddock every 12 hours and then dry cows and heifers are placed in that paddock for another 12 hours to eat the remaining forage, paddocks must be small enough so that all or most forage above 1.5 inches in each paddock can be eaten within the total 24-hour occupation period. On the other hand, under less intensive management, milking cows can graze a paddock for 1 to 2 days, followed by dry cows and heifers that graze the paddock for another 1 to 2 days to eat the remaining forage, giving a total occupation period of 2 to 4 days. If the herd is kept as one group of milking cows, dry cows, and heifers, then paddocks must be small enough so that all the forage is eaten within the 0.5-, 1-, or 2-day total occupation periods, depending on how often the animals are to be moved. Paddocks usually should be less than 2 acres in size, depending on pasture productivity and herd size. One-acre paddocks or smaller may be needed on excellent pasture for the highest pasture and animal productivity.

## Paddock Number

Farms that feed livestock on pasture need 10 to 80 paddocks, depending on how frequently the animals are moved. For example, if animals are in each paddock for a total of 4 days per rotation, 10 paddocks will eventually be needed to provide 36- to 42-day recovery periods. If animals are in paddocks for a total of 12 hours, about 80 paddocks are needed.

## Fencing

Electric fencing is the preferred method of dividing pasture into paddocks. Only one strand of smooth wire about 30 inches from the soil surface is needed to control dairy animals. Three strands of smooth wire or flexible net fencing is needed for sheep. Perimeter fence for sheep must have at least five strands to keep predators out. Cedar posts with insulators at the corners and gates and round fiberglass posts every 50 to 60 feet are commonly used. A better way is to use treated, self-insulating hardwood posts of various sizes to support high-tensile, spring-tightened smooth wire for perimeter fence and permanent paddock divisions. Portable fencing can be used for internal subdivisions to decrease the amount of permanent paddock fences that need to be built.

Ordinary fence chargers short out very easily and may not control livestock under intensive grazing management. The best way to charge an electric fence is to use a New Zealand-type energizer; they provide the dependable high shocking power that is needed. A smooth-wire fence is a psychological barrier, not a physical one. Animals must know that they will always be shocked if they touch the fence.

## Water

Ideally, drinking water should be available in the paddock where animals are grazing. Animals can then remain in the paddock, rather than having to walk back to the barn or other distant location to drink. Having adequate water in the paddock increases production efficiency. Another benefit is that manure is kept in the paddock, where it is needed to recycle nutrients for sustained plant production.

## CONCLUSION

The use of Voisin controlled grazing management can make farming profitable again for many financially stressed dairy farmers in the northeastern United States. Increased profits combined with the labor and time savings that result from use of this method can give farmers the money, time, and energy to enjoy life more.

## REFERENCES

Jones, C., and P. Burns. 1988. Economic Effects of Adoption of Rational Grazing. Orono, Maine: Soil Conservation Service.

Korte, C. J., A. C. P. Chu, and T. R. O. Field. 1987. Pasture production. Pp. 7–20 in Feeding Livestock on Pasture, A. M. Nicol, ed. Occasional Publication No. 10. Hamilton, New Zealand: New Zealand Society of Animal Production.

Murphy, W. M. 1987. Greener Pastures on Your Side of the Fence: Better Farming with Voisin Grazing Management. Colchester, Vt.: Arriba Publishing.

Murphy, B., B. Brigham, J. Brigham, A. Cleaves, D. Lockhar, and B. Lockhart. 1989. Voisin controlled grazing management: A better way to farm. A 33-minute video. Charlotte, Vt.: Perceptions. (Available from the Department of Plant and Soil Science, University of Vermont, Burlington, Vermont 05405.)

Pillsbury, B., and P. Burns. 1989. Economics of Adopting Voisin Grazing Management on a Vermont Dairy Farm. Winooski, Vt.: Soil Conservation Service.

Voisin, A. 1959. Grass Productivity. New York: Philosophical Library.

# Appendix B

# Expert Systems: An Aid to the Adoption of Sustainable Agriculture Systems

*Edwin G. Rajotte and Timothy Bowser*

Agricultural production has evolved into a complex business requiring the accumulation and integration of knowledge and information from many diverse sources, including marketing, horticulture, insect management, disease management, weed management, accounting, and tax laws. This is especially true of emerging sustainable practices that require even more information (to substitute for purchased inputs) for implementation. Very seldom do farm managers have all available information in a usable form at their disposal when major management decisions must be made. Increasingly, the modern grower must become expert in the acquisition of information for decision making to remain competitive. However, integration and interpretation of information from many sources may be beyond the means of individual growers, so they use the expertise of agricultural specialists. Unfortunately, agricultural specialist assistance is becoming relatively scarce at the same time that the complexity of agriculture is increasing. To alleviate this problem, it is essential that current information be structured and organized into a system for easy access by growers and agricultural specialists. No organized structure is currently available for information storage and retrieval; consequently, technical information, both experimental and experiential, is often lost or unavailable to potential users. One way to make this information readily available is through the use of electronic decision support systems.

The development of an electronic decision support system requires the combined efforts of specialists from many fields of agriculture and must be developed with the cooperation of the growers who will use them. Specialists tend to be trained in rather narrow domains and are best at solving problems within that domain. However, there is a growing realization that the complex problems faced by growers go beyond the abilities of individual specialists. Interdisciplinary teams of specialists must work in unison to formulate solutions to agricultural problems. Agriculture must be viewed as a system of interacting parts where the perturbation of one part affects many others.

The acquisition and utilization of information can be considered a means of reducing the amount of uncertainty in a given decision problem (Hey, 1979). Because high-quality information has not been easily accessible to growers when they are faced with important management decisions, decision making on the farm has been surrounded by a high degree of uncertainty. To compensate for the large degree of uncertainty, farm managers have increased inputs of chemical pesticides and fertilizers in an effort to minimize the variability in yield and quality that can occur from year to year. The price of this strategy, however, is reduction in potential profit and an increased threat to the environment because of the overuse of fertilizers and pesticides.

One way to alleviate these problems in agriculture is to substitute high-quality interpreted information for purchased production inputs such as fertilizer, labor, and pesticides. By providing farm business managers with up-to-date, interpreted information, the risk of decision making is reduced, the application of unnecessary inputs is eliminated, and profits are increased. The problem faced by land-grant colleges of agriculture and other providers of agricultural information has been how to deliver accurate information to farm managers rapidly in an integrated, interpreted fashion. Fortunately, several technologies are now available that can help overcome this problem: (1) data bases that include geographic information systems, (2) expert systems, (3) decision analysis tools, and (4) electronic communication through computer systems and telephone lines. A complication of this solution, however, is the fact that the adoption of computer technology by growers is predicated on a linkage between a particular farm operation and the access conditions of the particular technology (Audirac and Beaulieu, 1986). These access conditions are determined, in part, by the development of the technology and by private and public diffusion infrastructures. The development of diffusion strategies that consider growers' needs and capabilities relative to specific access conditions will accelerate the adoption of these new technologies.

## DESCRIPTION OF EXPERT SYSTEMS

This discussion concentrates on defining expert systems, describing the development of an apple production expert system, and reporting some of the reactions of commercial apple growers to this new information delivery technology. An expert system(s) is a computer program designed to simulate the combined problem-solving capabilities of a number of people who are experts in specialized disciplines or domains (Coulson and Saunders, 1987; Denning, 1986). Expert systems are able to draw and store inferences from information and are thus often called knowledge-based systems. A form of artificial intelligence, expert systems are capable of integrating and delivering quantitative information, much of which has been developed through basic and applied research, as well as heuristics (experientially based rules of thumb) to interpret quantitatively derived values, or for use when quantitative values do not exist.

Expert systems technology can be used as a delivery mechanism in a larger decision support system. By computing sequences of symbols that represent different levels in the solution of a problem, the expert system attempts to represent a common problem-solving pattern: "if conditions, then consequences" (Denning, 1986; Rajotte, 1987). Moreover, because an expert system remembers its logical chain of reasoning, a user may query the system about why a particular recommendation was given.

In agriculture, expert systems can be used to integrate the perspectives of individual disciplines (e.g., agronomy, horticulture, entomology, ecology, and economics) in a fashion that addresses the day-to-day, ad hoc decision-making processes required of modern farmers. Developed correctly, expert systems can become a powerful tool for providing farmers with the readily accessible, highly integrated decision support they need to practice a sustainable system of farming.

Unlike many industrial applications, most expert systems for agricultural production management are still in the developmental and testing phases (Schmisseur and Doluschitz, 1987). This chapter describes the creation of an expert system for apple production and provides the results of the first widespread field testing of expert systems by growers. Unlike most studies, this research has implemented an evaluation plan simultaneously with the beginning of adoption of the system. Thus, some of the problems with earlier research, such as lack of baseline data and the potential confounding of management ability and adoption (Wetzstein et al., 1985) can be avoided. The purpose of this study is to document the socioeconomic impact of expert systems in terms of changes in knowledge, skills, attitudes, and practices.

## DECISION MAKING FOR APPLE ORCHARDS

Apple orchards are highly diversified and complex ecological, economic, and social systems. Apple production is affected by a wide variety of insect, mite, disease, weed, and mammalian pests and is subject to the same economic and social constraints as any agricultural business enterprise. Moreover, orchardists are experiencing increased pressure from environmental and consumer groups to reduce their chemical use, particularly pesticides.

Apple producers have a need to utilize various sources of state-of-the-art agricultural knowledge as well as site-specific, on-farm information in a highly integrated fashion to reduce pesticide use and improve farm productivity and profitability. Alternative methods of pest management in apple production are needed in the face of increasing pesticide resistance and concerns about food safety and human health. The case for implementing integrated pest management (IPM) programs in apple production as one strategy to meet these requirements has been made previously (Rajotte et al., 1987).

However, the best means for effectively implementing IPM programs and other sustainable agriculture practices for widespread adoption are still being discovered. To overcome the initial complexities of converting to IPM, growers require more education, experience, and technical expertise. In addition, orchardists are confronted with an overwhelming amount of information that they need to assimilate in order to make decisions about production, harvesting, and the control of insects, diseases, and weeds. Traditional agricultural information and decision support delivery systems are discipline-oriented packages. Thus, growers must often integrate various disciplinary information and data for application to their own orchards (Rajotte et al., 1987). Rarely, if ever, do apple growers have the time or resources to compile and effectively assimilate all the required information involved in the daily decision-making process. An apple production expert system can provide an improved level of decision support in a timely and integrated fashion whenever and wherever growers require it.

## THE PENN STATE APPLE ORCHARD CONSULTANT

An expert system known as the Penn State Apple Orchard Consultant (PSAOC) has been developed to help apple growers make better decisions about production and pest management. After 4 years of development and testing (including 2 years supported in part by a U.S. Department of Agriculture [USDA] low-input sustainable agriculture [LISA] grant), this system has recently been made available for sale to fruit growers in Pennsylvania through Penn State Cooperative Extension (Travis et al., 1990). The system integrates various facets of apple production. It gives the apple grower the information necessary to reduce some purchased inputs by substituting high-quality, integrated, information derived from three sources (state-of-the-art apple production and IPM knowledge;

site specific, farm level data; and weather records). A primary emphasis of the PSAOC expert system is to decrease the detrimental environmental impacts associated with pesticide and fertilizer use as well as input costs, thereby improving farm profitability and reducing economic risk.

PSAOC was designed to view the apple orchard from an ecological perspective as a complex and highly interdependent system where the alteration of one component results in changes in the entire system. The system mimics the way in which growers must approach problem-solving in their orchards. The goal is to consider the orchard as a whole organism, and to make management recommendations in a holistic fashion, rather than making individual recommendations based on independent components (Heinemann et al., 1989).

Two unique characteristics of the PSAOC program are (1) the relative user friendliness of the system, and (2) a built-in user feedback loop that facilitates the incorporation of grower and user suggestions for improving the system into updated versions of the program (Heinemann et al., 1989). The two versions of the PSAOC system, Macintosh (Apple Computers, currently available) and DOS (available in 1991), were designed so that a person who has never used a computer may operate it. Operation of the system can be accomplished without using the keyboard in the Macintosh version. Growers' use of the system is being continuously monitored and evaluated, which allows them to have direct input into how the system is being developed. The software shell being used (PennShell) allows modifications to be made quickly so that updated versions can be rapidly distributed to growers.

Developers of PSAOC felt that these two components (user friendliness and user feedback loop) were critical to attaining the goals (Bowser, 1990; Heinemann et al., 1989). These two components contribute prominently to the ability of growers to input into the system data specific to their own orchards as well as up-to-the-minute weather data. With these baseline data in the system, growers may query PSAOC about specific problems of pest management, soil fertility, and orchard planting. They may also request in-depth supplementary information (including pictures) about an individual insect, disease, or weed. The user may ask the system to explain the logic behind a given recommendation (Bowser, 1990; Crassweller et al., 1989; Heinemann et al., 1989).

Recommendations are usually given with a range of alternatives (where alternatives exist), thus allowing growers to combine their own preferences and experiences with the recommendation being offered by the system. This combined package of information is then used to support the decision-making process of the grower in planning a pest management or other strategy.

## Structure of the Penn State Apple Orchard Consultant

When the PSAOC expert system was first introduced to growers in the field test it consisted of three main components: insects, diseases, and horticulture. Since each program fits onto one disk, a top level calling module provided a main menu to call each of the three main modules. In the most recent version, the insect and disease module were fully integrated into an IPM module. PSAOC is further divided into profiles (long- and short-term memory of an orchard block) and various decision modules that utilize recent orchard observations.

*Profiles*

The apple producer's orchard management program is based on orchard blocks. A block is the largest unit of an orchard within which consistent decisions are made (generically known as a management unit). A typical orchard may consist of several blocks that are each managed differently. Information about the block is stored in two separate files, called long-term and short-term profiles, and each block has its own profiles. The use of profiles eliminates the need for the grower to repeatedly enter information about the orchard that changes infrequently. The long-term profile consists of details about the orchard block that would not change from day to day. For example, the location of the block will not change at all. The tree varieties in each block, the ages of the trees, and the history of insect problems remain fixed for an entire growing season. Projected harvest dates usually remain fixed until the end of the growing season, when they may be adjusted. The short-term profile contains information that either needs updating on a more frequent basis or else has the potential for changing. For example, weather history data that need daily updating are kept in this profile. Crop load and market destination may change because of a number of environmental factors that alter the quantity and quality of the crop.

Information (besides weather) that changes from hour to hour within a day must be entered by the user at the beginning of a new session and is not stored in a profile. For instance, disease incidence and insect and mite population changes may be assessed as often as once a day.

The management program either can be initiated directly from the profile, in which case all profile information will automatically be loaded into the program, or else the user will be asked if a profile should be loaded. The user either can choose a previously defined profile or the user can create a new one.

*The Integrated Pest Management Module*

The user has the option of requesting a recommendation about an individual pest problem or running the IPM module, which considers all the orchard and pest characteristics as an integrated system when the management of each component will affect other components.

By considering site characteristics, horticultural parameters, weather conditions, pest severity, and predator density, for example, the program determines whether the insect and mite populations are over thresholds that signal the need for action to control these pests. It then calls a pesticide management module to establish pesticide application priorities. With the help of the expert system, the user then builds a recommendation by considering pesticide efficacy and appropriateness, timing, days to harvest, and tank compatibility. For instance, if the mite population is over the threshold level and predators are not sufficient to control the mites, miticide rates are determined. These rates will vary depending on the severity of the problem. Insecticide rates are then determined for the primary insect over the threshold level (i.e., most damaging). If the primary insect control is effective for all secondary insects, no more insecticide compounds will be considered. Otherwise, the module will determine other compounds and rates to control the secondary insects. Steps similar to those described in the preceding paragraph are taken to determine the disease control recommendations.

The program has now determined an array of miticides, insecticides, and fungicides that will control the pest problems in the orchard block. The array of pesticides is then checked against the days-to-harvest rules. Certain pesticides cannot be applied within a certain period of time before harvest, and that period varies between materials. The program checks the current date and the estimated harvest date and then eliminates any materials that are illegal to use during that time. Most growers mix pesticides into a single tank application. The final filter for the pesticide array is to determine tank mix compatibility between pesticides. Any incompatible chemicals are removed from the array. The user is given a choice of selecting from a list of the remaining pesticides.

Rates for the chosen pesticides are displayed on the computer screen. The program generally recommends a tank mix of a fungicide to control diseases, a miticide to control mites, a primary insecticide to control the most damaging insects, and a secondary insecticide to control any insects that are over threshold but that are not controlled by the primary insecticide. After reviewing the pesticides and rates, the user has the option of asking for a different combination of pesticides for the same pest problems. This option is offered because there are many pesticide combinations that may be suitable.

## PSAOC as a Tool for Sustainable Apple Production

Effective use of PSAOC provides growers with specific, IPM-oriented information that they may not have had in a usable form previously. This information may tell the grower that certain insect pests are present, but not at economically threatening levels that require application of a pesticide, or that conditions for a disease infection period have not been met, even though it is the proper season for disease infections. This information is substituted for the routine spraying practices that might have occurred without this knowledge. Thus, the ecosystem is spared the application of unnecessary pesticides, while the grower realizes an economic savings derived from not applying pesticides. Moreover, the yield and quality of the crop is maintained because pest problems are managed with a profitability objective.

PSAOC is a potentially effective tool for sustainable apple production for six reasons:

1. it delivers IPM-derived information and solutions to pest management problems, the benefits of which are outlined above;

2. it provides this information in a very up-to-date and site-specific fashion that is unattainable by traditional information delivery systems;

3. this information is always readily available to any grower with access to a computer and the software, relieving dependence on the accessibility of literature or human experts, thus enabling the grower to make critical, timely decisions whenever necessary;

4. when used effectively, it provides the apple grower with the opportunity to reduce the usage of chemical pesticides, thus reducing the negative impacts of apple production on the ecosystem and human health;

5. it can increase grower profits; profitability is an essential condition for sustainable agriculture; and

6. as additional low-input sustainable methods of production are developed, these can be easily incorporated into PSAOC.

It remains to be seen whether apple producers will successfully adopt this new agricultural innovation on a widespread basis. To address this question, a field test and evaluation of the expert system was conducted during 8 months of the 1988 and 1989 growing seasons. Some of the results of this field research are presented below.

## Field Testing the System

During regular extension educational meetings in 1988, apple growers were asked to volunteer for on-farm field testing of the expert system. Over 140 growers volunteered to participate in the first phase of the evaluation.

Of those volunteers, 26 apple growers were selected as a pilot test group. These growers were carefully selected to represent the spectrum of apple production characteristics in Pennsylvania, including farm size, geographic location, and experience with computers. These pilot test participants met with the study organizers for 1 day and were given instructional training and software. Fourteen growers who did not own computers were loaned Macintosh computers.

Growers agreed to use the system and to record their experiences with the system, suggestions for its improvement, and their usage patterns. A monthly telephone survey was used to collect the data being generated by the pilot group. Some results are discussed here. For a more complete discussion see Bowser (1990).

### Grower Surveys:  System Use and Practice Change

In this section grower usage of the PSAOC expert system is discussed, as are changes in farming practices resulting from this use of the system.

*General System Usage Patterns*

The number of times a grower uses the PSAOC expert system and the amount of time it is used during each session are indicators of the degree of adoption of the expert system. Table B-1 displays two measures of the

**TABLE B-1**   Penn State Apple Orchard Consultant
Expert System Use Characteristics of Growers

| System Use Characteristics | Percentage of Growers ($n = 26$) |
|---|---|
| Total no. of times system accessed by growers in 8 months | |
| 0 | 7.7 |
| 1–9 | 19.2 |
| 10–15 | 34.6 |
| 16–29 | 15.4 |
| 30–110 | 23.1 |
| Total no. of hours system used by growers in 8 months | |
| 0 | 7.7 |
| 1–3 | 26.9 |
| 4–6 | 15.4 |
| 7–9 | 23.1 |
| 10–40 | 26.9 |

frequency of use of the expert system: the total number of times that individual growers accessed PSAOC, and the total number of hours they used the system. Both measures are summations of the data from an 8-month period in 1988 and 1989 during which the study data were collected.

The first measure, the number of times that the growers accessed the system, represents the number of times an individual grower actually turned on and used the system, regardless of the duration of the session. This measure shows that 7.7 percent of the growers did not use the system at all, 53.8 percent of the growers used the system less than 16 times in 8 months (2 times per month), and 23.1 percent used it 4 times or more each month.

The second measure in Table B-1 represents the total number of hours that the system was actually used by the growers during those 8 months. Again, 7.7 percent did not use the system at all, 42.3 percent of the growers used it for less than 6 hours, and 26.9 percent used it 10 hours or more.

Total use of the system varied widely by year and time of year. Figure B-1 shows the percentage of growers who accessed the system each month. This variation is explained in two ways. The growers did not receive the system for use until late July 1988. A very high percentage of growers accessed the system during August 1988 (73.3 percent) because they were trying it for the first time. Use of the system in August 1989 was 3 1.8 percent, which more accurately reflects the need for information a grower would have just prior to harvest. The percentage of growers who

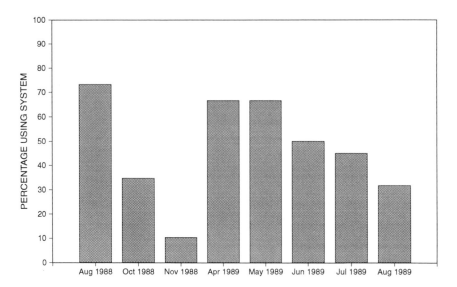

FIGURE B-1  Percentage of growers who accessed the Penn State Apple Orchard Consultant expert system each month.

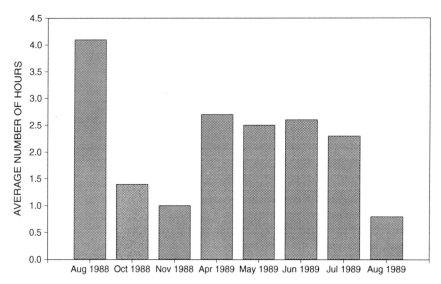

FIGURE B-2 Average number of hours per month the Penn State Apple Orchard Consultant expert system was accessed by growers.

used the system fell precipitously during October (34.8 percent) and November (10.3 percent) of 1988. During 1989, after the growers had the opportunity to review the system throughout the winter months, system use was high in the spring months. The spring is traditionally an intensive period of pest management because of favorable conditions for fungal and bacterial diseases caused by wet conditions. In addition, insect and mite populations begin to increase in the spring and are therefore more vulnerable to management actions. System use gradually decreases throughout the summer, as would be expected based on the declining informational needs of the growers.

Figure B-2 shows the average number of hours of use per month by growers who accessed the system. A pattern of variation similar to that described above occurred. On average, growers used the system fewer times but for longer durations earlier in the growing season than they did later in the growing season. This may be explained by the differences in types of information needed at different points in the growing season. Earlier in the season, growers were more involved in planning and scheduling for the season's work, which required more intensive and in-depth use of information sources. More importantly, pest problems (especially diseases) are much more complex in the spring than in the summer, requiring more time on the computer to extract a recommendation. During the summer months, growers are more involved in crop maintenance and troubleshooting and may be doing more of the double checking of their own knowledge mentioned above.

These measures of system use taken together indicate one aspect of adoption: use of the innovation. While the number of times that the PSAOC system is accessed shows how frequently the system is being used, the actual amount of time spent using the system may be a more significant indicator of adoption of the innovation. Some growers reported that they used the system primarily as a quick validation of their own knowledge regarding a decision. These growers reported a relatively high number of accesses and a low number of hours used. Conversely, the growers who reported that they used the system for many hours were presumably more fully engaging the logic of the system in their decision-making process.

*General Practice Change Characteristics*

The degree to which growers follow the recommendations presented by the expert system is a second aspect of adoption. Table B-2 displays two measures of the frequencies of changes induced by use of the system: (1) any change in growers' production practices and (2) increased pest monitoring. Both measures were derived from the eight monthly surveys.

The first measure is a sum of the number of times that growers indicated that use of the expert system stimulated some change in their production practices. Over the course of the 8 survey months, 65.2 percent of the growers indicated that they had changed standard production practices in some way during at least 1 month. Of these growers, 17.4 percent indicated some change during 3 different months of the 8 survey months.

A significant number of those sampled (65.2 percent) engaged a new and untried technology and were stimulated to change production practices as a result.

The second of the practice change characteristics displayed in Table B-2 is a sum of the number of times that a grower was stimulated by the expert system to go to the orchard and scout for a pest (monitoring). Pest monitoring is seminal to any IPM program. A large majority of growers (82.6 percent) reported that the system stimulated them to increase their monitoring at least once. A total of 30.3 percent of growers were stimulated to monitor their orchards four or more times. As the majority of pest monitoring occurs during April, May, and June, these numbers take on more significance when viewed as a subset of the eight monthly observations.

## Weekly Time Monitoring and Basic Economic Questionnaires

During the field test and evaluation process in the 1989 season, the economic impact of the apple expert system on cooperators' operations and net

**TABLE B-2**   Penn State Apple Orchard Consultant
Expert System Adoption Characteristics of Growers

| Production Practice Change Characteristic | Percentage of Growers ($n = 23$) |
|---|---|
| No. of times growers reported some change in practices, per grower | |
| 0 | 34.8 |
| 1 | 21.7 |
| 2 | 26.1 |
| 3 | 17.4 |
| No. of times system stimulated increased pest monitoring, per grower | |
| 0 | 17.4 |
| 1 | 26.0 |
| 2 | 4.4 |
| 3 | 21.7 |
| 4 | 21.7 |
| 6 | 4.4 |
| 7 | 4.4 |

income was estimated. Many growers already maintain pesticide logs that
contain most of the data needed for development of an apple enterprise
budget. A basic economic survey questionnaire was developed from the
pesticide record and crop history logsheet of a major commercial apple
processor to collect orchard characteristics, apple yields, and prices received. Additional information to aid in the comparison between expert
systems users and a control group of nonusers was incorporated into the
questionnaire. A weekly time monitoring survey was designed to gather
information on the amount of time each grower spent scouting (monitoring) his or her orchard each week as well as what pest problem was
being looked for. Pesticide application records were also collected to provide information on the chemicals and rates that the chemicals were applied to each orchard. The survey questionnaire was subjected to three
reviews: first, by the research team; next, by all the county agents involved
in the project; and finally, by selected growers who had expressed interest
in its development. This feedback was particularly helpful for developing the yield and price components of the questionnaire, which was a two-
part format that was collected in the spring and the fall.

Results from the monitoring surveys are still being analyzed. While the
findings reported here are preliminary and subject to change, they, too,
indicate that the expert system is an effective teaching tool. In the past,

extension information has encouraged growers to monitor for mites at the time of bloom and thereafter (week 8 of the growing season). Both PSAOC users and nonusers performed scouting at similar frequencies in the post-bloom period. However, a new prebloom monitoring practice is recommended by the expert system as an effective mite control strategy that may reduce pesticide usage later in the season. The nonusers of PSAOC were not as aware of this prebloom method. Figure B-3 shows that more PSAOC system users tended to monitor for European red mites earlier in the season than did the comparison group of nonusers. Similar behavior has been seen in PSAOC users who ended their monitoring processes sooner than did the control group, thus making more efficient use of limited time. This constitutes direct evidence that use of an expert system can stimulate measurable changes in farming practices.

A preliminary comparison of the farm-gate economics of expert system users versus those of expert system nonusers shows some trends. Even though Pennsylvania suffered through a poor apple-growing season in 1989, the preliminary results of the survey show that yields of PSAOC users and nonusers were roughly similar.

The cost of time spent monitoring the orchard for pests and using the expert system is also a component of the economic impact being examined. Specifically, the team is looking to answer the question of whether savings

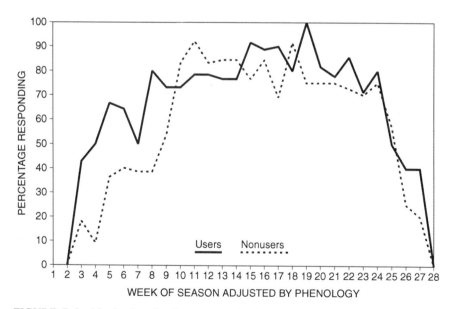

FIGURE B-3 Monitoring for European red mites (ERM) by users and nonusers of the Penn State Apple Orchard Consultant expert system.

on pesticide applications were being offset by greater costs in management. A weekly time-monitoring survey was developed that provides a checklist for most of the common items monitored. Primarily, it asks how much time was spent monitoring each block and using the expert system. This checklist went through the same review process the basic economics survey did.

No clear results have yet been obtained from the pesticide records analyzed thus far, but some interesting trends have been noted. There is some indication that system users may have applied lower amounts of some insecticides than nonusers did. Further analysis of this information may indicate whether or not the expert system is changing growers' practices regarding pesticide use and will provide the basis for partial budget analysis.

### Further Mechanisms to Obtain Grower Evaluation, Feedback, and Training

*Cooperators' Planning and Review Meetings*

The experiences with the PSAOC expert system during the 1988 and 1989 growing seasons were summarized during facilitated meetings of cooperating growers, researchers, and extension personnel in February 1989 and March 1990, respectively. The primary purposes of the meetings were to review the system's performance over the year to date, provide the growers with an opportunity for in-depth input and discussion about improvements in the program, and collectively plan for the upcoming year. In addition, a major benefit was to bring growers from 13 counties in Pennsylvania and researchers and extension agents from three states together to interact for the first time.

The nominal group technique was employed during working sessions with the growers group to solicit any suggestions that they had for improving either the software itself or the field evaluation process. Recommendations were distilled and ranked by growers according to importance during a later session.

Growers and extension agents also strongly suggested the inclusion of more economic information into the PSAOC expert system. A session devoted to procedures for collecting relevant budget data yielded an additional step in the proposed analysis of farm-level economic impacts.

Researchers and extension specialists from The Pennsylvania State University (University Park), University of Massachusetts (Amherst), University of Vermont (Burlington), and the Rodale Research Center (Maxatawney, Pennsylvania) also met for 1.5 days to plan and coordinate the following year's program. Additional responsibilities for expert systems development and evaluation were outlined for the second- and third-year plans of work.

## Midseason Grower Training Sessions

Based on feedback from growers as well as trends in the survey data, small-group training sessions were held at the Biglerville Fruit Laboratory and the Berks County Agriculture Center during the summer of 1989. It was determined that the newest version of PSAOC was not being comprehended adequately and therefore was not being used to its fullest efficiency. These training sessions sought to correct this problem by familiarizing the growers in-depth with the new aspects of the software.

## Electronic Mail Network Among Growers and Researchers

Also in response to feedback from growers, an electronic mail users group was formed to improve communications between cooperating growers, researchers, and extension personnel. Using The Pennsylvania State University's PenMail system, the growers are able to communicate with each other, with county extension agents, and with specialists on campus via electronic mail. This communications link has helped to make growers more comfortable with the computer and the information they receive.

The electronic mail system was set up in March 1989. Grower communications have included questions about insects, pest trapping, use of the computer, and information on the new version of PSAOC. The project's evaluation coordinator has sent out numerous informational and update bulletins. The growers are also receiving their own copy of the state horticultural newsletter by electronic mail. Half of the growers have accessed the system (for messages, responses, PenMail) roughly once a week, and the others have accessed the system about once a month. This system has worked well so far, and it is expected that usage will continue to grow.

## Site Visits to Cooperating Orchards

Visits to field test sites were made by evaluation staff at various points during the growing season, to observe orchard management and expert systems use by grower. These visits also provided more opportunity for the growers to give input into the development and improvement of PSAOC. It was noted that the expert system was more often found in the business office of the orchard, residing on the computer the grower used for accounting.

## Grower Panel at Professional Meetings

Three pilot study growers and the cooperating regional tree fruit extension agent presented a panel discussion on the Penn State Apple Orchard Consultant system to over 300 apple growers at a meeting of the Pennsylvania Horticultural Association in January 1990. Discussants provided insights into their experiences with testing of the system, citing both the problems and potentials of using the expert system in orchard management. Panelists were mostly supportive of the new technology, citing increased responsibility on the part of the grower to reduce environmental inputs and improve food safety while still maintaining profitability.

### Involvement with Cooperative Extension Agents

Cooperative extension agents were directly involved in the organization and implementation of the project. In addition to consulting on the structure and content of the survey process, agents were primarily responsible for the selection of cooperating growers for the project.

#### *County Extension Agents Survey on Expert Systems for Fruit Growers*

A survey was distributed by electronic mail in January 1989 to measure the familiarity of county extension agents with fruit expert systems and to solicit feedback on the overall expert systems program. The survey was necessary for two reasons: (1) many extension personnel were not informed about expert systems development, thus indicating some training sessions were necessary; and (2) feedback was received that indicated agents in cooperating counties could be better served and utilized by the evaluation process.

The survey was sent by PenMail to agents with horticultural responsibilities in all 67 county extension offices in Pennsylvania. Additional questions were asked of agents in the counties where growers were cooperating in the pilot study to solicit feedback on improvements to the evaluation process.

A vast majority (84 percent) of county extension agents were at most only somewhat familiar with expert systems for fruit production. Seventy-six percent of agents indicated that they would attend an in-service training program on how to use this technology in their programs.

#### *Extension Agent Expert System Training Session*

In response to feedback from county extension agents, training sessions for county extension personnel were scheduled during the March extension in-service training programs at The Pennsylvania State University. Agents participated in a lecture and discussion of what expert systems are and how they work. In another session, participants received hands-on experience with expert systems in a computer laboratory. This training was provided to help familiarize agents with expert systems and to lay the groundwork for the future diffusion of agricultural expert systems.

## Local Experts Network

A proposal has been made to extension administration to initiate a network of extension agents to serve as local experts to support expert systems users within a specified region. The local expert is a person who learns a new technology quickly and is motivated to help others learn it (Landy et al., 1987). Scharer (1983) suggests that the individual is central to the ultimate success of the training effort. This process, which is often used in the diffusion of software technologies, provides a more rapid response to user problems and educational needs than is currently available through Cooperative Extension programs. It is expected that this network will facilitate a more efficient and effective adoption process.

## CONCLUSIONS

The project reported here is the first in the literature of an agriculture-oriented expert systems being tested in the field with comparisons of user and nonuser practices. Evidence from this study supports the thesis of Audirac and Beaulieu (1986) that the access conditions of a technology need to be considered in the diffusion process. Those access conditions of the expert system derived from its technological development as well as its intrinsic characteristics are important variables in the diffusion process. In particular, two characteristics seem noteworthy based on the results of this study.

First, the Penn State Apple Orchard Consultant expert system is primarily an information delivery technology. While it contains data base production information (such as weather), it also requires the input of reliable, site-specific information in order to formulate recommendations for the user. The information requested as well as the resultant recommendations require the apple producer to form questions and to look at problems in a manner different from that of previous information delivery systems used in apple production. That this transition will not occur automatically is reflected by the fact that the test group exhibited various levels of use and that almost none of the changes in practices occurred until growers had sufficient time to develop some familiarity with the system's logic. Some growers indicated that they still do not trust the system to make decisions for them. This attitude is appropriate. PSAOC is not intended as a substitute for good management but as a source of information to guide and enlighten growers' decisions. Distrust of the PSAOC expert system could also be the result of incongruence between growers' perceptions of the system versus those of their apple orchards. The expert system is an information technology that is intrinsically different from most information technologies that have previously been used by apple producers. The kinds of practical and educational experience a grower or user has may affect how well the system is understood and, thus, adopted.

Second, the expert system is a technology that is inherently connected to microcomputers. For a grower to make use of the decision support capabilities of PSAOC, they must (1) have access to a microcomputer capable of running the system and (2) be able to operate the computer proficiently. While the software was designed and developed to be used by people with little or no computer experience, results of the study indicate that growers with the least amounts of computer experience also had the lowest rates of system use. This would appear to be an example of the access conditions of the technology not being congruent with the farming operation. This technology is inherently computer based, and a farming operation must have access to a computer and a person who can operate it before the technology will be adopted.

By substituting information for some chemical inputs, the Penn State Apple Orchard Consultant expert system has the potential to contribute to the generation of more sustainable apple production systems in the northeastern United States. This trend can accelerate through the introduction of more information-intensive, low-input IPM practices into the farm production system. This study has provided some preliminary evidence that changes in usual production practices occur as growers and users substitute information for purchased inputs, in this case, pesticides. It was also demonstrated that the substitution of information for inputs was stimulated by the expert system, which enabled the grower or user to collect, integrate, and interpret the information rapidly. However, based on other evidence produced by the study, it appears that the potential for sustainable agriculture that this technology holds will be diminished without some attention to better linking of the access conditions of the technology to the farming operation.

## RECOMMENDATIONS

More work will need to be done at the first stage of the diffusion process if the Penn State Apple Orchard Consultant is to become an effective tool for sustainable agriculture. This first stage concerns the set of activities which provide for the "establishment of diffusion agencies or a network of outlets from which the innovation is distributed to potential adopters" (Audirac and Beaulieu, 1986, p. 63).

In the present case, it is planned that this diffusion network will be the traditional Cooperative Extension Service network of university and county extension offices and personnel. In addition to acting as the distributive agent for this innovation, this network must also provide new educational training programs in key areas identified by this research, if the effective adoption of this innovation and its potential for sustainable agriculture are to be realized. Some growers are not using the system very often, and others are not being stimulated to change production practices based on their use of the system. In some of these instances, perhaps no change is necessary or advisable. In other instances, change would be highly beneficial in terms of grower profits and reduced pesticide use. In the latter case, effective adoption is not occurring and the potential to reduce the amount of pesticide inputs being used is diminished.

To correct this situation when the system is offered for general use by growers, it is recommended that the diffusion agency provide new educational programming in the following areas:

1. training in and basic orientation to computer use for farming operations in general and agricultural expert systems in particular; these training sessions should be held on a very localized basis and taught by people who are familiar with expert systems software and the cropping system being discussed;

2. training that provides an overview of the gradual modification of existing production systems to incorporate reduced-input methods; this training should focus on societal-level needs and responsibilities for reducing pesticide use as well as the long-term farm-level benefits for doing so;

3. establishment of a network of local experts to provide a resource for growers experiencing difficulties with the computer or expert system;

4. continual updating of system capabilities, so that recommendations remain scientifically current and appropriate;

5. training of extension specialists and agents to familiarize them with the possibilities and potentials of the system; and

6. beginning the process by delineating the criteria and goals for sustainable agriculture attainable with expert systems as a tool. In this way scientists will be better able to begin to design production systems for agricultural operations of all sizes that provide more flexibility in responding to dynamic production conditions, thus enabling time and spatially specific recommendations for the expert system to be better implemented. In the long run this may be the greatest contribution of agricultural expert systems development toward a more sustainable system of global agriculture.

## ACKNOWLEDGMENTS

The Penn State Apple Orchard Consultant expert system described here was developed by J. Travis and K. Hickey, Department of Plant Pathology;

E. Rajotte and L. Hull, Department of Entomology; R. Crassweller, Department of Horticulture; P. Heinemann, Department of Agricultural Engineering; and R. Bankert, V. Esh, J. Kelley, and C. Jung, Integrated Pest Management computer programmers. Program evaluation was conducted by J. McClure and T. Bowser, Department of Entomology; C. Sachs, W. Musser, and D. Laughland, Department of Agricultural Economics and Rural Sociology; and W. Kleiner, Pennsylvania State University Cooperative Extension. Cooperators from other institutions include L. Berkett, Department of Plant Pathology, University of Vermont; D. Cooley, Department of Plant Pathology, University of Massachusetts; and S. Wolfgang, orchard leader, Rodale Research Center. Partial support for this work was provided by LISA project LNE88-8, "Implementation of Electronic Decision Support System for Apple Production."

## REFERENCES

Audirac, I., and L. J. Beaulieu. 1986. Microcomputers in agriculture: A proposed model to study their diffusion/adoption. Rural Sociology 51(1):60–77.

Bowser, T. 1990. Adoption of Expert Systems by Apple Growers: A Test of a New Model. Unpublished master's thesis. Pennsylvania State University, University Park, Pa.

Coulson, R. N., and M. C. Saunders. 1987. Computer-assisted decision-making as applied to entomology. Annual Review of Entomology 32:415–437.

Crassweller, R. M., P. H. Heinemann, and E. G. Rajotte. 1989. An expert system for determining apple tree spacing. Hortscience 24(1):148.

Denning, P.J. 1986. The science of computing: Expert systems. American Scientist 71:18–20.

Heinemann, P. H., E. G. Rajotte, J. W. Travis, and T. Bowser. 1989. An expert system for apple orchard management. Paper presented at the 1989 International Meeting of the American Society of Agricultural Engineers and the Canadian Society of Agricultural Engineering.

Hey, J. D. 1979. Uncertainty in Microeconomics. New York: New York University Press.

Landy, F. J., H. Rastegary, and S. Motowidlo. 1987. Human-computer interactions in the workplace: Psychosocial aspects of VDT use. In Psychological Issues of Human Computer Interaction in the Work Place. Amsterdam: Elsevier/North-Holland Science Publishers B.V.

Rajotte, E. G. 1987. A reflective decision support system for Pennsylvania agriculture: Merging electronic information sources, artificial intelligence, and field experience. Agricultural Economics and Rural Sociology Staff Paper No. 144. University Park, Pa.: The Pennsylvania State University.

Rajotte, E. G., R. F. Kazmierczak, Jr., G. W. Norton, M. T. Lambur, and W. A. Allen. 1987. The national evaluation of extension integrated pest management (IPM) programs. Virginia Cooperative Extension Service Publication No. 491-010. Blacksburg, Va.: Virginia Cooperative Extension Service.

Scharer, L. L. 1983. User training: Less is more. Datamation 175–236.

Schmisseur, E., and R. Doluschitz. 1987. Expert systems insights: Future decision tools for farm managers. Journal of the American Society of Farm Managers and Rural Appraisers 51(2):51–57.

Travis, J., K. Hickey, E. Rajotte, L. Hull, R. Crassweller, R. Bankert, P. Heinemann, V. Esh, and C. Jung. 1990. Penn State Orchard Consultant. University Park, Pa.: The Pennsylvania State University.

Wetzstein, M. E., W. N. Musser, D. K. Linder, and G. K. Douse. 1985. An evaluation of integrated pest management with heterogeneous participation. Western Journal of Agricultural Economics 10(2):344–353.

# Authors

**VIVIEN GORE ALLEN** is associate professor of agronomy at Virginia Polytechnic Institute and State University. Her research areas include forage management and forage systems with emphasis on soil-plant-animal interrelationships and use of high-forage diets in ruminant and equine nutrition. She earned a Ph.D. degree in agronomy from Louisiana State University.

**VERNON L. ANDERSON** Since 1979 Anderson has been the animal scientist at the North Dakota State University Carrington Research Extension Center located near Carrington. His research includes drylot beef cow/calf management with a special emphasis on ruminant animal and crop production that is both integrated and complementary. He earned his M.S. degree in animal science from South Dakota State University.

**CHARLES M. BENBROOK** is president of Benbrook Consulting Services, Dickerson, Maryland. From 1984 to 1990, he was executive director of the Board on Agriculture at the National Academy of Sciences. Since 1979, he has worked in Washington, D.C., on agricultural science, technology, food safety, and policy issues. He has a Ph.D. degree from the University of Wisconsin-Madison.

**DAVID A. BENDER** is associate professor of horticulture at the Texas Agricultural Experiment Station. His research interest is the culture and physiology of vegetables. He has a Ph.D. degree in horticulture from Virginia Polytechnic Institute and State University.

**MARK BÖHLKE** From 1987 to 1990, Böhlke was a research technician at Rodale Research Center. He plans to enter a graduate program in pharmacognosy to study active compounds obtained from medicinal plants. He earned an M.S. degree in biology from the University of Michigan.

**TIMOTHY BOWSER** is a senior research technologist in the Department of Entomology at the Pennsylvania State University (PSU). He earned an M.S. degree in rural sociology from PSU and has research interests in social impact assessment, technology transfer, and sustainable agriculture.

**MIKE BRUORTON** joined the University of Georgia Extension Service in July 1987 where he worked as county extension agent until July 1991. He is currently county extension director of Clinch County, Homerville, Georgia. He has a B.S. degree in agronomy from Clemson University.

**PATRICK M. CARR** is research associate at the Carrington Research Center in North Dakota. He has a Ph.D. degree in crop and plant science from Montana State University. His areas of research include multiple cropping and novel crop development.

**RAYMOND I. CARRUTHERS** Since 1985 Carruthers has been a re-search ecologist/entomologist with the U.S. Department of Agriculture at Cornell University, where he serves as a lead scientist. His research addresses the use of parasites, pathogens, and predators as biological control agents of insect pests in agriculture. He received his Ph.D. degree from Michigan State University where he began his studies on the interactions of insects and their fungal pathogens.

**R. JAMES COOK** is adjunct professor of plant pathology at Washington State University and project leader of the Regional Cereal Disease Research Laboratory, Agricultural Research Service, U.S. Department of Agriculture. His research interests are the biological control of soil-borne plant pathogens, water relations of soil microorganisms, and cereal root rots. He earned his Ph.D. degree in phytopathology from the University of California, Berkeley.

**RANDALL A. CULPEPPER** is a graduate of the University of Georgia with a B.S. degree in agricultural journalism. He has travelled extensively over the Southeast reporting on the state of agriculture. Most recently, he has been with the University of Georgia Department of Extension Horti-culture.

**W. LEE DANIELS** Since 1987 Daniels has been a professor of soil and environmental sciences at Virginia Polytechnic Institute and State University, where he earned his Ph.D. degree in soil geomorphology.

His research program focuses on soil disturbance, land restoration, plant-soil relationships, and soil organic matter interactions.

**DALE R. DARLING**  For 28 years, Darling has worked for E. I. DuPont de Nemours and Company, Inc. He is manager of Agricultural Associations. He holds an M.S. degree in plant pathology from Texas A&M University and has undertaken additional studies in plant physiology and biochemistry while working on a U.S. Air Force grant studying "Effects of Missile Fuels and Components on Plant Growth and Development, Soils, Soil Structures, and Aquatic Life."

**THOMAS L. DOBBS** is professor of agricultural economics at South Dakota State University where his research focuses on production economics and public policy aspects of sustainable agriculture. His Ph.D. degree is from the University of Maryland. Dobbs has written extensively on issues of economic development, resource economics, and farm management.

**LAURIE E. DRINKWATER** is a postdoctoral scholar in the Department of Vegetable Crops, University of California, Davis. She received a Ph.D. degree from the Department of Zoology at Davis. Her areas of research interest include agroecology, nutrient cycling in agricultural systems, and design of ecologically sound cropping systems.

**MICHAEL DUFFY** is an associate professor of economics and extension economist at Iowa State University where he works in farm management, natural resources—especially land—and sustainable agriculture. He holds a Ph.D. degree from the Pennsylvania State University.

**JOSEPH P. FONTENOT** is John W. Hancock, Jr., Professor of Animal Science at Virginia Polytechnic Institute and State University. His primary areas of research are forage utilization by ruminants, especially as related to sustainable agriculture, and use of biodegradable animal wastes as livestock feed.

**RAYMOND E. FRISBIE**  Since 1980, Frisbie has been professor of entomology and integrated pest management (IPM) coordinator at Texas A&M University. He has a Ph.D. in entomology from the University of California at Riverside. His areas of research and education are in IPM in cotton and other crops, insecticide resistance management, and farm-level decision support systems.

**JOHN C. GARDNER**  A research agronomist, Gardner is also superintendent of North Dakota State University's Carrington Research Extension Center located in central North Dakota near Carrington. As part of a university-nonprofit-farmer partnership, he is studying ecological aspects of crop and livestock production in the Northern Great Plains. Gardner earned his Ph.D. degree in agronomy from the University of Nebraska.

**STEVEN J. GULDAN** is a research agronomist with Michael Fields Agricultural Institute, East Troy, Wisconsin. He is presently working cooperatively with, and is based at, the Carrington Research Extension Center in North Dakota. Guldan is researching the incorporation of legumes into cropping systems. He was awarded a Ph.D. degree in agronomy from the University of Minnesota.

**E. SCOTT HAGOOD** is associate professor, extension weed scientist, and extension project leader in the Department of Plant Pathology, Physiology, and Weed Science at Virginia Polytechnic Institute and State University. Hagood is doing applied research for weed control in agronomic crops. He received his Ph.D. degree from Purdue University.

**JONATHAN H. HARSCH** This *FarmFutures* magazine contributing editor is a partner in the research and consulting firm of Hudson & Harsch. He earned his Ph.D. degree in literature at the University of Dublin, Ireland, and his current research projects include new industrial uses for crops—status of technology and commercial adoption.

**FLOYD F. HENDRIX, JR.** Since 1973, Hendrix has been professor of plant pathology with a joint research-extension appointment. His Ph.D. degree in plant pathology was awarded by the University of California, Berkeley. His research interests include developing information for integrated pest management (IPM) control of fruit diseases. His extension interests include combining this information with that from entomologists and horticulturists into an IPM production program.

**CHARLES E. HESS** Sworn in on May 22, 1989, Hess is assistant secretary for science and education, U.S. Department of Agriculture (USDA). He is responsible for USDA research and education programs in the food and agricultural sciences, including general supervision of the Agricultural Research Service, the Cooperative State Research Service, the Extension Service, and the National Agricultural Library. Hess earned his Ph.D. degree in horticulture from Purdue University.

**JEFF B. HILLARD** is assistant professor of plant sciences at Louisiana Tech University. He has a Ph.D. degree in soil science from Texas A&M University and conducts soil fertility research studies on forages and small fruits.

**DAN L. HORTON** is an associate professor in the extension entomology department at the University of Georgia. He earned his Ph.D. in entomology from the University of Arkansas. His primary interest is the application of integrated pest management principles to commercial fruit culture.

**KIRSTEN HURAL** is a doctoral candidate in entomology at Cornell University. Her research interests are in the ecological genetics of host-pathogen interactions.

**JOHN E. IKERD** is extension professor and coordinator of the Center for Sustainable Agriculture Systems, University of Missouri, Columbia. He has a Ph.D. degree in agricultural economics from the University of Missouri. His research and extension work emphasizes a holistic, systems approach to farm management and rural resource development.

**RHONDA R. JANKE** An agronomy coordinator at Rodale Research Center, Janke is also adjunct assistant professor of agronomy, Pennsylvania State University. She supervises cropping systems research, collaborates with farmer cooperators in the Rodale Institute On-Farm Research program, and conducts research in the area of weed ecology. She holds a Ph.D. degree in agronomy from Cornell University.

**ROBERT A. KLICKER** For 44 years, Klicker has been an active partner in Klicker Brothers & Sons farming operation. In addition to managing 15,000 acres, he has worked with agricultural researchers for 37 years. On-farm testing of his crops for the past 36 years has resulted—during the last 9 years—in the successful balancing of his soil for crops, tilth, and microbial life. Klicker introduced and implemented an artificial insemination program for beef cattle that is still being used after 29 years. He has been instrumental in forming a new organization called Progressive Farmers Inland Northwest.

**TERRY KLOPFENSTEIN** is Kermit Wagner Professor of Animal Science at the University of Nebraska. He earned a Ph.D. degree from the Ohio State University and has been on the University of Nebraska-Lincoln faculty since 1965. He has been conducting research on forage and crop residue utilization and protein nutrition of cattle. Recent emphasis has been on systems of beef production.

**GERARD W. KREWER** Since 1983, Krewer has been a University of Georgia extension horticulturist advising fruit growers and county agents. He earned his Ph.D. in plant physiology/horticulture from Clemson University. His areas of interest are the culture and management of blueberries, stone fruits, muscadine grapes, Asian pears, and Mayhaws.

**DAVID R. LANCE** Since 1985, Lance has been a research entomologist with the U.S. Department of Agriculture at the Northern Grain Insects Research Laboratory in Brookings, South Dakota. He earned his Ph.D. degree in entomology from the University of Massachusetts at Amherst. His interests include insect behavior and ecology and their application to pest management.

**LES E. LANYON** is an associate professor of soil fertility in the Department of Agronomy at the Pennsylvania State University. He earned a Ph.D. degree in agronomy from the Ohio State University. His area of interest is the plant nutrient dynamics of crops, farms, and agriculture.

**CURTIS A. LAUB** is a researcher in the entomology department at Virginia Polytechnic Institute and State University (VPI&SU). He is currently involved in low-input sustainable agriculture research, with emphasis on natural controls of field crop pests. He received his M.S. degree from VPI&SU.

**DEBORAH K. LETOURNEAU** is an associate professor of ecology in the Board of Environmental Studies, University of California at Santa Cruz. She earned her Ph.D. degree in entomology from the University of California at Berkeley. Her research interest is tritrophic interactions among plants, herbivores, and their natural enemies in natural and managed systems.

**JOHN M. LUNA** is an assistant professor in the Department of Entomology at Virginia Polytechnic Institute and State University (VPI&SU) and serves as extension coordinator for sustainable agriculture. He also is director of an interdisciplinary farming systems project. His research interests include habitat manipulation for managing weeds and insect pests, and the integration of cover crops into low-external input conservation tillage systems. Luna received his Ph.D. degree in entomology from VPI&SU.

**CLARENCE MENDS** Since 1980, Mends has been a research associate and marketing analyst in the economics department at South Dakota State University (SDSU). He has an M.S. degree in economics from SDSU and is a member of a multidisciplinary research and extension team involved in farming systems studies at SDSU.

**JANE MT. PLEASANT** is an assistant professor in the Department of Soil, Crop and Atmospheric Sciences at Cornell University. She earned a Ph.D. degree in soil science from North Carolina State University. Her area of specialization is in soil and cropping systems management.

**WILLIAM M. MURPHY** is a professor of agronomy in the Department of Plant and Soil Science at the University of Vermont. He has a Ph.D. degree in agronomy from the University of Wisconsin. His main research interest concerns detailed measurements and dynamics of grass-legume swards under controlled grazing management.

**ELIZABETH W. NEUENDORFF** Since 1982, Neuendorff has been a research associate at the Texas A&M University Agricultural Research and Extension Center, Overton. She has an M.S. degree in horticulture from Texas A&M University. Her areas of specialization include temperate and small fruit physiology and management systems.

**GARY NIMR** is a horticultural field technician for the Texas A&M University Agricultural Research and Extension Center at Overton. He earned his B.S. degree from Stephen F. Austin State University and is currently working on low-input sustainable management systems for strawberries and blueberries.

**CLAYTON W. OGG** is with the Office of Policy Analysis at the Environmental Protection Agency. He has a Ph.D. degree from the Department of Agricultural Economics at the University of Minnesota. Much of his recent research focuses on integration of farm policy and environmental policy.

**ROBERT I. PAPENDICK**   Since 1972, Papendick has been research leader of the U.S. Department of Agriculture Agricultural Research Service's Land Management and Water Conservation research unit at Pullman, Washington. He earned a Ph.D. degree in soil physics from South Dakota State University. His research interests include dryland soil and water conservation and alternative agriculture.

**JAMES F. PARR** is national program leader for Dryland Agriculture and Soil Fertility with the U.S. Department of Agriculture (USDA) at Beltsville, Maryland. Since 1984, he has served as coordinator of a USDA/U.S. Agency for International Development cooperative project to improve the sustainability of farming systems in the semiarid tropics. Parr earned his Ph.D. degree in soil microbiology from Purdue University. He has broad experience in soil and water conservation and management of organic wastes to improve soil productivity.

**KIM PATTEN** conducted research for Texas A&M on small fruit production systems from 1985 to 1990. He is presently a cranberry research-extension specialist for Washington State University. He has a Ph.D. degree in horticulture from Washington State University.

**NEIL H. PELSUE, JR.,** is chair of the Agricultural and Resource Economics Department at the University of Vermont, with extension, teaching, and research responsibilities in agricultural marketing and policy. He has conducted research in alternative agricultural enterprises and currently teaches a course in the economics of sustainable agriculture. In 1989–1990 he served as interim coordinator of the Low-Input Sustainable Agriculture program in the Northeast region.

**STEVEN E. PETERS**   Since 1986, Peters has been a project leader conducting farming systems research at the Rodale Institute Research Center. He has an M.S. degree in vegetable crops and soil science from Cornell University. From 1974 to 1981, he was a commercial market gardener.

**DOUGLAS G. PFEIFFER**   Since 1981, Pfeiffer has been a faculty member in the Department of Entomology at Virginia Polytechnic Institute and

State University. He obtained his Ph.D. degree in entomology from Washington State University. Pfeiffer is responsible for developing integrated pest management programs for trees and small fruits. Most of his research activity focuses on mating disruption and other alternative tactics and the development of action thresholds for foliar feeding arthropods.

**EDWIN G. RAJOTTE** is associate professor of entomology at the Pennsylvania State University. He has a Ph.D. degree in entomology from Rutgers University. He is a specialist in fruit integrated pest management and information delivery systems.

**HAROLD F. REETZ, JR.** Since 1982, Reetz has served as regional director, Potash & Phosphate Institute, responsible for education and research programs in crop and soil management to promote development and implementation of production practices that are agronomically sound, economically viable, and environmentally responsible. Reetz earned his Ph.D. degree from Purdue University in crop physiology and ecology. He has been involved in numerous national and international symposia and committee work related to developing and implementing best management practices.

**GAIL RICHARDSON** is a senior research consultant to INFORM, a national environmental research and education organization. She has a Ph.D. in political science from the University of Wisconsin and has directed INFORM's agricultural water conservation program since 1982.

**ALAN J. SAWYER** is an ecologist with the Plant Protection Research Unit of the U.S. Department of Agriculture's Agricultural Research Service in Ithaca, New York. His research involves the use of systems analysis and modeling to study the spatial dynamics of insect populations and their fungal pathogens. He earned his Ph.D. degree in entomology from Michigan State University.

**NEILL SCHALLER** Since September 1990, Schaller has been associate director of the Institute for Alternative Agriculture, Greenbelt, Maryland. For 2 years prior to that he was director of the Low-Input Sustainable Agriculture (LISA) research and education program in the U.S. Department of Agriculture (USDA). Schaller has a Ph.D. degree in agricultural economics from the University of California-Berkeley.

**BLAINE G. SCHATZ** As associate agronomist at the North Dakota State University Carrington Research Center, Schatz directs the field crops research effort in the area of traditional crop production studies. His interests are crop rotations, alternative crops and production systems, and field crop management. He earned an M.S. degree in agronomy at North Dakota State University.

**CAROL SHENNAN**  Since 1985 Shennan has been an assistant professor in the vegetable crops department, University of California, Davis.  She has a Ph.D. in botany from the University of Cambridge, United Kingdom.  Her research interests include resource use and nutrient cycling in cropping systems and plant responses to environmental stress.

**BARBARA J. SMITH** is affiliated with the Small Fruit Research Station, Agricultural Research Service, U.S. Department of Agriculture.  Her areas of research interest are diseases of small fruit and breeding for disease resistance.  She earned a Ph.D. degree in plant pathology from Louisiana State University.

**JAMES D. SMOLIK** is a professor in the Plant Science Department at South Dakota State University (SDSU), where he earned a Ph.D. degree in plant pathology.  His research studies include long-term comparisons of alternative and conventional farming systems.  He is also investigating the effects of nematodes on growth of several crops.  Smolik manages SDSU's Northeast Research Station.

**JAMES L. STARR** is a member of the faculty of plant pathology and microbiology at Texas A&M University.  He has a Ph.D. degree in plant pathology from Cornell University.  His research interests include nematode population dynamics, determination of damage functions for different nematode species-crop interactions, and the development and deployment of host resistance.

**NICHOLAS D. STONE**  In 1988, Stone came to Virginia Polytechnic Institute and State University (VPI&SU) to help form the Center for Computer-Aided Decision Making in the College of Agriculture.  He received his Ph.D. degree from the University of California at Berkeley, and is an associate professor of entomology at VPI&SU.  His research interests include population ecology, mathematical modeling, decision theory, and artificial intelligence.

**PRESTON G. SULLIVAN** is a technical specialist for ATTRA (Appropriate Technology Transfer for Rural Areas), a nonprofit sustainable agricultural organization.  He holds a Ph.D. degree from Virginia Polytechnic Institute and State University in crop and soil environmental sciences.

**GERALD R. SUTTER**  After graduating from Iowa State University with a Ph.D. degree in entomology, Sutter conducted research on the development of new control technology for soil-inhibiting arthropod pests of cereals.  He served as research leader for the entomology projects at the U.S. Department of Agriculture-Agricultural Research Service Northern Grain Insects Research Laboratory from 1972 to 1989.

**DANIEL B. TAYLOR** is an associate professor in the Department of Agricultural Economics at Virginia Polytechnic Institute and State University. He earned a Ph.D. degree in agricultural economics from Washington State University. His current areas of research include the economics of low-input agriculture, international agricultural development, environmental impacts of agriculture, and public policy analysis.

**ARIENA H. C. VAN BRUGGEN** Since 1986, van Bruggen has been an assistant professor of vegetable diseases in the Department of Plant Pathology at the University of California at Davis. She received her Ph.D. degree from Cornell University. Her research interests include agricultural systems analysis and epidemiology of bacterial and fungal plant pathogens.

**DAVID H. VAUGHAN** is a professor in the Department of Agricultural Engineering at Virginia Polytechnic Institute and State University. He earned a Ph.D. degree in biological and agricultural engineering from North Carolina State University. His areas of research include sustainable agricultural systems, conservation tillage/planting systems, agricultural energy analysis, and renewable energy resources, including biofuels.

**FEKEDE WORKNEH** is a graduate research assistant in plant pathology in the Department of Plant Pathology, University of California at Davis.